INTEGRAL EQUATIONS, BOUNDARY VALUE PROBLEMS AND RELATED PROBLEMS

INTEGRAL EQUATIONS, BOUNDARY VALUE PROBLEMS AND RELATED PROBLEMS

Dedicated to Professor Chien-Ke Lu
on the occasion of his 90th birthday

Yinchuan, Ningxia, China 19 – 23 August 2012

Editor

Xing Li

Ningxia University, China

NEW JERSEY • LONDON • SINGAPORE • BEIJING • SHANGHAI • HONG KONG • TAIPEI • CHENNAI

Published by

World Scientific Publishing Co. Pte. Ltd.
5 Toh Tuck Link, Singapore 596224
USA office: 27 Warren Street, Suite 401-402, Hackensack, NJ 07601
UK office: 57 Shelton Street, Covent Garden, London WC2H 9HE

British Library Cataloguing-in-Publication Data
A catalogue record for this book is available from the British Library.

INTEGRAL EQUATIONS, BOUNDARY VALUE PROBLEMS AND RELATED PROBLEMS

ISBN 978-981-4452-87-8

Printed in Singapore.

PREFACE

This proceedings volume is a collection of 26 selected papers of the 15th conference of integral equations,boundary value problems and related problems which held at Ningxia University during August 19-23, 2012.

The conference was co-sponsored by Chinese Society of Mathematics, Association for Science and Technology of Ningxia, and Mathematics Society of Ningxia. The conference organizers were Ningxia University, Wuhan University, Peking University, Fudan University, Sun Yat-sen University, Beijing Normal University, Hebei University, Hebei Normal University and Renmin University of China. The conference brought active researchers together to reassess recent advances on integral equations, boundary value problems and their applications, and to provide guidance for relevant research topics for the future.

The theory and application of integral equations and boundary value problems is important subject within applied mathematics. Integral equations and boundary value problems are used as mathematical models for many and varied physical situations, and they also occur as reformulations of other mathematical problems. The conference topics focus on the theory and analysis of integral equations,integral operators and boundary value problems; Applications of integral equations and boundary value problems to mechanics and physics; Approximate solutions of integral equations and boundary value problems; Boundary value problems for partial differential equations and functional equations; And some problems in Clifford analysis.

As editors we express our gratitude to the contributors of the volume for submitting their manuscripts. Thanks also to Prof. Huili Han, Dr. Shenghu Ding and Mr. Pengpeng Shi from the School of Mathematics and Computer Science at Ningxia University for having organized the conference and unified the manuscripts for this volume. We also thank Dena of World Scientific Publishing Company for making the publication of this volume possible. Finally,the chief editor would like to thank the financial support of 211 fund of Ningxia University and the National Natural Science Foundation of China (10962008, 51061015).

ORGANIZING COMMITTEES

EDITORIAL BOARD

for the 15th conference of integral equations, boundary value problems and related problems

CONTENTS

SOME PROPERTIES OF A KIND OF SINGULAR INTEGRAL OPERATOR FOR K-MONOGENIC FUNCTION IN CLIFFORD ANALYSIS*

LIPING WANG[a,b], ZUOLIANG XU[a], YUYING QIAO[b†]

[a]*School of Information, Renmin University of China, Beijing 100872, China*

This paper has two parts. In the first part, we study the boundedness for a kind of singular integral operator, which is related to the Cauchy-type integral of k-monogenic function in Clifford Analysis. In the second part, we show the Hölder continuous of the singular integral operator.

Keywords: Clifford Analysis, k-monogenic function, singular integral operator, Hölder continuous.

AMS No: 30G30, 30G35.

1. Introduction

Clifford analysis is a kind of function theory which was established in order to study the solution of Dirac equation in spinor fields. And it is as such a direct generation to higher dimensions of the classical function theory of holomorphic function in the complex analysis. Clifford analysis is an important branch of modern analysis that studies functions defined on R^n with their values in Clifford algebra space, which is an associative and incommutable algebra structure. It possesses both theoretical and applicable values to many fields, such as elastic mechanics, fluid mechanics and so on. Since last century, a number of mathematicians have made great efforts in real and complex Clifford analysis. For example Brack, Relanghe, Sommen[1], Wen[2], Huang, Qiao[3,4]etc.

In addition, the solutions of partial differential equation and their boundary value problems have been applied in many subjects such as

*This research is supported by NSFC (No.10971224, No.11171349) and Natural Science Foundation of Hebei Province(No.A2010000346).

†Corresponding author, E-mail: yuyingqiao@163.com.

physics, chemistry and engineering technology. And singular integral operators are one of the basic tools to solve partial differential equation and their boundary value problems. Hence for many years studying the properties of singular integral operators have been a hot and significative topics. References [5-9] studied some properties and application of Cauchy type integral operator and singular integral operator in higher dimensional. And when we solve some problem of degenerate equations and degenerate system in higher dimensional space, we need define and study some singular generalized integral operator in ordered to give a solution of degenerate equations and degenerate system.

In this paper, we define a kind of singular integral operator. Firstly, we discuss the boundedness for the singular integral operators in an nonempty open bounded connected set in R^{n+}, which is related to the Cauchy-type integral of k-monogenic function in Clifford Analysis. Secondly, we discuss the Hölder continuous of the singular integral operator.

2. Preliminaries

Let $e_1, \cdots, e_n$ be an orthogonal basis of the Euclidean space R^n, and let $\mathcal{A}_n(R)$ be the 2^n-dimensional Clifford algebra with basis $\{e_A : e_A = e_{\alpha_1} \cdots e_{\alpha_h}\}$, where $A = \{\alpha_1, \cdots, \alpha_h\} \subseteq \{1, \cdots, n\}, 1 \leq \alpha_1 < \alpha_2 < \cdots < \alpha_h \leq n$. and $e_\emptyset = e_0 = 1$. Hence the real Clifford algebra is composed of elements having the type $a = \sum\limits_A x_A e_A,\ x_A \in R$. Its module is defined as $|a| = \left(\sum\limits_A |x_A|^2\right)^{\frac{1}{2}}$.

The associative and noncommutative multiplication of the basis in $\mathcal{A}_n(R)$ is governed by the rules

$$\begin{cases} e_i^2 = -1, \ \ i = 1, 2, \cdots, n, \\ e_i e_j = -e_j e_i, \ \ 1 \leq i, j \leq n, \ i \neq j, \\ e_{\alpha_1} e_{\alpha_2} \cdots e_{\alpha_h} = e_{\alpha_1 \alpha_2 \cdots \alpha_h}, \ \ 1 \leq \alpha_1 < \cdots < \alpha_h \leq n. \end{cases}$$

In addition, Suppose Ω is an nonempty open bounded connected set in $R^{n+} = \{x = (x_1, x_2, \cdots, x_n) \in R^n \,|x_n \geq 0\}$ and the boundary $\partial\Omega$ is a differentiable, oriented and compact Liapunov surface in R^{n+}. The function f which is defined in Ω with values in $\mathcal{A}_n(R)$ can be expressed as $f(x) = \sum\limits_A f_A(x) e_A$, where the $f_A(x)$ is a real-valued function. In the paper, let $f(x) \in C^{(r)}(\Omega, \mathcal{A}_n(R)) = \{f | f : \Omega \to \mathcal{A}_n(R),\ f(x) = \sum\limits_A f_A(x) e_A\}$, where $f_A(x)$ has continuous $r-$times differentials.

And the Dirac operator is defined as
$Df=\sum_{i}^{n}e_i\frac{\partial f}{\partial x_i}=\sum_{i}^{n}\sum_{A}e_ie_A\frac{\partial f_A}{\partial x_i},\ fD=\sum_{i}^{n}\frac{\partial f}{\partial x_i}e_i=\sum_{i}^{n}\sum_{A}e_Ae_i\frac{\partial f_A}{\partial x_i}$.
f is called a left(right) k-monogenic function if $D^kf(x)=0(f(x)D^k=0)(r\geq k,k<n,k\in Z)$. $f(x)\in L^p(\Omega)$, which means $D^jf(x)\in L^p(\Omega),j=0,1,\cdots,k-1$ and $L_p[f(x),\Omega]=\sum_{j=0}^{k-1}L_p[D^jf(x),\Omega]$.

In the following, we define a kind of singular integral operator and give some necessary lemma. Then we discuss some properties of the singular integral operator.

Definition 2.1 *Let Ω be as stated above and $f(x)\in C^{(r)}(\Omega,\mathcal{A}_n(R))$, we define a kind of singular integral operator*

$$(T[f])(y)=\sum_{j=0}^{k-1}(-1)^j\int_{\Omega}\frac{A_{j+1}}{\omega_n}\frac{(x-y)^{j+1}}{x_n^{\alpha}|x-y|^n}dxD^jf(x),$$

where x_n is the $n-th$ component of x, $f(x)\in L^p(\Omega)$, $0<\alpha<1$ is a positive constant and ω_n is the area of the unit sphere in R^n, A_j is as stated the reference [10], it is irrelevant to the x,y.

Remark 1: When $y\in\Omega^-=R^n-\overline{\Omega}$, $T[f]$ is a normal generalized integral. When $y\in\overline{\Omega}$, $T[f]$ is its Cauchy principal value.

Remark 2: When $\alpha=0,n=2,k=1$, the singular integral operator is a normal Teodorescu operator.

Lemma 2.1[10] *For any $x,y_1,y_2\in R^n$, when $j\geq0$, we have*

$$|\frac{(x-y_1)^{j+1}}{|x-y_1|^n}-\frac{(x-y_2)^{j+1}}{|x-y_2|^n}|$$

$$\leq[\sum_{i=1}^{n-1}\frac{(|x|+|y_2|)^j}{|x-y_1|^i|x-y_2|^{n-i}}+\sum_{i=1}^{j}\frac{(|x|+|y_2|)^{j-i}}{|x-y_1|^{n-i}}]|y_1-y_2|.$$

Remark: When $j=0$, The second part is vanishing.

Lemma 2.2[11] *Suppose Ω is a bounded domain in $R^n,n\geq2$, and let α and β satisfy $0<\alpha,\ \beta<n,\ \alpha+\beta>n$, then for all $x_1,x_2\in R^n$ such that*

$x_1 \neq x_2$, *we have*

$$\int_\Omega |t-x_1|^{-\alpha}|t-x_2|^{-\beta}dt \leq M(\alpha,\beta)|x_1-x_2|^{n-\alpha-\beta}.$$

3. The main results

Theorem 3.1 *Let* $\Omega \subset R^n (n>2)$ *be as stated above,* $f(x) \in L^p(\Omega)$, $p > \dfrac{n+2}{1-\alpha}$, *Then we have*

$$|T[f]| \leq M(n,p,\Omega)L_p[f,\Omega],$$

where p, q, β_0, *are positive constants and satisfy* $\dfrac{1}{p}+\dfrac{1}{q}=1$, $\beta_0 = \dfrac{2(1-\alpha)}{n} - \dfrac{4(q-1)}{nq}$.

Proof *By Hölder formula, we have*

$$|T[f]| \leq J_1\sum_{j=0}^{k-1}\int_\Omega \frac{|D^j f(x)|}{|x_n|^\alpha |x-y|^{n-j-1}}|dx|$$

$$\leq J_2\sum_{j=0}^{k-1}[\int_\Omega |D^j f(x)|^p|dx|]^{\frac{1}{p}} \cdot [\int_\Omega \frac{1}{|x_n|^{\alpha q}|x-y|^{(n-j-1)q}}]^{\frac{1}{q}}$$

$$\leq J_2 L_p[f,\Omega]\sum_{j=0}^{k-1}[\int_\Omega \frac{|dx|}{|x_n|^{\alpha q}|x-y|^{(n-j-1)q}}]^{\frac{1}{q}}$$

$$\leq J_2 L_p[f,\Omega]\sum_{j=0}^{k-1}\int_0^d \frac{|dx|}{|x_n|^{\alpha q}|x_n-y_n|^{q(n-j-1)\frac{\beta_0}{2}}}dx_n$$

$$\cdot\int_\Omega \frac{1}{|\underline{x}-\underline{y}|^{q(n-j-1)(1-\frac{\beta_0}{2})}}d\underline{x},$$

where $d = \max_{x\in\Omega} x_n$, $\underline{x} = (x_1, x_2, \cdots, x_{n-1})$, $\underline{y} = (y_1, y_2, \cdots, y_{n-1})$. $\beta_0 = \dfrac{2(1-\alpha)}{n} - \dfrac{4(q-1)}{nq}$.

For $p > \dfrac{n+2}{1-\alpha}$, *we have*

$$\begin{aligned} 1 < q < \frac{n+2}{n+1+\alpha} &= 1 + \frac{1-\alpha}{n+1+\alpha} \\ < \frac{n}{n-1+\alpha} &= 1 + \frac{1-\alpha}{n-1+\alpha} < 2. \end{aligned} \tag{3.1.1}$$

$$0 < \beta_0 = \frac{2(1-\alpha)}{n} - \frac{4(q-1)}{nq} < 1.$$

Then

$$q(n-j-1)\frac{\beta_0}{2} + \alpha q \leq q(n-1)\frac{\beta_0}{2} + \alpha q < 2 - q < 1, (j = 0, 1, \cdots, k-1).$$

Hence the integral $\displaystyle\int_0^d \frac{1}{|x_n|^{\alpha q}|x_n - y_n|^{q(n-j-1)\frac{\beta_0}{2}}} dx_n$ *is bounded.*

Again from $q(n-j-1)(1-\dfrac{\beta_0}{2}) \leq (n-1)q(1-\dfrac{\beta_0}{2}), (j = 0, 1, \cdots, k-1)$, *and*

$$\begin{aligned} (n-1)q(1-\frac{\beta_0}{2}) &= (n-1)q\left[1 - \frac{1-\alpha}{n} + \frac{2(q-1)}{nq}\right] \\ &= (n-1)\left[\frac{(n-1+\alpha)q}{n} + \frac{2q}{n} - \frac{2}{n}\right] \\ &< (n-1)\left[\frac{n+\alpha+1}{n}(1 + \frac{1-\alpha}{n+1+\alpha}) - \frac{2}{n}\right] \\ &= (n-1)(\frac{n+2}{n} - \frac{2}{n}) = n-1. \end{aligned} \tag{3.1.2}$$

We obtain $\displaystyle\int_\Omega \frac{1}{|\underline{x} - \underline{y}|^{q(n-j-1)(1-\frac{\beta_0}{2})}} d\underline{x}$ *is bounded too.*

To sum up, we have $|T[f]| \leq M(n, p, \Omega) L_p[f, \Omega]$.

Theorem 3.2 *Let* $\Omega \subset R^n (n > 2)$ *be as stated above,* $f(x) \in L^p(\Omega)$, $p > \dfrac{n+2}{1-\alpha}$, *Then exist* $J > 0$, *for all* $y_1, y_2 \in \Omega$, *we have*

$$|T[f](y_1) - T[f](y_2)| \leq J|y_1 - y_2|^\beta,$$

where p, q, β_0, β *are positive constants and satisfy* $\dfrac{1}{p}+\dfrac{1}{q}=1$, $\beta_0=\dfrac{2(1-\alpha)}{n}-\dfrac{4(q-1)}{nq}$, $0<\beta=\dfrac{n-nq(1-\frac{\beta_0}{2})}{q}<1$.

Proof *From Lemma 2.1, we have*

$$\begin{aligned}
&|T[f](y_1)-T[f](y_2)|\\
&\le J_3\sum_{j=0}^{k-1}\int_\Omega \frac{1}{|x_n|^\alpha}\Big|\frac{(x-y_1)^{j+1}}{|x-y_1|^n}-\frac{(x-y_2)^{j+1}}{|x-y_2|^n}\Big||dx||D^jf(x)|\\
&\le J_3[\sum_{j=0}^{k-1}\int_\Omega \frac{1}{|x_n|^\alpha}\sum_{i=1}^{n-1}\frac{(|x|+|y_2|)^j}{|x-y_1|^i|x-y_2|^{n-i}}|y_1-y_2||dx||D^jf(x)| \qquad (3.2.1)\\
&+\sum_{j=0}^{k-1}\int_\Omega \frac{1}{|x_n|^\alpha}\sum_{i=1}^{j}\frac{(|x|+|y_2|)^{j-i}}{|x-y_1|^{n-i}}|y_1-y_2||dx||D^jf(x)|]\\
&=J_3[I_1+I_2].
\end{aligned}$$

In the following, firstly we evaluate I_2.

$$\begin{aligned}
I_2 &\le J_4|y_1-y_2|\sum_{j=0}^{k-1}\sum_{i=1}^{j}\int_\Omega \frac{1}{|x_n|^\alpha|x-y_1|^{n-i}}|dx||D^jf(x)|\\
&\le J_4|y_1-y_2|\{[\int_\Omega |Df(x)|^p|dx|]^{\frac{1}{p}}[\int_\Omega \frac{1}{|x_n|^{\alpha q}|x-y_1|^{(n-1)q}}|dx|]^{\frac{1}{q}}\\
&+\sum_{i=1}^{2}[\int_\Omega |D^2f(x)|^p|dx|]^{\frac{1}{p}}[\int_\Omega \frac{1}{|x_n|^{\alpha q}|x-y_1|^{(n-i)q}}|dx|]^{\frac{1}{q}}\\
&+\cdots\\
&+\sum_{i=1}^{k-1}[\int_\Omega |D^{k-1}f(x)|^p|dx|]^{\frac{1}{p}}[\int_\Omega \frac{1}{|x_n|^{\alpha q}|x-y_1|^{(n-i)q}}|dx|]^{\frac{1}{q}}\}.
\end{aligned}$$

Obviously, we know from the above that all integrals are bounded by the similar proof as Theorem 3.1. Again for $0<\beta<1$, Ω *is bounded, So*

$$I_2 \le J_5|y_1-y_2| = J_5|y_1-y_2|^{1-\beta}|y_1-y_2|^{\beta} \le J_6|y_1-y_2|^{\beta}. \tag{3.2.2}$$

Secondly

$$I_1 \le J_7|y_1-y_2|\sum_{j=0}^{k-1}\sum_{i=1}^{n-1}\int_\Omega \frac{1}{|x_n|^\alpha}\frac{1}{|x-y_1|^i|x-y_2|^{n-i}}|dx||D^j f(x)|$$

$$\le J_8|y_1-y_2|\sum_{i=1}^{n-1}[\int_\Omega \frac{1}{|x_n|^{\alpha q}}\frac{1}{|x-y_1|^{iq}|x-y_2|^{(n-i)q}}|dx|]^{\frac{1}{q}}.$$

Transforming $\int_\Omega \frac{1}{|x_n|^{\alpha q}|x-y_1|^{iq}|x-y_2|^{(n-i)q}}|dx|$ *as follows*

$$\int_\Omega \frac{1}{|x_n|^{\alpha q}|x-y_1|^{iq}|x-y_2|^{(n-i)q}}|dx|$$

$$= \int_\Omega \frac{1}{|x_n|^{\alpha q}|x-y_1|^{iq\frac{\beta_0}{2}}|x-y_1|^{iq(1-\frac{\beta_0}{2})}}$$

$$\cdot\frac{1}{|x-y_2|^{(n-i)q\frac{\beta_0}{2}}|x-y_2|^{(n-i)q(1-\frac{\beta_0}{2})}}|dx|$$

$$\le \int_0^d \frac{1}{|x_n|^{\alpha q}|x_n-y_{1,n}|^{iq\frac{\beta_0}{2}}|x_n-y_{2,n}|^{(n-i)q\frac{\beta_0}{2}}}|dx_n|$$

$$\cdot\int_{\underline{\Omega}} \frac{1}{|\underline{x}-\underline{y_1}|^{iq(1-\frac{\beta_0}{2})}|\underline{x}-\underline{y_2}|^{(n-i)q(1-\frac{\beta_0}{2})}}|d\underline{x}| = I_{11}\cdot I_{12},$$

where $\underline{y_1}=(y_{1,1},y_{1,2},\cdots,y_{1,n-1})$, $\underline{y_2}=(y_{2,1},y_{2,2},\cdots,y_{2,n-1})$.

Again for $p>\frac{n+2}{1-\alpha}$, *thus by (3.1.1), we know* $1<q<2$, $0<\beta_0<1$, *Hence*

$$\alpha q + iq\frac{\beta_0}{2} + (n-i)q\frac{\beta_0}{2} = \alpha q + nq\frac{\beta_0}{2} = 2-q<1.$$

So I_{11} *is bounded. Thus there is no harm to suppose* $I_{11}\le J_9$.

In the following, we evaluate I_{12}.

Firstly, when $\underline{y_1}\ne\underline{y_2}$. *Suppose* $\alpha' = iq(1-\frac{\beta_0}{2}) \le (n-1)q(1-\frac{\beta_0}{2})$, $\beta' = (n-i)q(1-\frac{\beta_0}{2}) \le (n-1)q(1-\frac{\beta_0}{2})$, $(i=1,2,\cdots,n-1)$. *By (3.1.2), we know* $0 < \alpha', \ \beta' < n-1$.

Again for

$$\begin{aligned} \alpha' + \beta' &= qn(1 - \frac{\beta_0}{2}) \\ &= (n+\alpha+1)q - 2 > (n+\alpha+1) - 2 \\ &> n+1-2 = n-1. \end{aligned} \tag{3.2.3}$$

Thus by Lemma 2.2, we know when $\underline{y_1} \neq \underline{y_2}$, *we have*

$$I_{12} \le J_{10}|\underline{y_1} - \underline{y_2}|^{n-1-nq(1-\frac{\beta_0}{2})}.$$

Hence, again for $|\underline{y_1} - \underline{y_2}| \le |y_1 - y_2|$, *we have*

$$\begin{aligned} I_1 &\le J_8|y_1 - y_2| \sum_{i=1}^{n-1} J_9^{\frac{1}{q}} J_{10}^{\frac{1}{q}} |\underline{y_1} - \underline{y_2}|^{\frac{1}{q}[n-1-nq(1-\frac{\beta_0}{2})]} \\ &= J_{11}|y_1 - y_2||\underline{y_1} - \underline{y_2}|^{\frac{n-1-nq(1-\frac{\beta_0}{2})}{q}} \\ &\le J_{11}|y_1 - y_2|^{1+\frac{n-1-nq(1-\frac{\beta_0}{2})}{q}} \\ &= J_{11}|y_1 - y_2|^{\frac{q-1}{q}}|y_1 - y_2|^{\frac{n-nq(1-\frac{\beta_0}{2})}{q}} \\ &\le J_{12}|y_1 - y_2|^{\beta}, \end{aligned} \tag{3.2.4}$$

where $\beta = \dfrac{n - nq(1-\frac{\beta_0}{2})}{q}$.

And by equations (3.1.2) and (3.2.3), we know

$$-1 < n-1-nq(1-\frac{\beta_0}{2}) < 0.$$

Thus

$$-1 < -\frac{1}{q} < \frac{n-1-nq(1-\frac{\beta_0}{2})}{q} = \frac{n-nq(1-\frac{\beta_0}{2})}{q} - \frac{1}{q} < 0.$$

Namely

$$0 < \beta = \frac{n-nq(1-\frac{\beta_0}{2})}{q} < \frac{1}{q} < 1.$$

Therefore, when $\underline{y_1} \neq \underline{y_2}$, *by (3.2.1),(3.2.2) and (3.2.4), we obtain*

$$|T[f](y_1) - T[f](y_2)| \leq J_3[J_{12} + J_6]|y_1 - y_2|^\beta = J_{13}|y_1 - y_2|^\beta.$$

Secondly, when $\underline{y_1} = \underline{y_2}$. *Now* $y_{1,n} \neq y_{2,n}$, *otherwise* $y_1 = y_2$. *Thus there is no harm to suppose* $\underline{y_1} = \underline{y_2} = \underline{y}$, *namely* $y_1 = (\underline{y}, y_{1,n})$, $y_2 = (\underline{y}, y_{2,n})$.

Then if we take $y_3 = \underline{y} + \dfrac{|y_{2,n} - y_{1,n}|}{2\sqrt{3}} e_1 + \dfrac{y_{2,n} + y_{1,n}}{2} e_n$, *obviously we have* $\underline{y_3} \neq \underline{y_1}$, $\underline{y_3} \neq \underline{y_2}$. *and*

$$|y_1 - y_3| \leq \frac{1}{\sqrt{3}}|y_1 - y_2|, \ |y_2 - y_3| \leq \frac{1}{\sqrt{3}}|y_1 - y_2|. \tag{3.2.5}$$

Let

$$I_1' = \sum_{j=0}^{k-1} \int_\Omega \frac{1}{|x_n|^\alpha} \sum_{i=1}^{n-1} \frac{(|x| + |y_3|)^j}{|x - y_1|^i |x - y_3|^{n-i}} |y_1 - y_3||dx||D^j f(x)|,$$

$$I_1'' = \sum_{j=0}^{k-1} \int_\Omega \frac{1}{|x_n|^\alpha} \sum_{i=1}^{n-1} \frac{(|x| + |y_3|)^j}{|x - y_2|^i |x - y_3|^{n-i}} |y_2 - y_3||dx||D^j f(x)|.$$

Similar to the proof earlier, we know

$$I_1' \leq J_{14}|y_1 - y_3|^\beta, \quad I_1'' \leq J_{15}|y_2 - y_3|^\beta.$$

Let

$$I_2' = \sum_{j=0}^{k-1} \int_\Omega \frac{1}{|x_n|^\alpha} \sum_{i=1}^{j} \frac{(|x| + |y_3|)^{j-i}}{|x - y_1|^{n-i}} |y_1 - y_3||dx||D^j f(x)|,$$

$$I_2'' = \sum_{j=0}^{k-1} \int_\Omega \frac{1}{|x_n|^\alpha} \sum_{i=1}^{j} \frac{(|x| + |y_3|)^{j-i}}{|x - y_2|^{n-i}} |y_2 - y_3||dx||D^j f(x)|.$$

Then similar to the proof of equation (3.2.2), we have

$$I_2' \leq J_{16}|y_1 - y_3|^\beta, \quad I_2'' \leq J_{17}|y_2 - y_3|^\beta.$$

Therefore by the above conclusion and equation (3.2.5), we have

$$
\begin{aligned}
&|T[f](y_1) - T[f](y_2)| \\
&\leq |T[f](y_1) - T[f](y_3)| + |T[f](y_2) - T[f](y_3)| \\
&\leq J_{18}[I_1' + I_2'] + J_{19}[I_1'' + I_2''] \\
&\leq J_{18}[J_{14}(\tfrac{1}{\sqrt{3}})^{\beta}|y_1 - y_2|^{\beta} + J_{16}(\tfrac{1}{\sqrt{3}})^{\beta}|y_1 - y_2|^{\beta}] \\
&\quad + J_{19}[J_{15}(\tfrac{1}{\sqrt{3}})^{\beta}|y_1 - y_2|^{\beta} + J_{17}(\tfrac{1}{\sqrt{3}})^{\beta}|y_1 - y_2|^{\beta}] \\
&= J_{20}|y_1 - y_2|^{\beta}.
\end{aligned}
$$

Therefore, to sum up, for all $y_1, y_2 \in \Omega$, *we obtain*

$$|T[f](y_1) - T[f](y_2)| \leq J|y_1 - y_2|^{\beta},$$

where $J = \max\{J_{13},\ J_{20}\}$.

References

[1] F. Brack, R. Delanghe, F. Sommen. *Clifford analysis (Research Notes in Mathematics 76).* Pitman Books Ltd, London , 1982.

[2] Wen Guochun. *Clifford Analysis and Elliptic Systems, Hyperbolic Systems of First Order Equations.* World Sci. Publ., Teaneck, NJ, 1991.

[3] Huang Sha, Qiao Yuying, Wen Guochun. *Real and complex Clifford analysis.* USA: Springer, 2006.

[4] Qiao Yuying. *A boundary value problem for hypermonogenic function in Clifford analysis.* Science in China Ser.A Mathematics, 2005, 48:324-332.

[5] H. Begher. *Iterations of Pompeiu operators.* Meum Diff Equ math phys, 12(1997): 13-21.

[6] R.P. Gilbert, Z.Y.Hou, X.W. Meng . *Vekua Theory in higher Dimensional complex space: The* Π*-operator in* C^n. Complex Var, 21(1993): 99-103.

[7] Bian Xiaoli, S.L. Eriksson, Li Junxia, Qiao Yuying. *Cauchy integral formula and Plemelj formula of bihypermonogenic functions in real Clifford analysis.* Complex Variables and Elliptic Equations, 54(2009), 10: 957C976.

[8] Yang Peiwen. *The operators T in Quaternionic Analysis and two classes of boundary value problems.* Acta Math Sinica, 44(2001), 2: 343-350.

[9] Wang Liping, Qiao Yuying, Qiao Haiying. *The fixed point theorem and iterative approximation of some class Quasi-Cauchy integral operator in real Clifford Analysis.* Advances in Mathematics, 38(2009), 4:385-396 (in chinese).

[10] Li Xiaoling, Qiao Yuying, Hu Weiwei, Wang Haiyan. *Some Properties of the k-regular Functions and Boundary Behavior of the Higher Order Cauchy-type Integral in Clifford Analysis.* Advances in Mathematics,39(2010),1:31-48 (in chinese).

[11] R. Gilbert, J. Buchanan. *First Order Elliptic Systems: A Function Theoretic Approach.* Academic Press, New York, 1983.

SOME RESULTS RELATED WITH MÖBIUS TRANSFORMATION IN CLIFFORD ANALYSIS

ZHONGXIANG ZHANG

School of Mathematics and Statistics, Wuhan University,
Wuhan 430072, P. R. China
E-mail: zhangzx9@126.com

In this paper, the hyper-complex forms of equations for the plane and the sphere in $\mathcal{R}^3$ are given. Then, the concept of symmetric points with respect to the plane and the sphere in $\mathcal{R}^3$ is introduced, equivalent equations for symmetric points respect to the plane and the sphere are given. Some special Möbius transformations in hyper-complex space Cl_3 are considered. Some properties of the mentioned Möbius transformation are given. For example: The Möbius transformation maps the plane and the sphere in $\mathcal{R}^3$ one-to-one onto the plane or the sphere; The Möbius transformation preserves the symmetric points with respect to the plane and the sphere in Cl_3; The cross ratio of four distinct points is also invariant under the Möbius transformation. A relation between the monogenic function and Möbius transformation is shown. Finally, based on the properties of Möbius transformations, integral formulas for change of variables under Möbius transformations are constructed.

Keywords: Clifford algebra, Möbius transformation, Poisson integral representation.

2000 MR Subject Classification 30G35.

1. Introduction and Preliminaries

Möbius transformations of the complex plane play an important role in a variety of mathematical and physical problems. Naturally, Möbius transformation in higher dimension is a very interesting topic, it may be fully described by using Clifford algebras. In [9-11], Clifford analysis and quaternionic analysis were systematically studied. In [8, 11, 14, 16-18 etc.], Möbius transformations in Clifford analysis were well studied, many interesting results related with Möbius transformations were shown. Most of these theories were built for Dirac operator $\sum_{i=1}^{n} e_i \frac{\partial}{\partial x_i}$. So, the corresponding theories for operator $\sum\limits_{i=0}^{n} e_i \frac{\partial}{\partial x_i}$ are also interesting. Since there are a large amount of practical problems in physics and mechanics in three dimension. So, in contrast to the theories in complex plane, similar results in the so called three dimensional hyper-complex space Cl_3 would be also interesting. For this purpose, some results for differential operator $\sum\limits_{i=0}^{2} e_i \frac{\partial}{\partial x_i}$ are presented in this paper.

Let $V_{n,0}$ be an n–dimensional ($n \geq 1$) real linear space with basis $\{e_1, e_2, \cdots, e_n\}$, $C(V_{n,0})$ be the 2^n–dimensional real linear space with basis

$$\{e_A, A = \{h_1, \cdots, h_r\} \in \mathcal{P}N, 1 \leq h_1 < \cdots < h_r \leq n\},$$

where N stands for the set $\{1, \cdots, n\}$ and $\mathcal{P}N$ denotes the family of all order-preserving subsets of N in the above way. we denote $e_\emptyset$ as e_0 and e_A as $e_{h_1 \cdots h_r}$ for $A = \{h_1, \cdots, h_r\} \in \mathcal{P}N$. The product on $C(V_{n,0})$ is defined by

$$\begin{cases} e_A e_B = (-1)^{\#(A\cap B)}(-1)^{P(A,B)} e_{A\triangle B}, \text{ if } \ A, B \in \mathcal{P}N, \\ \lambda\mu = \sum\limits_{A\in \mathcal{P}N} \sum\limits_{B\in \mathcal{P}N} \lambda_A \mu_B e_A e_B, \qquad \text{if } \ \lambda = \sum\limits_{A\in \mathcal{P}N} \lambda_A e_A, \ \mu = \sum\limits_{B\in \mathcal{P}N} \mu_B e_B. \end{cases} \tag{1}$$

where $\#(A)$ is the cardinal number of the set A, the number $P(A, B) = \sum\limits_{j\in B} P(A, j)$, $P(A, j) = \#\{i, i \in A, i > j\}$, the symmetric difference set $A\triangle B$ is also order-preserving in the above way, and $\lambda_A \in \mathcal{R}$ is the coefficient of the e_A–component of the Clifford number λ. We also denote λ_A as $[\lambda]_A$, $\lambda_{\{i\}}$ as $[\lambda]_i$ and λ_0 as $\mathrm{Re}\lambda$. It follows at once from the multiplication rule (1) that e_0 is the identity element written now as 1 and in particular,

$$\begin{cases} e_i^2 = -1, & \text{if } \ i = 1, \cdots, n, \\ e_i e_j = -e_j e_i, & \text{if } \ 1 \leq i < j \leq n, \\ e_{h_1} e_{h_2} \cdots e_{h_r} = e_{h_1 h_2 \cdots h_r}, & \text{if } \ 1 \leq h_1 < h_2 \cdots, < h_r \leq n. \end{cases} \tag{2}$$

Thus $C(V_{n,0})$ is a real linear, associative, but non-commutative algebra and it is called the Clifford algebra over $V_{n,0}$.

The real linear space with basis $\{e_0, e_1, e_2\}$ is a subspace of $C(V_{2,0})$, it is so-called three dimensional hyper-complex space and denoted by Cl_3. For $\forall \mathbf{x} = x_0 e_0 + x_1 e_1 + x_2 e_2 \in Cl_3$, $|\mathbf{x}| = \sqrt{x_0^2 + x_1^2 + x_2^2}$. The operator D which is written as

$$D = \sum_{k=0}^{2} e_k \frac{\partial}{\partial x_k} : C^{(r)}(\Omega, C(V_{2,0})) \to C^{(r-1)}(\Omega, C(V_{2,0})).$$

Let f be a function with value in $C(V_{2,0})$ defined in Ω, where Ω is an open non empty subset in $\mathcal{R}^3$. The operator D acts on the function f from the left and from the right being governed by

$$D[f] = \sum_{k=0}^{2}\sum_{A} e_k e_A \frac{\partial f_A}{\partial x_k}, \quad [f]D = \sum_{k=0}^{2}\sum_{A} e_A e_k \frac{\partial f_A}{\partial x_k}.$$

2. Hyper-complex Forms of Equations for Plane and Sphere in $\mathcal{R}^3$

Suppose $\mathbf{y} = y_0 + y_1e_1 + y_2e_2$, $\mathbf{B} = B_0 + B_1e_1 + B_2e_2$, then

Theorem 2.1. *The hyper-complex form of equations for the plane* Π *with the normal vector* (B_0, B_1, B_2) *in* $\mathcal{R}^3$ *is expressed as*

$$\boldsymbol{B}\overline{\boldsymbol{y}} + \boldsymbol{y}\overline{\boldsymbol{B}} + C = 0, \tag{3}$$

where $\boldsymbol{B} \neq 0, C \in \mathcal{R}$.

Theorem 2.2. *The hyper-complex form of equations for the sphere* S *in* $\mathcal{R}^3$ *is expressed as*

$$A\boldsymbol{y}\overline{\boldsymbol{y}} + \boldsymbol{B}\overline{\boldsymbol{y}} + \boldsymbol{y}\overline{\boldsymbol{B}} + C = 0, \tag{4}$$

where $A, C \in \mathcal{R}$, $A \neq 0$ *and* $|\boldsymbol{B}|^2 - AC > 0$.

3. Hyper-complex Forms of Equations for Symmetric Points with respect to Plane and Sphere in $\mathcal{R}^3$

In this section, suppose the hyper-complex equation of the plane Π be as: $\mathbf{B}\overline{\mathbf{y}} + \mathbf{y}\overline{\mathbf{B}} + C = 0$, where $\mathbf{B} \neq 0$. Suppose the hyper-complex equation of the sphere be as: $A\mathbf{y}\overline{\mathbf{y}} + \mathbf{B}\overline{\mathbf{y}} + \mathbf{y}\overline{\mathbf{B}} + C = 0$, where $A, C \in \mathcal{R}$, $A \neq 0$ and $|\mathbf{B}|^2 - AC > 0$.

Definition 3.1. $\mathbf{y}^*$ and $\mathbf{y}^{**}$ are called symmetric points with respect to Π if $\mathbf{y}^*$, $\mathbf{y}^{**}$ satisfy the following condition: (I) $\mathbf{y}^* - \mathbf{y}^{**}$ is the normal vector of Π; (II) $\dfrac{\mathbf{y}^* + \mathbf{y}^{**}}{2} \in \Pi$.

Theorem 3.1. $\boldsymbol{y}^*$ *and* $\boldsymbol{y}^{**}$ *are symmetric with respect to* Π *is equivalent to the following equation:*

$$\boldsymbol{B}\overline{\boldsymbol{y}^*} + \boldsymbol{y}^{**}\overline{\boldsymbol{B}} + C = 0. \tag{5}$$

Definition 3.2. Let Σ be as above, $\forall \mathbf{y}^* = y_0e_0 + y_1e_1 + y_2e_2$, $\mathbf{y}^{**}$ is called the symmetric point of $\mathbf{y}^*$ with respect to Σ if

$$\mathbf{y}^{**} - \mathbf{x} = \frac{R^2}{|\mathbf{y}^* - \mathbf{x}|^2}(\mathbf{y}^* - \mathbf{x}). \tag{6}$$

Theorem 3.2. *Suppose the hyper-complex equation of the sphere Σ be as: $A\boldsymbol{y}\overline{\boldsymbol{y}} + \boldsymbol{B}\overline{\boldsymbol{y}} + \boldsymbol{y}\overline{\boldsymbol{B}} + C = 0$, where $A, C \in \mathcal{R}$, $A \neq 0$ and $|\boldsymbol{B}|^2 - AC > 0$, then $\boldsymbol{y}^{**}$ be the symmetric point of $\boldsymbol{y}^*$ with respect to Σ is equivalent to the following equation:*

$$A\boldsymbol{y}^{**}\overline{\boldsymbol{y}^*} + \boldsymbol{B}\overline{\boldsymbol{y}^*} + \boldsymbol{y}^{**}\overline{\boldsymbol{B}} + C = 0. \tag{7}$$

4. Möbius Transformation

In this section, we shall consider the fractional linear transformation:

$$T(\mathbf{x}) = \frac{a\mathbf{x} + b}{c\mathbf{x} + d}, \tag{8}$$

where $\mathbf{x} = x_0 + x_1e_1 + x_2e_2$, $a, b, c, d \in \mathcal{R}$, and $ad - bc \neq 0$. (8) is also called Möbius transformation.

Theorem 4.1. *Suppose $T(\boldsymbol{x}) = \dfrac{a\boldsymbol{x} + b}{c\boldsymbol{x} + d}$, where $\boldsymbol{x} \in Cl_3$, $a, b, c, d \in \mathcal{R}$, and $ad - bc \neq 0$. T is not identity transformation I, Denote $\triangle = (d - a)^2 + 4bc$. Then:*

(I) If $c = 0$, then the set of the fixed point of T consists of at most one real number. Moreover, if $a \neq d$, then the fixed point is $\boldsymbol{x} = \dfrac{b}{d - a}$.

(II) If $c \neq 0$ and $\triangle > 0$, then the set of the fixed point of T consists of two different real numbers. The fixed points are $\boldsymbol{x}_1 = \dfrac{a - d + \sqrt{\triangle}}{2c}$ and $\boldsymbol{x}_2 = \dfrac{a - d - \sqrt{\triangle}}{2c}$.

(III) If $c \neq 0$ and $\triangle = 0$, then the set of the fixed point of T consists of one real number and the fixed point is $\boldsymbol{x} = \dfrac{a - d}{2c}$.

(IV) If $c \neq 0$ and $\triangle < 0$, then the set of the fixed points of T can be written as

$$\left\{ \boldsymbol{x} | \boldsymbol{x} \in Cl_3, x_0 = \frac{a - d}{2c}, x_1^2 + x_2^2 = -\frac{\triangle}{4c^2} \right\}. \tag{9}$$

Theorem 4.2. *Let T_1 and T_2 be two Möbius transformations. If T_1 and T_2 satisfy any of the following condition:*

(I) $T_1(\boldsymbol{x}_k)=T_2(\boldsymbol{x}_k)$, $k=1,2,3$, $\boldsymbol{x}_k$ *are distinct points, and* $\boldsymbol{x}_k \in \mathcal{R}$ *for* $k=1,2,3$.

(II) $T_1(\boldsymbol{x}_k)=T_2(\boldsymbol{x}_k)$, $k=1,2$, $\boldsymbol{x}_1 \in \mathcal{R}$ *and* $\boldsymbol{x}_2 \notin \mathcal{R}$.

(II) $T_1(\boldsymbol{x}_k)=T_2(\boldsymbol{x}_k)$, $k=1,2$, $\boldsymbol{x}_1, \boldsymbol{x}_2 \notin \mathcal{R}$ *and* $Re\{\boldsymbol{x}_1\} \neq Re\{\boldsymbol{x}_2\}$.

Then T_1 *and* T_2 *are identical throughout* Cl_3.

Three elementary Möbius transformations are expressed as follows: (I) $T(\mathbf{x}) = \mathbf{x} + b$, $b \in \mathcal{R}$. (II) $T(\mathbf{x}) = a\mathbf{x}$, $a \in \mathcal{R}$. (III) $T(\mathbf{x}) = \frac{1}{\mathbf{x}} = \frac{\overline{\mathbf{x}}}{|\mathbf{x}|^2}$.

Theorem 4.3. *Every Möbius transformation* T *can be written as a product of three elementary Möbius transformations.*

Theorem 4.4. *The image of plane or surface in* Cl_3 *under any Möbius transformation* T *is still plane or surface in* Cl_3.

Theorem 4.5. *Let* T *be a Möbius transformation and let* Σ *be a plane or sphere in* Cl_3. *Denote* Σ^* *as the transformed plane or sphere of* Σ. *Then for* $\forall \boldsymbol{x} \in Cl_3$,

$$S_{\Sigma^*}(T(\boldsymbol{x})) = T(S_{\Sigma}(\boldsymbol{x})), \tag{10}$$

where $S_{\Sigma}(\boldsymbol{x})$ *denotes the symmetric point of* $\boldsymbol{x}$ *with respect to* Σ.

Definition 4.1. Let $\mathbf{y} \in Cl_3$, z_2, z_3, z_4 be distinct points and $z_2, z_3, z_4 \in \mathcal{R}$. ($\infty$ is allowed as one of the above points)

$$\frac{(\mathbf{y} - z_3)(z_2 - z_4)}{(\mathbf{y} - z_4)(z_2 - z_3)} \tag{11}$$

is called the cross ratio of $\mathbf{y}, z_2, z_3, z_4$, and denoted by $(\mathbf{y}, z_2, z_3, z_4)$.

Definition 4.2. (∞, z_2, z_3, z_4) is defined as follows: $(\infty, z_2, z_3, z_4) = \lim_{|\mathbf{y}| \to \infty} (\mathbf{y}, z_2, z_3, z_4)$. Similarly, $(\mathbf{y}, \infty, z_3, z_4) = \lim_{z_2 \to \infty} (\mathbf{y}, z_2, z_3, z_4)$, $(\mathbf{y}, z_2, \infty, z_4) = \lim_{z_3 \to \infty} (\mathbf{y}, z_2, z_3, z_4)$, $(\mathbf{y}, z_2, z_3, \infty) = \lim_{z_4 \to \infty} (\mathbf{y}, z_2, z_3, z_4)$.

Theorem 4.6. *The cross ratio of four distinct points* $\boldsymbol{y}, z_2, z_3, z_4$ *is invariant under any Möbius transformation, where* $\boldsymbol{y} \in Cl_3$, $z_2, z_3, z_4 \in \mathcal{R}$.

Theorem 4.7. *Let* $\{z_2, z_3, z_4\}$ *and* $\{z_2^*, z_3^*, z_4^*\}$ *be given with each consisting of three distinct real numbers (*∞ *is allowed as one of the above points in each triple), then there is a uniquely determined Möbius transformation* T *that maps the first set of points to the second:*

$$T(z_2) = z_2^*, \ T(z_3) = z_3^*, \ T(z_4) = z_4^*.$$

For Möbius transformation $\mathbf{y} = T(\mathbf{x})$, Denote $J(T(\mathbf{x})) \triangleq \dfrac{\partial(y_0, y_1, y_2)}{\partial(x_0, x_1, x_2)}$.

Theorem 4.8. *Let T be a Möbius transformation, then*

$$J(T(\boldsymbol{x})) = \frac{(ad - bc)^3}{|c\boldsymbol{x} + d|^6}. \tag{12}$$

Theorem 4.9. *Let Ω_* be an open no-empty subset of $\mathcal{R}^3$ and T be a Möbius transformation, $T(\boldsymbol{x}) = \dfrac{a\boldsymbol{x} + b}{c\boldsymbol{x} + d}$, denoting $\Omega \triangleq T(\Omega_*)$, then $D[f] = 0$ in Ω is equivalent to $D\left[\dfrac{c\overline{\boldsymbol{x}} + d}{|c\boldsymbol{x} + d|^3} f(T(\boldsymbol{x}))\right] = 0$ in Ω_*.*

Theorem 4.10. *Let Ω_* be an open no-empty subset of $\mathcal{R}^3$ and T be a Möbius transformation, $\boldsymbol{y} = T(\boldsymbol{x}) = \dfrac{a\boldsymbol{x} + b}{c\boldsymbol{x} + d}$, denoting $\Omega \triangleq T(\Omega_*)$, then*

$$D_x\left[\frac{c\overline{\boldsymbol{x}} + d}{|c\boldsymbol{x} + d|^3} f(T(\boldsymbol{x}))\right] = (ad - bc)\frac{c\boldsymbol{x} + d}{|c\boldsymbol{x} + d|^5} D_y\left[f(\boldsymbol{y})\right], \tag{13}$$

where $\boldsymbol{x} \in \Omega_*$, $\boldsymbol{y} \in \Omega$.

Theorem 4.11. *Let Ω_* be an open no-empty subset of $\mathcal{R}^3$ and T be a Möbius transformation, $\boldsymbol{y} = T(\boldsymbol{x}) = \dfrac{a\boldsymbol{x} + b}{c\boldsymbol{x} + d}$, denoting $\Omega \triangleq T(\Omega_*)$, then*

$$\frac{c\overline{\boldsymbol{x}} + d}{|c\boldsymbol{x} + d|^3} D_x\left[\frac{c\overline{\boldsymbol{x}} + d}{|c\boldsymbol{x} + d|^3} f(T(\boldsymbol{x}))\right] = \frac{1}{(ad - bc)^2} J(T(\boldsymbol{x})) D_y\left[f(\boldsymbol{y})\right], \tag{14}$$

where $\boldsymbol{x} \in \Omega_*$, $\boldsymbol{y} \in \Omega$.

Theorem 4.12. *Let Ω_* be an open no-empty subset of $\mathcal{R}^3$ and T be a Möbius transformation, $\boldsymbol{y} = T(\boldsymbol{x}) = \dfrac{a\boldsymbol{x} + b}{c\boldsymbol{x} + d}$, denoting $\Omega \triangleq T(\Omega_*)$, then*

$$\left[g(T(\boldsymbol{x}))\frac{c\overline{\boldsymbol{x}} + d}{|c\boldsymbol{x} + d|^3}\right] D_x = (ad - bc)\left[g(\boldsymbol{y})\right] D_y \cdot \frac{c\boldsymbol{x} + d}{|c\boldsymbol{x} + d|^5}, \tag{15}$$

where $\boldsymbol{x} \in \Omega_*$, $\boldsymbol{y} \in \Omega$.

Theorem 4.13. *Let Ω_* be an open no-empty subset of $\mathcal{R}^3$ and T be a Möbius transformation, $T(\boldsymbol{x}) = \dfrac{a\boldsymbol{x} + b}{c\boldsymbol{x} + d}$, denoting $\Omega \triangleq T(\Omega_*)$, then $[g]D = 0$ in Ω is equivalent to $\left[g(T(\boldsymbol{x}))\dfrac{c\overline{\boldsymbol{x}} + d}{|c\boldsymbol{x} + d|^3}\right] D = 0$ in Ω_*.*

Theorem 4.14. *Let Ω_* be an open no-empty subset of $\mathcal{R}^3$ and T be a Möbius transformation, $\boldsymbol{y} = T(\boldsymbol{x}) = \dfrac{a\boldsymbol{x}+b}{c\boldsymbol{x}+d}$, denoting $\Omega \triangleq T(\Omega_*)$, then*

$$\left[g(T(\boldsymbol{x}))\frac{c\overline{\boldsymbol{x}}+d}{|c\boldsymbol{x}+d|^3}\right] D_x \cdot \frac{c\overline{\boldsymbol{x}}+d}{|c\boldsymbol{x}+d|^3} = \frac{1}{(ad-bc)^2} J(T(\boldsymbol{x})) \cdot [g(\boldsymbol{y})]\, D_y, \quad (16)$$

where $\boldsymbol{x} \in \Omega_*$, $\boldsymbol{y} \in \Omega$.

5. Integral Formulas for Change of Variables under Möbius Transformation

In this section, we shall give integral formulas for change of variables under Möbius transformation, these formulas will help to calculate the surface integral for Clifford valued functions by choosing the suitable domain or suitable functions. Analogous results can be also found in [16, 18].

Suppose U be a domain in Cl_3, Ω be a bounded domain with smooth boundary $\partial\Omega$ satisfying $\Omega \subset U$, T be any Möbius transformation as in section 4 and $T(\mathbf{x}) = \dfrac{a\mathbf{x}+b}{c\mathbf{x}+d}$. Denote $\Omega^* \triangleq T^{-1}(\Omega)$.

Theorem 5.1. *Let $f, g \in C^1(U, C(V_{2,0}))$, U, Ω and Ω^* be as above, then*

$$\int_{\partial\Omega} g(\boldsymbol{y})\, d\sigma_y f(\boldsymbol{y}) = \pm(ad-bc)^2 \int_{\partial\Omega^*} g(T(\boldsymbol{x}))\frac{c\overline{\boldsymbol{x}}+d}{|c\boldsymbol{x}+d|^3}\, d\sigma_x \frac{c\overline{\boldsymbol{x}}+d}{|c\boldsymbol{x}+d|^3} f(T(\boldsymbol{x})), \quad (17)$$

where the right side of (17) is taken $+$ in case of $J(T(\boldsymbol{x})) > 0$ and taken $-$ in case of $J(T(\boldsymbol{x})) < 0$.

Theorem 5.2. *Let $f, g \in C^1(B(\boldsymbol{y}_0, R), C(V_{2,0})) \bigcap C(\overline{B(\boldsymbol{y}_0, R)}, C(V_{2,0}))$, then*

$$\int_{\partial B(\boldsymbol{y}_0,R)} g(\boldsymbol{y})\, d\sigma_y f(\boldsymbol{y}) = \pm(ad-bc)^2 \int_{\partial\Omega^*} g(T(\boldsymbol{x}))\frac{c\overline{\boldsymbol{x}}+d}{|c\boldsymbol{x}+d|^3}\, d\sigma_x \frac{c\overline{\boldsymbol{x}}+d}{|c\boldsymbol{x}+d|^3} f(T(\boldsymbol{x})), \quad (18)$$

where the right side of (18) is taken $+$ in case of $J(T(\boldsymbol{x})) > 0$ and taken $-$ in case of $J(T(\boldsymbol{x})) < 0$, $\Omega^* \triangleq T^{-1}(B(\boldsymbol{y}_0, R))$.

Acknowledgements

This work is supported by a DAAD-K.C. Wong Education Foundation and NNSF for Young Scholars of China (No. 11001206).

References

1. H. Begehr, Iterations of Pompeiu operators. Mem. Diff. Eq. Math. Phys. 12 (1997), 3–21.
2. H. Begehr, Iterated integral operators in Clifford analysis. Journal for Analysis and its Applications 18 (1999), 361–377.
3. H. Begehr, Representation formulas in Clifford analysis. Acoustics, Mechanics,and the Related Topics of Mathematical Analysis. World Scientific Singapore, 2002, 8–13.
4. H. Begehr, Zhang Zhongxiang, Du Jinyuan, On Cauchy-Pompeiu formula for functions with values in a universal Clifford algebra. Acta Mathematica Scientia 23B(1) (2003), 95–103.
5. F. Brack, R. Delanghe and F. Sommen, Clifford Analysis. Research Notes in Mathematics 76. Pitman Books Ltd, London, 1982.
6. R. Delanghe, On regular analytic functions with values in a Clifford algebra. Math. Ann. 185 (1970), 91–111.
7. R. Delanghe, On the singularities of functions with values in a Clifford algebra. Math. Ann. 196 (1972), 293–319.
8. R. Delanghe, Clifford analysis: History and perspective. Computational Methods and Function Theory. 1 2001, 107–153.
9. R. Delanghe, F. Brackx, Hypercomplex function theory and Hilbert modules with reproducing kernel. Proc. London Math. Soc. 37 1978, 545–576.
10. R. Delanghe, F. Sommen, V. Soucek, Clifford algebra and spinor-valued functions. Kluwer, Dordrecht, 1992.
11. S. L. Eriksson, H. Leutwiler, Hypermonogenic functions and Möbius transformations. Advances in Applied Clifford Algebras, 11(s2), 2001, 67–76.
12. K. Gürlebeck, W. Sprössig. Quaternionic analysis and elliptic boundary value problems. Akademie-Verlag, Berlin, 1989.
13. V. Iftimie, Functions hypercomplex. Bull. Math. Soc. Sci. Math. R. S. Romania. 9(57), 1965, 279–332.
14. R. Kraußhar, A characterization of conformal mappings in $\mathcal{R}^4$ by a formal differentiability condition. Bulletin de la Societe Royale des Sciences de Liege. 70, 2001, 35–49.
15. E. Obolashvili, Higher order partial differential equations in Clifford analysis. Birkhauser, Boston, Basel, Berlin, 2002.
16. E. M. Olea, Morera type problems in Clifford analysis. Rev. Mat. Iberoam. 17, 2001, 559–585.
17. J. Peetre, T. Qian, Möbius covariance of iterated Dirac operators. J. Austral. Math. Soc. (Series A) 56, 1994, 403–414.
18. T. Qian, J. Ryan, Conformal transformations and Hardy spaces arising in Clifford analysis. J. Operator Theory. 35, 1996, 349–372.
19. Xu Zhenyuan, Boundary value problems and function theory for spin-invariant differential operators. Ph.D. thesis, Ghent State University, 1989.
20. Zhang Zhongxiang, On k-regular functions with values in a universal Clifford algebra. J. Math. Anal. Appl. 315(2), 2006, 491-505.
21. Zhang Zhongxiang, Some properties of operators in Clifford analysis. Complex Var., elliptic Eq. 6(52), 2007, 455-473.

THE SCATTERING OF SH WAVE ON THE ARRAY OF PERIODIC CRACKS IN A PIEZOELECTRIC SUBSTRATE BONDED A HALF-PLANE OF FUNCTIONALLY GRADED MATERIALS

JUNQIAO LIU and XING LI*

School of civil and water conservancy engineering, Ningxia University, Yinchuan, 750021, P.R. of China
**E-mail: lix666@yeah.net*
www.nxu.edu.cn

JUNQIAO LIU†, SHUZHUAN DONG, XIYAN YAO and CAIFENG WANG

Faculty of Applied Mathematics, Yuncheng University, Yuncheng, 044000, P.R. of China
†*E-mail: liujunqiao@yeah.net*

The interaction of SH wave with the array of periodic cracks in a piezoelectric substrate bonded to a functionally graded materials (FGM) coatings is considered. The governing equations along with permeable crack boundary, regularity and continuity conditions across the interface are reduced to a coupled set of Hilbert singular integral equations which are solved approximately by applying Cheyshev polynomials. Numerical results for the normalized dynamic stress intensive factors (NDSIF) and the normalized electric displacement intensive factors (NEDIF) are presented. The effects of geometric parameters and the physical parameters, and the effects of frequency and angles of SH wave on the NDSIF and the NEDIF are discussed.

Keywords: Functionally graded materials;Singular integral equations;The array of periodic cracks;Piezoelectric substrate;Intensity factors.

1 Introduction

In recent years, the piezoelectric materials (PZM) have been found to have wide applications in the smart systems of aerospace, automotive, medical and electronic fields due to the intrinsic coupling characteristic between their electric and mechanical fields [1]. Due to its brittleness, PZM have a tendency to develop critical cracks during the manufacturing and the poling process. These defects greatly affected the mechanical integrity and electromechanical performance of PZM [2,3]. Zhou studied the interaction of SH wave with two cracks in the same line of the piezoelectric media in 2001 [4]. Narita and Shindo studied the scattering of SH wave from the crack in composite plane with PZM, where the dual integral equation are obtained by using Fourier transform [5,6]. Wang and Meguid studied the interaction of SH wave with some cracks in PZM by using integral transform and Chebyshev polynomials [7,8]. Gu and Yu discussed the elastic wave scattering by an interface crack between a piezoelectric layer and an elastic substrate in 2002 [9]. Ueda.S studied diffraction of anti-plan shear waves in a piezoelectric laminated with a vertical crack by the integral transform and the singular integral equation method in 2003 [10]. The scattering of SH wave from a crack in the piezoelectric substrate bonded with a half-space of FGM is studied by Li X [11]. The scattering of SH wave by a vertical crack in a FGM coated the piezoelectric strip was considered in 2010 by Liu [12]. Liu studied the scattering of SH wave by the array of periodic cracks in magneto-electro-elastic composites in 2011 [13].

However, in this paper, the scattering of SH wave from the array of periodic cracks in a piezoelectric substrate bonded to a half-plane of functionally graded materials (FGM) is studied. Based on the Fourier integral transform and traditional permeable crack electrical boundary conditions, the Hilbert singular integral equation can be obtained by defining the dislocation density function. The singular integral equation is solved by the approximation of Chebyshev polynomials and the numerical calculations are carried out [14]. The effects of geometric parameters and the physical parameters, the effects of frequency and angles of SH wave on the NDSIF and the NEDIF are discussed.

2 Description of the problem

Consider FGM perfectly bonded to a piezoelectric substrate weakened by a periodic array of cracks, as shown in Fig.1. Assume that the crack of length $2c$ lies in the periodic array of $2L$. The thickness between the crack and the bonded line is h. Refer to a cartesian coordinate system (x, y, z) located at the center of the crack in which $z-$ axis is the poling axis of the piezoelectric media.

The shear modulus and the mass densities of the FGM are expressed by exponential form as

$$\mu(y) = c_{44} \exp[\gamma(y-h)], \rho(y) = \rho_0 \exp[\gamma(y-h)] \tag{1}$$

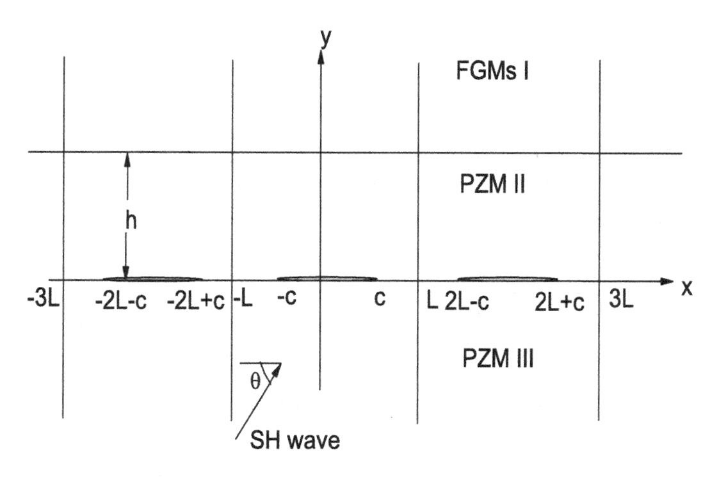

Fig. 1. Geometry of the crack problem

where c_{44}, ρ_0, γ are the elastic modulus, the mass density and the graded coefficient in the PZE respectively. $\mu(y)$ and $\rho(y)$ are the shear modulus and the mass densities of the FGM. Let time-harmonic SH wave originating at $y \to \infty$ be incident with an incidence angle θ on the cracks. The anti-plane shear stress act on the materials, so the displacement components are written as

$$u(x, y, t) = 0, v(x, y, t) = 0, w^{(t)}(x, y, t) = w^{(i)}(x, y, t) + w(x, y, t) \tag{2}$$

where $w^{(t)}(x, y, t)$, $w^{(i)}(x, y, t)$ and $w(x, y, t)$ are the total field, incidence field and scattering field, respectively. In the incidence field, the displacement and the electric potential are expressed as

$$w^{(i)}(x, y, t) = A_0 \exp\{-i[\alpha_2(x\cos\theta + y\sin\theta) + \omega t]\} \tag{3}$$

$$\phi^{(i)}(x, y, t) = \frac{e_{15}}{\varepsilon_{11}} w^{(i)}(x, y, t) \tag{4}$$

where A_0, θ, ω, e_{15}, ε_{11}, t are the amplitude, the incidence angle, the frequency, the piezoelectric coefficient, dielectric coefficient and time, respectively. $i = \sqrt{-1}$, $\alpha_2 = \frac{\omega}{c_2}$ (the wave number), $c_2 = \sqrt{\frac{c_{44}}{\rho_0}}$ (the wave velocity).

We regard the piezoelectric substrate as two layers, the part below the crack as the third layer and the part above the crack as the second layer. The first layer is the FGM. The constitutive relations of the piezoelectric substrate are expressed as

$$\tau_{31}(x,y,t) = e_{15}\phi_{,x}(x,y,t) + c_{44}w_{,x}(x,y,t) \tag{5}$$

$$\tau_{32}(x,y,t) = e_{15}\phi_{,y}(x,y,t) + c_{44}w_{,y}(x,y,t) \tag{6}$$

$$D_1 = -\varepsilon_{11}\phi_{,x}(x,y,t) + e_{15}w_{,x}(x,y,t) \tag{7}$$

$$D_2 = -\varepsilon_{11}\phi_{,y}(x,y,t) + e_{15}w_{,y}(x,y,t) \tag{8}$$

The constitutive relations of the FGM can be written as

$$\sigma_{31} = c_{44}\exp[\gamma(y-h)]w_{,x}(x,y,t); \sigma_{32} = c_{44}\exp[\gamma(y-h)]w_{,y}(x,y,t) \tag{9}$$

where τ_{3j}, σ_{3j} and $D_j (j = 1,2)$ represent the stress of PZM, the stress of FGM and the electric displacement of PZM, respectively. Substituting (3) and (4) into (6), the stress in the incidence field will be written as

$$\tau_{32}^{(i)}(x,y,t) = -iA_0\alpha_2 C^* \sin\theta \exp\{-i[\alpha_2(x\cos\theta + y\sin\theta) + \omega t]\} \tag{10}$$

where $C^* = \frac{e_{15}^2 + c_{44}\varepsilon_{11}}{\varepsilon_{11}}$.

Substituting (5) $\sim$ (9) into the following governing equations

$$\frac{\partial\tau_{31}(x,y,t)}{\partial x} + \frac{\partial\tau_{32}(x,y,t)}{\partial y} = \rho_0\frac{\partial^2 w(x,y,t)}{\partial t^2} \tag{11}$$

$$\frac{\partial D_1(x,y,t)}{\partial x} + \frac{\partial D_2(x,y,t)}{\partial y} = 0 \tag{12}$$

$$\frac{\partial\sigma_{31}(x,y,t)}{\partial x} + \frac{\partial\sigma_{32}(x,y,t)}{\partial y} = \rho_0\frac{\partial^2 w(x,y,t)}{\partial t^2} \tag{13}$$

We can obtain three equations as follows,

$$e_{15}(\phi_{,xx}(x,y,t)+\phi_{,yy}(x,y,t))+c_{44}(w_{,xx}(x,y,t)+w_{,yy}(x,y,t)) = \rho_0\frac{\partial^2 w(x,y,t)}{\partial t^2} \tag{14}$$

$$-\varepsilon_{11}\phi_{,xx}(x,y,t) + e_{15}w_{,xx}(x,y,t) - \varepsilon_{11}\phi_{,yy}(x,y,t) + e_{15}w_{,yy}(x,y,t) = 0 \tag{15}$$

$$w_{,xx}(x,y,t) + \gamma w_y(x,y,t) + w_{,yy}(x,y,t) = \frac{\rho_0}{c_{44}}\frac{\partial^2 w(x,y,t)}{\partial t^2} \tag{16}$$

The electric permeable boundary conditions are appropriate to a silt crack in PZM. Then the boundary conditions can be written as follows,

$$\tau_{32}^{+}(x,0)=\tau_{32}^{-}(x,0)=-\tau_{32}^{i}(x,0),D_2^{+}(x,0)=D_2^{-}(x,0)=D_2^{i}(x,0),$$

$$\phi^{+}(x,0)=\phi^{-}(x,0)(|x|\le c) \tag{17}$$

$$\tau_{32}^{+}(x,0)=\tau_{32}^{-}(x,0),D_2^{+}(x,0)=D_2^{-}(x,0),$$

$$\phi^{+}(x,0)=\phi^{-}(x,0),w(x,0^{+})=w(x,0^{-})(a<|x|<L) \tag{18}$$

$$\sigma_{32}^{+}(x,h)=\tau_{32}^{-}(x,h),D_2^{-}(x,h)=0,w^{+}(x,h)=w^{-}(x,h),(|x|<+\infty) \tag{19}$$

$$w(-L,y)=w(L,y);\quad \phi(-L,y)=\phi(L,y); \tag{20}$$

$$\delta_{31}(-L,y)=\delta_{31}(L,y);\quad \delta_{32}(-L,y)=\delta_{32}(L,y); \tag{21}$$

$$D_1(-L,y)=D_1(L,y);\quad D_2(-L,y)=D_2(L,y) \tag{22}$$

$$\delta_{32}(x,y)=\tau_{32}(x,y)=D_2(x,y)=0 \qquad |y|=\infty \tag{23}$$

3. Conversion of the problem

Because of the same time as in the incidence field, the displacement and electric potential in the scattering field are expressed as follows,

$$w(x,y,t)=w(x,y)\exp(-i\omega t),\phi(x,y,t)=\frac{e_{15}}{\varepsilon_{11}}w(x,y,t) \tag{24}$$

Substituting eq. (24) into eqs. (14) $\sim$ (16)

$$c_{44}\nabla^2 w(x,y)+e_{15}\nabla^2\phi(x,y)=-\rho_0\omega^2 w(x,y) \tag{25}$$

$$-\varepsilon_{11}\nabla^2\phi(x,y)+e_{15}\nabla^2 w(x,y)=0 \tag{26}$$

$$\nabla^2 w(x,y)+\gamma\frac{\partial w(x,y)}{\partial y}=-\frac{\rho_0\omega^2}{c_{44}}w(x,y) \tag{27}$$

Let $m_2^2=\frac{\rho_0\varepsilon_{11}\omega^2}{\varepsilon_{11}c_{44}+e_{15}^2}$, $f(x,y)=\phi(x,y)-\frac{e_{15}}{\varepsilon_{11}}w(x,y)$, $m_1^2=\frac{\rho_0\omega^2}{c_{44}}$. We can obtain

$$\nabla^2 w(x,y)+m_2^2 w(x,y)=0;\nabla^2 f(x,y)=0;$$

$$\nabla^2 w(x,y) + \gamma \frac{\partial w(x,y)}{\partial y} + m_1^2 w(x,y) = 0 \tag{28}$$

We introduce the finite Fourier integral transforms as follows

$$w_2(x,y) = \frac{1}{L}[\frac{1}{2}a_{2w0} + \sum_{n=1}^{\infty} a_{2wn} cos\frac{n\pi x}{L}] \tag{29}$$

$$f_2(x,y) = \frac{1}{L}[\frac{1}{2}a_{2f0} + \sum_{n=1}^{\infty} a_{2fn} cos\frac{n\pi x}{L}] \tag{30}$$

$$w_3(x,y) = \frac{1}{L}[\frac{1}{2}a_{3w0} + \sum_{n=1}^{\infty} a_{3wn} cos\frac{n\pi x}{L}] \tag{31}$$

$$f_3(x,y) = \frac{1}{L}[\frac{1}{2}a_{3f0} + \sum_{n=1}^{\infty} a_{3fn} cos\frac{n\pi x}{L}] \tag{32}$$

$$w_1(x,y) = \frac{1}{L}[\frac{1}{2}a_{1w0} + \sum_{n=1}^{\infty} a_{1wn} cos\frac{n\pi x}{L}] \tag{33}$$

eqs. (25) $\sim$ (27) can be rewritten as

$$a_{2w0} = A_{2w0}\cos m_2 y + B_{2w0}\sin m_2 y, a_{2wn} = A_{2wn}\exp(\lambda_0 y) + B_{2wn}\exp(-\lambda_0 y) \tag{34}$$

$$a_{3w0} = A_{3w0}\cos m_2 y + B_{3w0}\sin m_2 y, a_{3wn} = A_{3wn}\exp(\lambda_0 y) + B_{3wn}\exp(-\lambda_0 y) \tag{35}$$

$$a_{1w0} = A_{1w0}\exp(\lambda_3 y) + B_{1w0}\exp(\lambda_4 y), a_{1wn} = A_{1wn}\exp(\lambda_1 y) + B_{1wn}\exp(-\lambda_2 y) \tag{36}$$

$$a_{2f0} = A_{2f0}y + B_{2f0}, a_{2fn} = A_{2fn}\exp(\lambda_f y) + B_{2fn}\exp(-\lambda_f y) \tag{37}$$

$$a_{3f0} = A_{3f0}y + B_{3f0}, a_{3fn} = A_{3fn}\exp(\lambda_f y) + B_{3fn}\exp(-\lambda_f y) \tag{38}$$

where

$$\lambda_0 = \sqrt{\frac{n^2\pi^2}{L^2} - m_2^2}, \lambda_{1,2} = \frac{-\gamma \pm \sqrt{\gamma^2 - 4(m_1^2 - \frac{n^2\pi^2}{L^2})}}{2},$$

$$\lambda_{3,4} = \frac{-\gamma \pm \sqrt{\gamma^2 - 4m_1^2}}{2}, \lambda_f = \frac{n\pi}{L} \tag{39}$$

The stress and the electric displacement can be written

$$\delta_{32} = c_{44} \exp(\gamma(y-h)) \frac{1}{L} [\frac{1}{2} a'_{1w0} + \sum_{n=1}^{\infty} a'_{1wn} \cos \frac{n\pi x}{L}] \tag{40}$$

$$\tau_{32} = \frac{e_{15}}{L} (\frac{a'_{2f0}}{2} + \sum_{n=1}^{\infty} a'_{2fn} \cos \frac{n\pi x}{L}) + \frac{e_{15}^2 + c_{44}\varepsilon_{11}}{L\varepsilon_{11}} (\frac{a'_{2w0}}{2} + \sum_{n=1}^{\infty} a'_{2wn} \cos \frac{n\pi x}{L}) \tag{41}$$

$$D_2 = -\varepsilon_{11} [\frac{1}{2} a'_{2f0} + \sum_{n=1}^{\infty} a'_{2fn} \cos \frac{n\pi x}{L}] \tag{42}$$

$$\tau_{32} = \frac{e_{15}}{L} [\frac{1}{2} a'_{3f0} + \sum_{n=1}^{\infty} a'_{3fn} \cos \frac{n\pi x}{L}] + \frac{e_{15}^2 + c_{44}\varepsilon_{11}}{L\varepsilon_{11}} [\frac{1}{2} a'_{3w0} + \sum_{n=1}^{\infty} a'_{3wn} \cos \frac{n\pi x}{L}] \tag{43}$$

$$D_2 = -\varepsilon_{11} [\frac{1}{2} a'_{3f0} + \sum_{n=1}^{\infty} a'_{3fn} \cos \frac{n\pi x}{L}] \tag{44}$$

Applying the boundary conditions, we can obtain

$$A_{1w0} = A_{1wn} = A_{3f0} = A_{3w0} = A_{2f0} = A_{2w0} = 0 \tag{45}$$

$$B_{1w0} = B_{3fn} = B_{3w0} = B_{3wn} = B_{2w0} = B_{2f0} = B_{3f0} = 0 \tag{46}$$

$$A_{2wn} \exp(\lambda_0 h) + B_{2wn} \exp(-\lambda_0 h) = B_{1wn} \exp(-\lambda_2 h) \tag{47}$$

$$\exp(\lambda_f h) A_{2fn} - \exp(-\lambda_f h) B_{2fn} = 0 \tag{48}$$

$$-\varepsilon_{11} c_{44} \lambda_2 \exp(-\lambda_2 h) B_{1wn} = \lambda_0 (e_{15}^2 + c_{44}\varepsilon_{11}) [A_{2wn} \exp(\lambda_0 h) - B_{2wn} \exp(\lambda_0 h)] \tag{49}$$

$$A_{3fn} = A_{2fn} - B_{2fn}; A_{3wn} - A_{2wn} + B_{2wn} = 0 \tag{50}$$

$$e_{15}(A_{2wn} + B_{2wn}) + \varepsilon_{11}(A_{2fn} + B_{2fn}) = e_{15} A_{3wn} + \varepsilon_{11} A_{3fn} \tag{51}$$

We define the following dislocation density function

$$g(x) = \frac{\partial [w(x, 0^+) - w(x, 0^-)]}{\partial x}, |x| \leq a \tag{52}$$

We can obtain

$$A_{3wn} - A_{2fn} - B_{2fn} = \frac{1}{\lambda_f} \int_{-c}^{c} g(x) \sin \lambda_f u du \tag{53}$$

Applying the boundary condition (18)

$$\frac{1}{L}\int_{-c}^{+c}\sum_{n=0}^{\infty}\frac{G_1(n)}{\lambda_f}g(u)\sin\lambda_f u\cos\lambda_f x du=\tau^{(i)}(x) \tag{54}$$

Let

$$\alpha=\lim_{n\to\infty}\frac{G_1(n)}{\lambda_f}=\frac{e_{15}^4-2c_{44}^2\varepsilon_{11}^2}{e_{15}(e_{15}^2+2c_{44}\varepsilon_{11})-2c_{44}\varepsilon_{11}^2} \tag{55}$$

we obtain integral equation in which the unknown function is $g(x)$.

$$\frac{1}{4L}\int_{-c}^{+c}(\cot\frac{\pi(u-x)}{2L}+\cot\frac{\pi(u+x)}{2L})g(u)du+\frac{1}{L}\int_{-c}^{+c}k_1(x,u)g(u)du=\frac{\tau^{(i)}(x)}{\alpha} \tag{56}$$

where

$$k_1(x,u)=\sum_{n=0}^{\infty}(m-\alpha)\sin(\lambda_f u)\cos(\lambda_f x) \tag{57}$$

$$m=[\frac{e_{15}^2\lambda_f(1-\exp(2\lambda_f h))(\lambda_0(e_{15}^2+c_{44}\varepsilon_{11})-c_{44}\varepsilon_{11}\lambda_2)}{-2\varepsilon_{11}\lambda_0(e_{15}^2+c_{44}\varepsilon_{11})\exp(2\lambda_f h)}+c_{44}\lambda_2]$$

$$\frac{2\varepsilon_{11}\lambda_0(e_{15}^2+c_{44}\varepsilon_{11})\exp(2\lambda_f h)}{2\varepsilon_{11}^2c_{44}\lambda_2\exp(2\lambda_f h)+(1+\exp(2\lambda_f h))e_{15}(\lambda_0(e_{15}^2+c_{44}\varepsilon_{11})-\varepsilon_{11}c_{44}\lambda_2)}$$

4. Solution of the problem

To solve (56), the following normalized quantity are

$$(\overline{x},\overline{u},\overline{L},\overline{h})=(x,u,L,h)/c \tag{58}$$

$$(\overline{g(u)},\overline{k_1(u,x)},\overline{\tau_{32}^{(i)}(x)})=(g(u/c),k_1(u/c,x/c),\tau_{32}^{(i)}(x/c)) \tag{59}$$

For simplicity in what follows, the bar appearing with the dimensionless quantities is omitted. The eq.(56) can be written

$$\frac{1}{4L}\int_{-1}^{+1}(\cot\frac{\pi(u-x)}{2L}+\cot\frac{\pi(u+x)}{2L})g(u)du+\frac{1}{L}\int_{-1}^{+1}k_1(x,u)g(u)du=\frac{\tau^{(i)}(x)}{\alpha} \tag{60}$$

where $\tau^{(i)}(x)=\frac{e_{15}^2+c_{44}\varepsilon_{11}}{\varepsilon_{11}}(-i\alpha_2A_0)\sin\theta\exp(-i\alpha_2x\cos\theta)$.

Considering eqs. (19) and (54), we have

$$\int_{-1}^{1}g(u)du=0 \tag{61}$$

The solution to eq. (60) can be expressed

$$g(u) = \frac{\psi(u)}{\sqrt{1-u^2}}, \psi(u) = \sum_{j=0}^{\infty} B_j T_{2j+1}(u) \tag{62}$$

where $T_{2j+1}(u)$ is the first kind of Chebyshev polynomials, B_j is constant to be determined.

Substituting eq. (62) into eq. (60), we can obtain the equation as follows,

$$\sum_{l=0}^{n} \frac{\omega_l}{n}(\cot\frac{\pi(u_l-x_j)}{2L} + \cot\frac{\pi(u_l+x_j)}{2L})\psi(u_l) + \sum_{l=0}^{n} \frac{4\omega_l}{n} k_1(x_j,u_l)\psi(u_l) = \frac{4L\tau_0(x_j)}{\alpha} \tag{63}$$

$$\sum_{l=0}^{n} \omega_l \psi(u_l)) = 0 \tag{64}$$

where $u_l = \cos\frac{l\pi}{n}, l = 0,1,2...n; x_j = \cos\frac{(2j-1)\pi}{2n}, j = 1,2...n;\ \omega_0 = \omega_n = \frac{1}{2}, \omega_i = 1, i = 1,2...n-1$.

For $y = 0$, we note that

$$\tau_{32}(x,0) = \frac{\alpha}{4L}\int_{-1}^{+1}(\cot\frac{\pi(u-x)}{2L} + \cot\frac{\pi(u+x)}{2L})g(u)du + \frac{1}{L}\int_{-1}^{+1} k_1(x,u)g(u)du \tag{65}$$

$$D_2(x,0) = \frac{\beta}{4L}\int_{-1}^{+1}(\cot\frac{\pi(u-x)}{2L} + \cot\frac{\pi(u+x)}{2L})g(u)du + \frac{1}{L}\int_{-1}^{+1} k_2(x,u)g(u)du \tag{66}$$

where

$$\beta = \lim_{n\to\infty}\frac{G_2(n)}{\lambda_f} = \frac{-2c_{44}\varepsilon_{11}}{2\varepsilon_{11}^2 c_{44} + e_{15}^2} \tag{67}$$

Considering the singular part, the stress and electric displacement can be rewritten

$$\tau_{32}(x,0) = \frac{\alpha}{2}\int_{-1}^{+1}(\frac{1}{u-x} + \frac{1}{u+x})\frac{\psi(u)}{\sqrt{1-u^2}}du + \frac{1}{L}\int_{-1}^{+1} k_1^*(x,u)\frac{\psi(u)}{\sqrt{1-u^2}}du \tag{68}$$

$$D_2(x,0) = \frac{\beta}{2}\int_{-1}^{+1}(\frac{1}{u-x} + \frac{1}{u+x})\frac{\psi(u)}{\sqrt{1-u^2}}du + \frac{1}{L}\int_{-1}^{+1} k_2^*(x,u)\frac{\psi(u)}{\sqrt{1-u^2}}du \tag{69}$$

Where the second integral is finite when $|x| \to 1$.

Applying the following formula

$$\frac{1}{\pi}\int_{-1}^{+1}\frac{1}{u-x}\frac{T_j(u)}{\sqrt{1-u^2}}du=\frac{[(x^2-1)^{\frac{1}{2}}-x]^j}{(-1)^{j+1}(x^2-1)^{\frac{1}{2}}},|x|>1 \tag{70}$$

We define the dynamic stress intensive factor as follows

$$K_{III}=|\lim_{x\to 1}\sqrt{2\pi\,(x^2-1)}\tau_{32}(x,0)|=\sqrt{2\pi}\alpha\psi(1) \tag{71}$$

Let

$$K^*_{III}=\frac{K_{III}}{\sqrt{2\pi}\alpha} \tag{72}$$

K^*_{III} is the normalized dynamic stress intensity factor (NDSIF).

Define the electric displacement intensive factor as follows

$$K_D=|\lim_{x\to 1}\sqrt{2\pi\,(x^2-1)}D_2(x,0)|=\sqrt{2\pi}\beta\psi(1) \tag{73}$$

Let

$$K^*_D=\frac{\beta}{\alpha}K^*_{III} \tag{74}$$

where K^*_D is the normalized electric displacement intensity factor (NEDIF).

5 Numerical results and discussion

The integral transform and singular integral equation technique is used to obtain some analytical results. Since α and β doesn't appear ω, Eqs.(83) shows that the NDSIF and the NEDIF have the same varies when the frequency of SH wave increasing. To examine the effects of geometric parameters and the physical parameters, the effects of frequency and incident angles of SH wave on the NDSIF and the NEDIF, the approximate solution to the Hilbert singular integral equation has been valuated numerically. For numerical examples, the electro-elastic properties and the mass densities of these materials were $c_{44}=2.3\times 10^{10}N/m^2$, $e_{15}=17c/m^2$, $\varepsilon_{11}=150.3\times 10^{-10}c/\nu m$, $\rho_0=7500kg/m^2$, respectively. The result is shown in the Figs. 2 $\sim$ 6. Where x is refer to the non-dimensional frequency ($x=\omega/cc_2$), K^*_{III} ($K^*_{III}=\frac{K_{III}}{\sqrt{2\pi}\alpha}$)is referred to as the NDSIF.

Fig.2 displays the effects of periodic array $2L$ on the K^*_{III} when both the graded coefficientγ and the non-dimensional thickness h/c are 1. With the non-dimensional frequency ($x=\omega/cc_2$)of the SH wave increasing from $0.2\sim 2.0$, the value of NDSIF went up. The value of NDSIF fell when the non-dimensional array L/c increasing from 7 to 12. The value of NDSIF

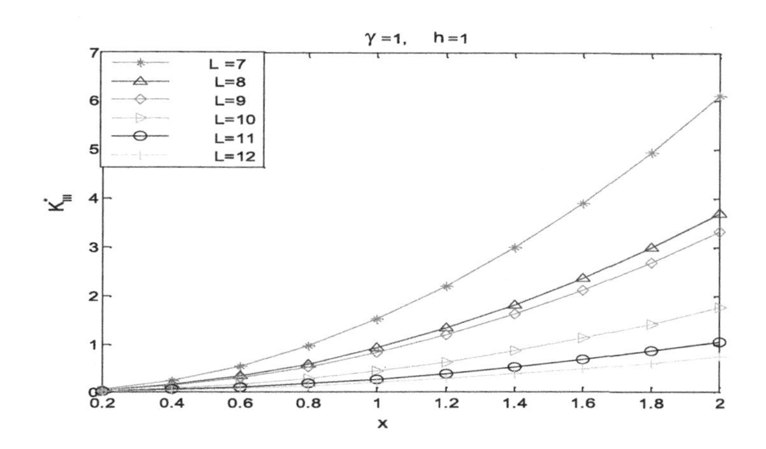

Fig. 2. The effects of the NDSF on ω, L

V

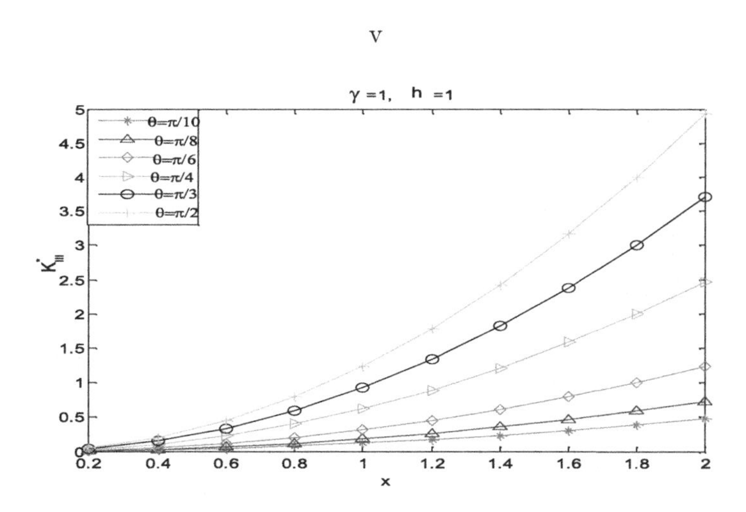

Fig. 3. The effects of the NDSF on ω, θ

increased much slower with the non-dimensional frequency increasing, the non-dimensional array L/c becomes bigger.

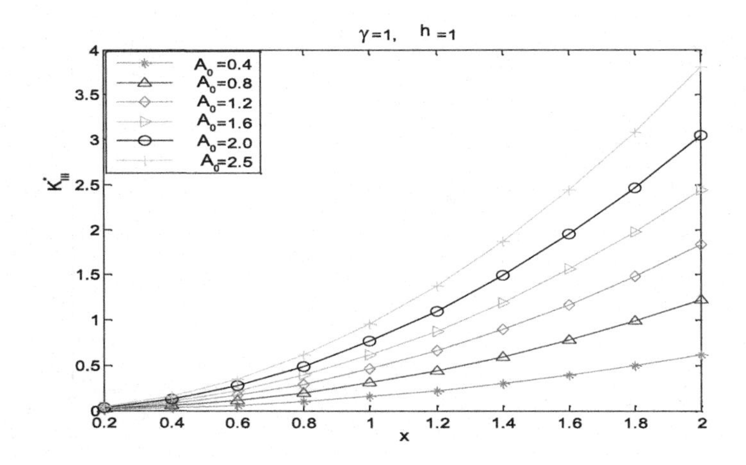

Fig. 4. The effects of the NDSF on ω, A_0

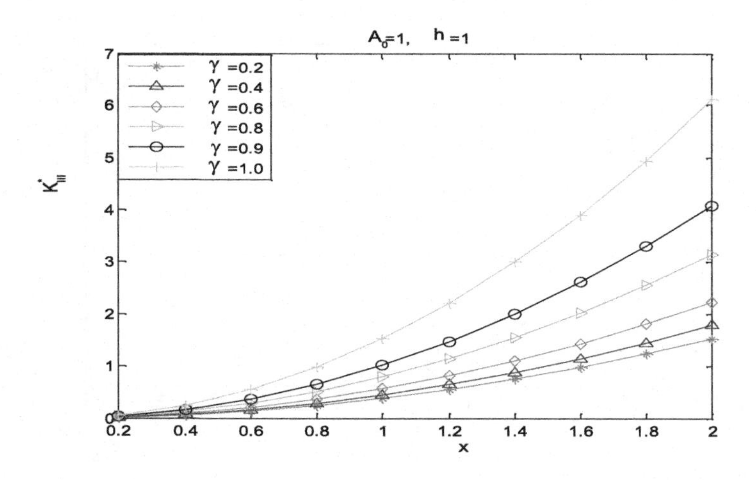

Fig. 5. The effects of the NDSF on ω, γ

Fig.3 presents the curve about effects of incident angles of SH wave on the NDSIF for the case $\gamma = 1, h/c = 1$. When θ increasing from $\frac{\pi}{10}$ to $\frac{\pi}{6}$,

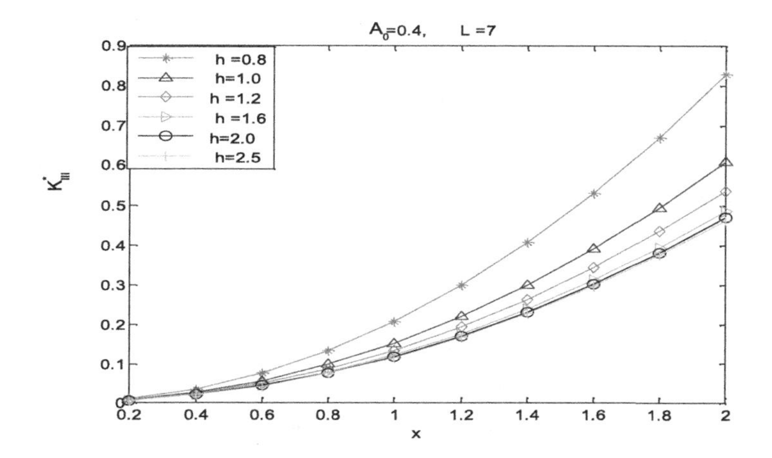

Fig. 6. The effects of the NDSF on ω, h

the variation of the NDSIF is not obviously. When the incidence angle is $\frac{\pi}{2}$, the NDSIF are maximum.

From Fig.3, the thickness h affects more the maximum of the NDSIF and the NEDIF. With h increasing, the maximum of the NDSIF and the NEDIF are increasing, too. That is the FGM have function to stop the NDSIF and the NEDIF increasing.

Seen the Fig.4, when the amplitude increasing, the NDSIF and the NEDIF are also increasing. At the same time the maximum become bigger.

Observed Fig.5, when the length of crack increasing, the NDSIF and the NEDIF are also increasing.

6 Conclusions

In the present paper, the scattering of SH wave on the array of periodic cracks in a piezoelectric substrate bonded a half-plane of functionally graded materials has been investigated. Using the Fourier transforms, the elasticity equations were converted analytically into a system of singular integral equations with Hilbert kernels. The singular integral equations are solved by the numerically using Chebyshev polynomials. From Figs.2-6, it can be concluded that for a fixed value of the parameter $\omega a/c_2$, the normalized dynamic stress intensity factor K_{III}^* increases for increasing values of the graded parameter γ, the angle of incidence θ, and the amplitude A_0, respectively.

Acknowledgements

This work was supported by the National Natural Science Foundation of China (Nos.10962008 and 51061015) and Research Fund for the Doctoral Program of Higher Education of China (No. 20116401110002).

References

1. Rao S. S., Sunar M.(1994). Piezoelectricity and ite use in disturbance sensing and control of flexible structures: a survey. Appl. Mech. Rev., 47: 113~123.
2. Chen Z. T., Meguid S. A.(2000). The transient response of a piezoelectric strip with a vertical crack under electromechnical impact load[J]. International Journal of Solids and Structures, 37: 6051~6062.
3. Meguid S. A., Wang X. D.(1998). Dynamic antiplane behavior of interacting cracks in a piezoelectric medium. International Journal Fracture, 91: 391~403.
4. Zhou Z. G., Li H. C., Wang B.(2001). Investigation of the scattering of antiplane shear waves by two conlinear cracks in a piezoelectric material using a new method[J]. Acta Mechanica, 147(1):87~97.
5. Narita F., Shindo Y.(1999). Scattering of antiplane shear waves by a finite crack in piezoelectric laminates[J]. Acta Mechnica, 134(1):27~43.
6. Narita F., Shindo Y. (1998). Scattering of love waves by a surface-beaking crack in piezoelectric layered media[J]. JSME International Journal series A, 41: 40~48.
7. Wang X. D., Meguid S. A.(2000). Modeling and analysis of the dynamic behavior of piezoelectric materials containing interacting cracks[J]. Mechanics of Materials, 32(7):723~737.
8. Wang X. D. (2001). On the dynamic behavior of interaction interfacial cracks in piezoelectric media[J]. International Journal of solids and Structures, 38(8): 831~851.
9. Gu B.,Yu S.W., Feng X. Q. .(2002). Elastic wave scattering by an interface crack between a piezoelectric layer and elastic substrate[J]. International Journal of Fracture, 116: L29~L34.
10. Sei Ueda. (2003).Diffraction of antiplane shear waves in a piezoelectric laminate with a vertical crack[J]. European Journal of Mechanica A/Solids, 22: 413~422.
11. Li X., Liu JQ.(2009). Scattering of the SH wave from a crack in a piezoelectric subsrate bonded to a half-space of functionally graded materials[J]. Acta Mechnica, 208:299~308.
12. Liu JQ., Duan HQ. Li X. (2010). The scattering of SH wave on a vertical crack in a coated piezoelectric strip[J]. Chinese Journal of Solid Mechanics, 31(4): 385~391.

13. Liu JQ., Li X. (2011). The scattering of SH wave on a Periodic Array of cracks in Magneto-electro-elastic Composites[J]. Chinese Journal of Engineering Mathematics, 28(4): 470~474.
14. F. Erdogan and G.D.Gupta.(1972). On the Numerical solution of singular integral equations[J]. Quarterly of Applied Mathematics, 1: 525~534.

ANTI-PLANE PROBLEM OF TWO COLLINEAR CRACKS IN A FUNCTIONALLY GRADED COATING–SUBSTRATE STRUCTURE

SHENG-HU DING*, XING LI

School of Mathematics and Computer Science, Ningxia University, Yinchuan, 750021, China

**Corresponding author (Tel: +86 951 2061405; fax: +86 951 2061405. Email: dshsjtu2009@163.com).*

A theoretical treatment of antiplane crack problem of two collinear cracks on the two sides of and perpendicular to the interface between a functionally graded orthotropic strip bonded to an orthotropic homogeneous substrate is put forward. The problem is formulated in terms of a singular integral equation with the crack face displacement as the unknown variable. Numerical calculations are carried out, and the influences of the orthotropy and nonhomogeneous parameters on the mode III stress intensity factors are investigated.

Keywords: Functionally graded orthotropic strip; Collinear cracks; Singular integral equation; Stress intensity factor

1. Introduction

With the application of functionally gradient materials in engineering, most of the current researches on the fracture analysis of the FGMs interface have been devoted to the FGMs interlayer and the interface between the functionally graded material (FGM) coating and the homogeneous substrate [1-5]. Ding et al. [6,7] studied the problem of the model III crack in a functionally graded piezoelectric layer bonded to a piezoelectric half-plane. The dynamic fracture problem of the weak-discontinuous interface between a FGM coating and a FGM substrate have been studied by Li et al.[8,9]. The mode III fracture problem of a cracked functionally graded piezoelectric surface layer bonded to a cracked functionally graded piezoelectric substrate are studied by Chen and Chue [10]. Introducing a bi-parameter exponential function to simulate the continuous variation of material properties, two bonded functionally graded finite strips with two collinear cracks are investigated by Ding and Li [11].

Their processing techniques mean that FGMs are seldom isotropic [12,13]. Therefore, the failure behaviors of orthotropic FGMs attracted some researchers' attention. Supposing the material properties to be one-dimensionally dependent, Chen and Liu [14] studied the transient response of an embedded crack and edge crack perpendicular to the boundary of an orthotropic functionally graded strip. Wang and Mai [15] studied the static and transient solutions for a periodic array of cracks in a functionally graded coating bonded to a homogeneous half-plane. A theoretical treatment of mode I crack problem is put forward for a functionally graded orthotropic strip. Guo et al.[16,17] considered the internal crack and edge crack perpendicular to the boundaries respectively. Assuming the variation of material properties is a power law function, Feng et al. [18] investigated the dynamic fracture problem of a mode III crack, which is also parallel to the boundary of an orthotropic FGM strip.

Though there are lots of papers related to the crack problem of FGMs, very few papers on the antiplane crack problem of two collinear cracks on the two sides of and perpendicular to the interface between a functionally graded orthotropic strip bonded to a homogeneous orthotropic substrate are published. Such a problem is important not only because a practical FGM may subject to an antiplane mechanical deformation, but also because it can provide a useful analogy to the most important in-plane crack problem. The material properties are assumed to vary continuously along the x-direction. The crack problem can be reduced into a system of singular integral equations after applying the integral transform technique. The influences of geometrical and physical parameters and crack interactions on the stress intensity factor were analyzed.

2. Description of the problem

Consider a graded coating bonded to a homogeneous substrate. There is two collinear cracks vertical to the interface between two mediums, as shown in Fig.1. Each strip contains an internal crack with crack length $(b_j - a_j)(j = 1, 2)$ perpendicular to the interface, respectively. The principal axes of the medium are along the x and y-axes. By adjusting the thickness, h_1 and h_2, of the two bonded materials, the model in Fig.1 can be generalized to represent a FGM coating with finite thickness bonded to a homogeneous substrate with finite or infinite thickness. In this paper we assume that the medium has orthotropic properties. To make the analysis tractable, it is further assumed that the shear moduli of the graded region is given by the following two-parameter expression

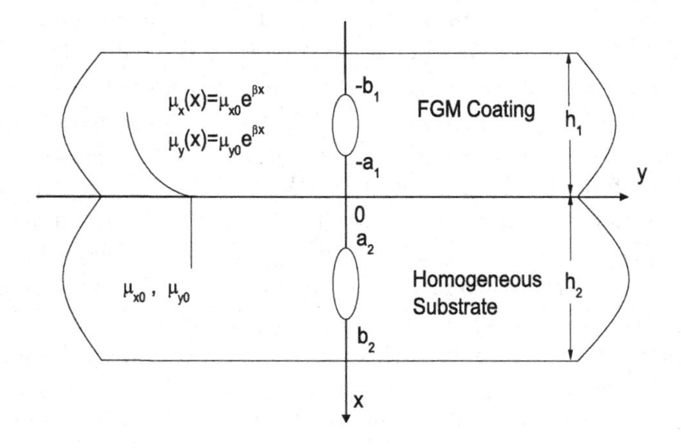

Fig. 1. The crack geometry in bonded materials.

$$\mu_x(x) = \mu_{x0} e^{\beta x}, \quad \mu_y(x) = \mu_{y0} e^{\beta x}, \tag{1}$$

where μ_{x0} and μ_{y0} are elastic constants on the surface $x = 0$. Many authors have used the assumptions shown in Eq.(1) for FGM applications [9,14,16,17].

The basic elasticity equations for nonhomogeneous materials undergoing antiplane shear deformations are

$$\frac{\partial \tau_{xz}}{\partial x} + \frac{\partial \tau_{yz}}{\partial y} = 0, \tag{2}$$

$$\tau_{xz} = \mu_x(x, y) \frac{\partial w}{\partial x}, \quad \tau_{yz} = \mu_y(x, y) \frac{\partial w}{\partial y}, \tag{3}$$

where w is the antiplane displacement, τ_{xz} and τ_{yz} are the shear stresses, and $\mu_x(x, y)$ and $\mu_y(x, y)$ are the shear moduli.

Substituting Eq.(3) to Eq.(2) becomes

$$\mu_{x0} \frac{\partial^2 w_1}{\partial x^2} + \mu_{y0} \frac{\partial^2 w_1}{\partial y^2} + \beta \mu_{x0} \frac{\partial w_1}{\partial x} = 0, \quad \text{for the graded layer,} \tag{4}$$

$$\mu_{x0} \frac{\partial^2 w_2}{\partial x^2} + \mu_{y0} \frac{\partial^2 w_2}{\partial y^2} = 0, \quad \text{for the substrate.} \tag{5}$$

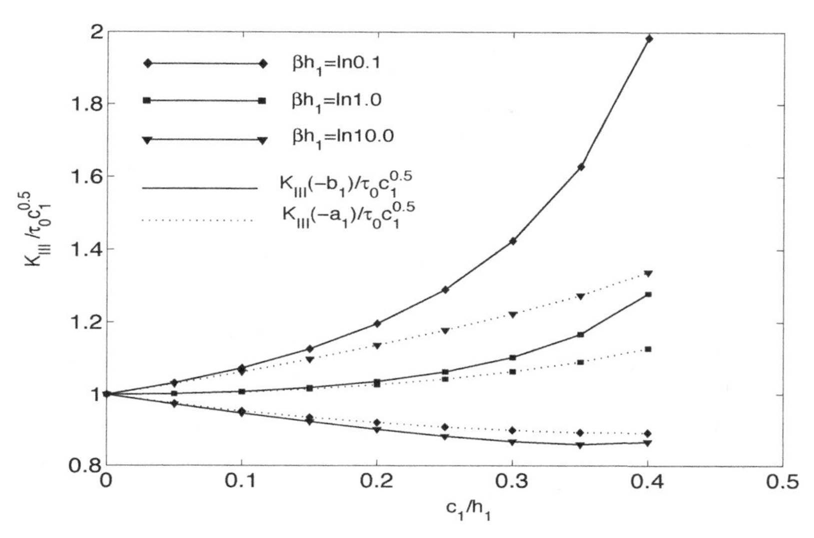

Fig. 2. Characteristic curves of the normalized SIFs vary with c_1/h_1 at different βh_1 ($h_2/h_1 = 1.0$, $c_2/h_2 = 0.0$).

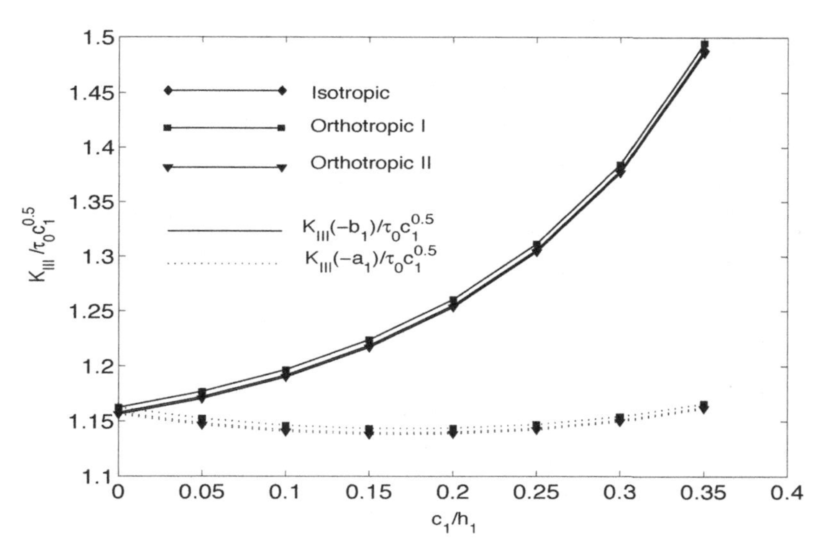

Fig. 3. Comparison of SIFs for collinear cracks between different FGM coating-substrate structures ($h_2/h_1 = 5.0$, βh_1 =ln0.5, $c_2/h_2 = 0.2$).

3. Derivation of the singular integral equations

Employing the Fourier transform on Eqs.(4) and (5), the solutions for w_1 and w_2 become

$$w_1(x,y) = \frac{1}{2\pi}\int_{-\infty}^{\infty} C_1(\alpha)e^{m_1 y}e^{-i\alpha x}d\alpha + \frac{2}{\pi}\int_0^{\infty}(A_1(\alpha)e^{n_1 x} + A_2(\alpha)e^{n_2 x})\sin\alpha y d\alpha, \tag{6}$$

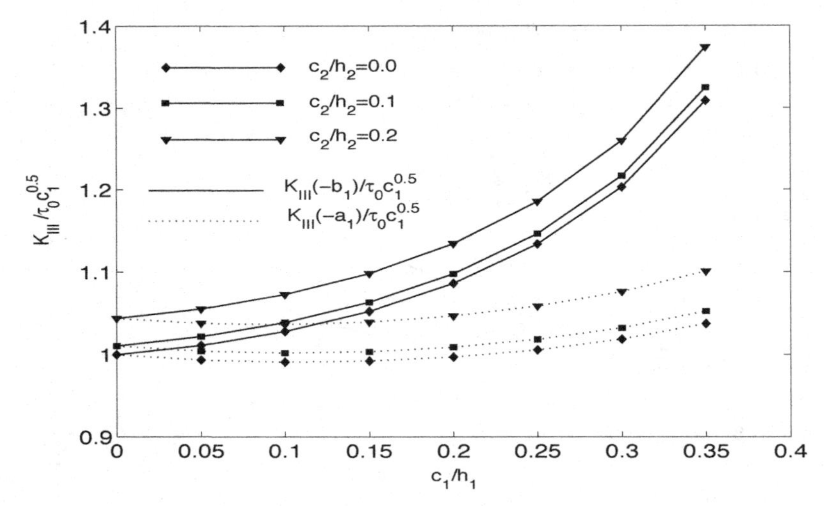

Fig. 4. Effect of c_1/h_1 on the SIFs for single carck and collinear cracks in an orthotropic I FGM coating-substrate structure ($h_2/h_1 = 1.0$, $\beta h_1 = \ln 0.5$).

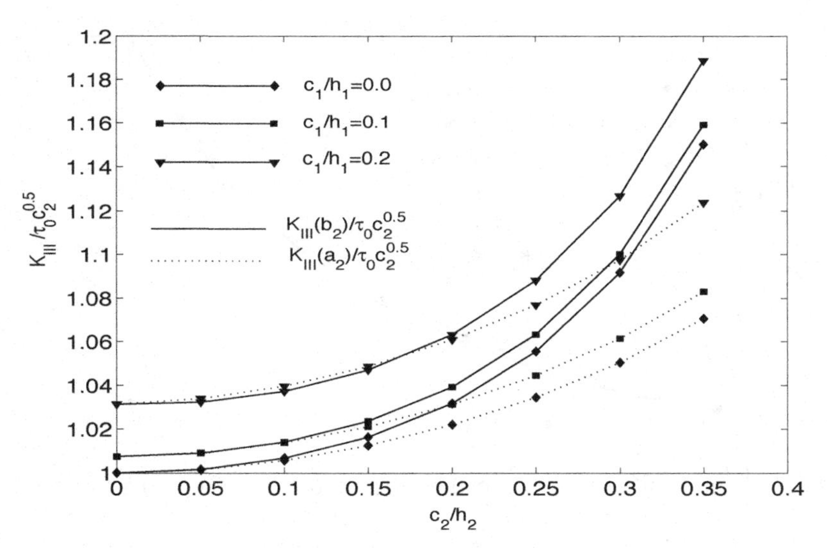

Fig. 5. Effect of c_2/h_2 on the SIFs for single crack and collinear cracks in an orthotropic I FGM coating-substrate structure ($h_2/h_1 = 1.0$, $\beta h_2 = \ln 0.5$).

$$w_2(x, y) = \frac{1}{2\pi}\int_{-\infty}^{\infty} C_2(\alpha)e^{p_1 y}e^{-i\alpha x}d\alpha + \frac{2}{\pi}\int_0^{\infty}(B_1(\alpha)e^{q_1 x} + B_2(\alpha)e^{q_2 x})\sin\alpha y d\alpha, \tag{7}$$

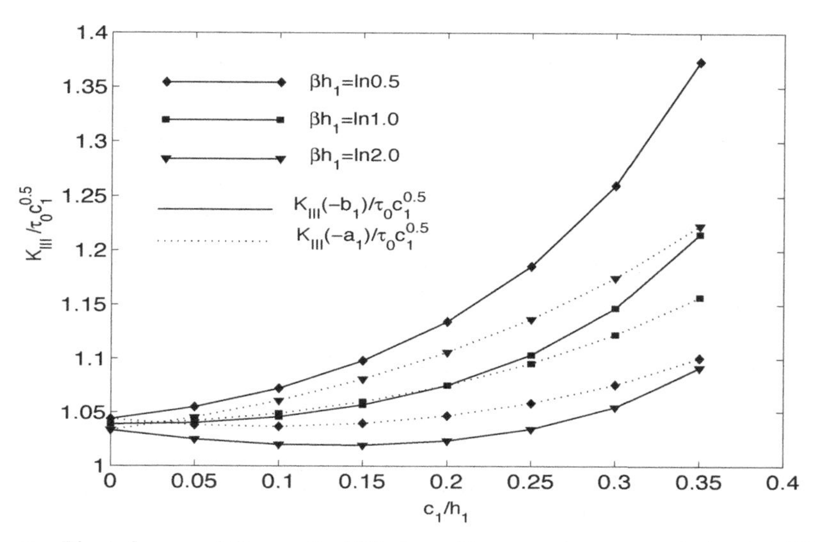

Fig. 6. The influence of βh_1 on the SIFs for collinear cracks in an orthotropic I FGM coating-substrate structure ($h_2/h_1 = 1.0$, $c_2/h_2 = 0.2$).

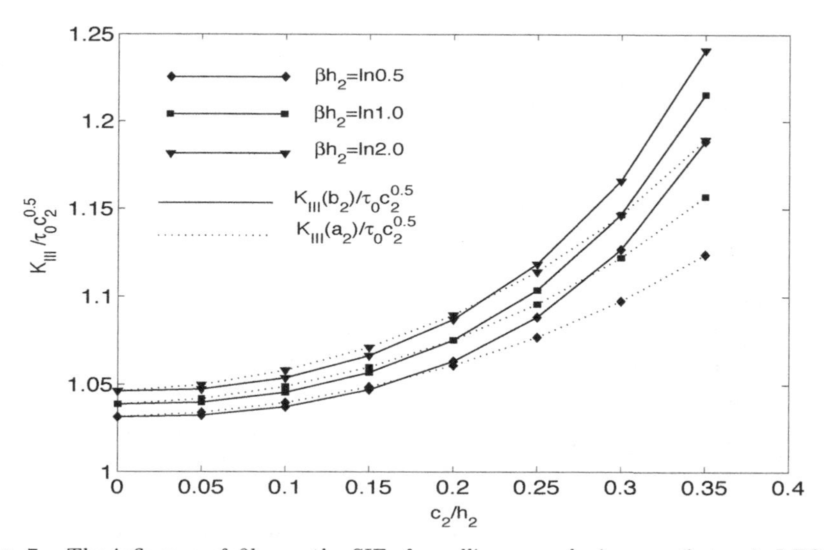

Fig. 7. The influence of βh_2 on the SIFs for collinear cracks in an orthotropic I FGM coating-substrate structure ($h_2/h_1 = 1.0$, $c_1/h_1 = 0.2$).

where

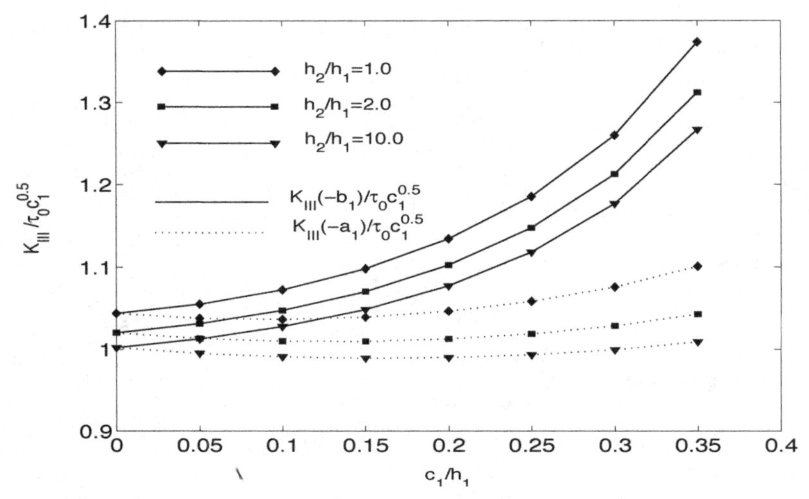

Fig. 8. The influence of h_2/h_1 on the SIFs for collinear cracks in an orthotropic I FGM coating-substrate structure (βh_1 =ln0.5, c_2/h_1 =0.2).

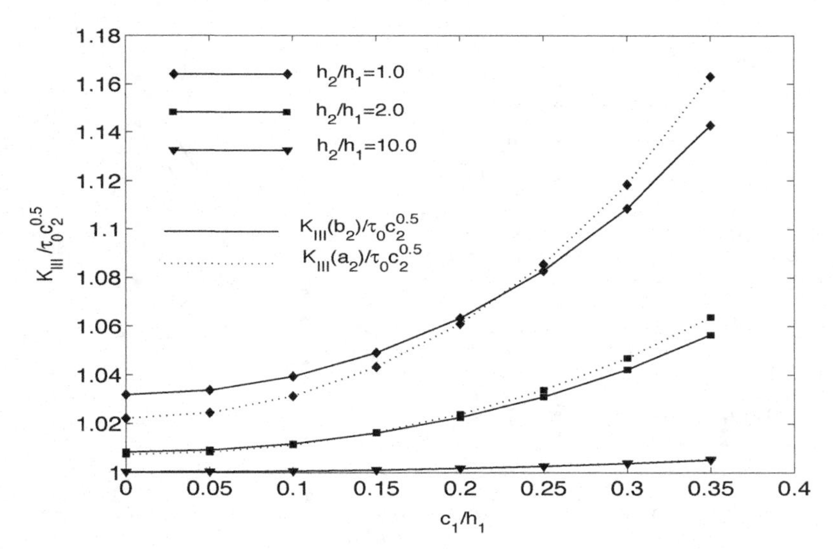

Fig. 9. The influence of h_2/h_1 on the SIFs for collinear cracks in an orthotropic I FGM coating-substrate structure (βh_1 =ln0.5, c_2/h_1 =0.2).

$$m_1 = -m_2 = -\mu_1\sqrt{\alpha^2 + i\alpha\beta}, \quad n_1 = -\frac{\beta}{2} + \frac{\sqrt{\beta^2 + 4\alpha^2\mu_2^2}}{2},$$

$$n_2 = -\frac{\beta}{2} - \frac{\sqrt{\beta^2 + 4\alpha^2\mu_2^2}}{2}, \tag{8}$$

$$p_1 = -p_2 = -\mid \alpha \mid \mu_1, \quad q_1 = -q_2 = -\mid \alpha \mid \mu_2, \quad \mu_1 = \sqrt{\frac{\mu_{x0}}{\mu_{y0}}}, \quad \mu_2 = 1/\mu_1, \tag{9}$$

and $C_1(\alpha), C_2(\alpha), ..., B_2(\alpha)$ are unknown functions to be obtained from the continuity and boundary conditions. The boundary and continuity conditions can be stated as

$$w_1(0, y) = w_2(0, y), \quad \tau_{xz1}(0, y) = \tau_{xz2}(0, y), \quad -\infty < y < \infty, \tag{10}$$

$$\tau_{xz1}(-h_1, y) = 0, \quad \tau_{xz2}(h_2, y) = 0, \quad -\infty < y < \infty, \tag{11}$$

$$w_1(x, 0) = 0, \quad -h_1 < x < -b_1, \quad -a_1 < x < 0, \tag{12}$$

$$w_2(x, 0) = 0, \quad 0 < x < a_2, \quad b_2 < x < h_2, \tag{13}$$

$$\tau_{zy1}(x, 0) = -\tau_1(x), \quad -b_1 < x < -a_1, \tag{14}$$

$$\tau_{zy2}(x, 0) = -\tau_2(x), \quad a_2 < x < b_2. \tag{15}$$

In order to derive the singular integral equation, the following two auxiliary functions are introduced

$$g_1(x_1) = \frac{\partial w_1(x_1, 0)}{\partial x_1}, \quad -b_1 < x_1 < -a_1, \tag{16}$$

$$g_2(x_2) = \frac{\partial w_2(x_2, 0)}{\partial x_2}, \quad a_2 < x_2 < b_2, \tag{17}$$

where $x = x_1$ for $x < 0$, and $x = x_2$ for $x > 0$.

By applying Fourier transform, the unknown functions $C_1(\alpha)$ and $C_2(\alpha)$ can be expressed in the form of dislocation functions as

$$C_1(\alpha) = \frac{i}{\alpha} \int_{-b_1}^{-a_1} g_1(t_1) e^{i\alpha t_1} dt_1, \quad C_2(\alpha) = \frac{i}{\alpha} \int_{a_2}^{b_2} g_2(t_2) e^{i\alpha t_2} dt_2. \tag{18}$$

After employing the continuity conditions Eqs. (10) and (11) and the Fourier inverse transform, the following can be obtained

$$A_1 + A_2 - B_1 - B_2 = \frac{1}{2\pi} \int_{-\infty}^{\infty} \left\{ \frac{\alpha}{p_1^2 + \alpha^2} C_2(\rho) - \frac{\alpha}{m_1^2 + \alpha^2} C_1(\rho) \right\} d\rho, \tag{19}$$

$$n_1A_1+n_2A_2-q_1B_1-q_2B_2=\frac{-1}{2\pi}\int_{-\infty}^{\infty}i\rho\left\{\frac{\alpha}{p_1^2+\alpha^2}C_2(\rho)-\frac{\alpha}{m_1^2+\alpha^2}C_1(\rho)\right\}d\rho, \tag{20}$$

$$n_1\exp(-n_1h_1)A_1+n_2\exp(-n_2h_1)A_2=\frac{1}{2\pi}\int_{-\infty}^{\infty}i\rho\frac{i\alpha\rho\exp(i\rho h_1)}{m_1^2+\alpha^2}C_1(\rho)d\rho, \tag{21}$$

$$q_1\exp(q_1h_2)B_1+q_2\exp(q_2h_2)B_2=\frac{1}{2\pi}\int_{-\infty}^{\infty}i\rho\frac{i\alpha\rho\exp(-i\rho h_2)}{p_1^2+\alpha^2}C_2(\rho)d\rho. \tag{22}$$

By solving Eqs. (19)–(22) for functions A_1, A_2, B_1 and B_2, we have

$$A_1(\alpha)=\frac{D_{11}}{|D|}R_1-\frac{D_{21}}{|D|}R_2+\frac{D_{31}}{|D|}R_3-\frac{D_{41}}{|D|}R_4, \tag{23}$$

$$A_2(\alpha)=\frac{-D_{12}}{|D|}R_1+\frac{D_{22}}{|D|}R_2-\frac{D_{32}}{|D|}R_3+\frac{D_{42}}{|D|}R_4, \tag{24}$$

$$B_1(\alpha)=\frac{D_{13}}{|D|}R_1-\frac{D_{23}}{|D|}R_2+\frac{D_{33}}{|D|}R_3-\frac{D_{43}}{|D|}R_4, \tag{25}$$

$$B_2(\alpha)=\frac{-D_{14}}{|D|}R_1+\frac{D_{24}}{|D|}R_2-\frac{D_{34}}{|D|}R_3+\frac{D_{44}}{|D|}R_4, \tag{26}$$

where

$$R_j(\alpha)=\int_{-b_1}^{-a_1}g_1(t_1)R_{j1}dt_1+\int_{a_2}^{b_2}g_2(t_2)R_{j2}dt_2,\quad j=1-4, \tag{27}$$

$$R_{11}=-\frac{\alpha e^{-n_2t_1}}{\mu_1^2n_2(n_1-n_2)}+\frac{\alpha}{2\mu_1^2n_1n_2},\quad R_{12}=\frac{\alpha e^{-q_2t_2}}{\mu_1^2q_2(q_2-q_1)}+\frac{\alpha}{2\mu_1^2q_1q_2}, \tag{28}$$

$$R_{21}=\frac{\alpha e^{-n_2t_1}}{\mu_1^2(n_1-n_2)},\quad R_{22}=\frac{\alpha e^{-q_2t_2}}{\mu_1^2(q_2-q_1)}, \tag{29}$$

$$R_{31}=\frac{\alpha e^{-n_1(t_1+h_1)}}{\mu_1^2(n_2-n_1)},\quad R_{32}=0,\quad R_{41}=0,\quad R_{42}=\frac{\alpha e^{-q_2(h_2-t_2)}}{\mu_1^2(q_1-q_2)}, \tag{30}$$

with $D=(d_{ij})(i,j=1-4)$ is a square coefficient matrix of order 4 in Eqs.(23)-(26), $|D|$ is the determinant of D, and D_{ij} are the cofactors of the elements $d_{ij}(i,j=1-4)$.

From the load conditions (14) and (15) on the crack surfaces, the following singular integral equations can be obtained

$$\frac{1}{\pi}\int_{-b_1}^{-a_1}\frac{g_1(t_1)}{t_1-x_1}dt_1+\int_{-b_1}^{-a_1}g_1(t_1)K_{11}(x_1,t_1)dt_1+\int_{-b_1}^{-a_1}g_1(t_1)K_{12}(x_1,t_1)dt_1 + \int_{a_2}^{b_2}g_2(t_2)K_{13}(x_1,t_2)dt_2=-\frac{\tau_1(x_1)}{\mu_1\mu_{y0}e^{\beta x_1}}, \tag{31}$$

$$\frac{1}{\pi}\int_{a_2}^{b_2}\frac{g_2(t_2)}{t_2-x_2}dt_2+\int_{a_2}^{b_2}g_2(t_2)K_{21}(x_2,t_2)dt_2+\int_{-b_1}^{-a_1}g_1(t_1)K_{22}(x_2,t_1)dt_1 + \int_{a_2}^{b_2}g_2(t_2)K_{23}(x_2,t_2)dt_2=-\frac{\tau_2(x_2)}{\mu_1\mu_{y0}}, \tag{32}$$

where

$$\begin{aligned} K_{11}(x_1,t_1)=\frac{1}{2\mu_1}\Bigg\{&\int_0^A\frac{i[m_1(\alpha)-m_1(-\alpha)]}{\alpha}\cos[\alpha(t_1-x_1)]d\alpha \\ &+\int_A^\infty\frac{i[m_1(\alpha)-m_1(-\alpha)+i\mu_1\beta]}{\alpha}\cos[\alpha(t_1-x_1)]d\alpha \\ &+\int_A^\infty\frac{\mu_1\beta}{\alpha}\cos[\alpha(t_1-x_1)]d\alpha \\ &+\int_0^\infty\frac{-[m_1(\alpha)+m_1(-\alpha)+2\mu_1\alpha]}{\alpha}\sin[\alpha(t_1-x_1)]d\alpha\Bigg\}, \end{aligned} \tag{33}$$

$$K_{12}(x_1,t_1)=\frac{2}{\mu_1}\int_0^\infty\sum_{j=1}^{4}\left\{\alpha\left(\frac{D_{j1}}{|D|}e^{n_1x_1}-\frac{D_{j2}}{|D|}e^{n_2x_1}\right)(-1)^{j+1}R_{j1}\right\}d\alpha, \tag{34}$$

$$K_{13}(x_1,t_2)=\frac{2}{\mu_1}\int_0^\infty\sum_{j=1}^{4}\left\{\alpha\left(\frac{D_{j1}}{|D|}e^{n_1x_1}-\frac{D_{j2}}{|D|}e^{n_2x_1}\right)(-1)^{j+1}R_{j2}\right\}d\alpha, \tag{35}$$

$$K_{21}(x_2,t_2)=0, \tag{36}$$

$$K_{22}(x_2,t_1)=\frac{2}{\mu_1}\int_0^\infty\sum_{j=1}^{4}\left\{\alpha\left(\frac{D_{j3}}{|D|}e^{q_1x_2}-\frac{D_{j4}}{|D|}e^{q_2x_2}\right)(-1)^{j+1}R_{j1}\right\}d\alpha, \tag{37}$$

$$K_{23}(x_2,t_2)=\frac{2}{\mu_1}\int_0^{\infty}\sum_{j=1}^{4}\left\{\alpha\left(\frac{D_{j3}}{|D|}e^{q_1x_2}-\frac{D_{j4}}{|D|}e^{q_2x_2}\right)(-1)^{j+1}R_{j2}\right\}d\alpha, \tag{38}$$

where A is an arbitrary positive constant.

The solutions of the singular integral equations Eqs. (31) and (32) with the Cauchy type kernel are

$$g_1(t_1)=\frac{G_1(t_1)}{\sqrt{(t_1+a_1)(t_1+b_1)}},\quad g_2(t_2)=\frac{G_2(t_2)}{\sqrt{(t_2-a_2)(t_2-b_2)}}. \tag{39}$$

Consider two collinear internal cracks in a FGM coating bonded to a homogeneous substrate (Fig.1). For this case, note that the singular term $\dfrac{1}{t-x}$ is associated with two embedded cracks in two materials and leads to the standard square-root singularity for the unknown function $g_1(t_1)$ and $g_2(t_2)$. It may easily be shown that for $-b_1>-h_1$ and $-a_1<0$, and $a_2>0$ and $b_2<h_2$, $K_{ij}(i=1,2,j=1-3)$ remain bounded in the closed interval $-b_1\leq(x_1,t_1)\leq -a_1$ and $a_2\leq(x_2,t_2)\leq b_2$, respectively.

In order to change Eqs. (31) and (32) into standard form, we define dimensionless quantities

$$\begin{gathered}t_1=u_1(b_1-a_1)/2-(b_1+a_1)/2,\quad x_1=r_1(b_1-a_1)/2-(b_1+a_1)/2,\\ t_2=u_2(b_2-a_2)/2+(b_2+a_2)/2,\quad x_2=r_2(b_2-a_2)/2+(b_2+a_2)/2,\\ f_j(u_j)=g_j(t_j),\quad F_j(u_j)=G_j(t_j),\quad j=1,2,\end{gathered} \tag{40}$$

then Eq.(39) becomes

$$f_j(u_j)=\frac{F_j(u_j)}{\sqrt{1-u_j^2}},\quad j=1,2, \tag{41}$$

where $F_j(u_j)(j=1,2)$ are bounded functions, which can be expressed as

$$F_1(u_1)=\sum_{n=0}^{\infty}A_nT_n(u_1),\quad F_2(u_2)=\sum_{n=0}^{\infty}B_nT_n(u_2), \tag{42}$$

where T_n is the Chebyshev polynomial of the first kind and A_n and B_n are unknown constants. $F_1(u_1)$ and $F_2(u_2)$ must fulfill the following single-valuedness as

$$\int_{-1}^{1} F_1(u_1)/\sqrt{1-u_1^2}\, du_1 = 0, \quad \int_{-1}^{1} F_2(u_2)/\sqrt{1-u_2^2}\, du_2 = 0. \tag{43}$$

The results of stress intensity factors (SIFs) become

$$K_{III}(-b_1) = \mu_1 \mu_{y0} e^{-\beta b_1} \sqrt{(b_1 - a_1)/2} \sum_{n=1}^{\infty} (-1)^n A_n, \tag{44}$$

$$K_{III}(-a_1) = -\mu_1 \mu_{y0} e^{-\beta a_1} \sqrt{(b_1 - a_1)/2} \sum_{n=1}^{\infty} A_n, \tag{45}$$

$$K_{III}(a_2) = \mu_1 \mu_{y0} \sqrt{(b_2 - a_2)/2} \sum_{n=1}^{\infty} (-1)^n B_n, \tag{46}$$

$$K_{III}(b_2) = -\mu_1 \mu_{y0} \sqrt{(b_2 - a_2)/2} \sum_{n=1}^{\infty} B_n. \tag{47}$$

4. Results and discussion

In the following numerical computations, the shear loads τ_1 and τ_2 applied on the crack surfaces are assumed to be equal when cracks in FGM coating and homogeneous substrate, respectively. τ_1 and τ_2 are constant tractions, which are also assumed to equal τ_0. Numerical calculations were carried out for three kinds of materials (i.e. isotropic, orthotropic I and II), whose mechanical properties can be found in [14]. The only difference between orthotropic I and II is that their reinforced directions are perpendicular to each other. For internal cracks problem, the stress intensity factors at the crack tips are normalized by $\tau_0\sqrt{c_1}$ or $\tau_0\sqrt{c_2}$, where $c_1 = (b_1 - a_1)/2, c_2 = (b_2 - a_2)/2$. First of all, let us verify the validity of the analytical solution procedure. We compare the normalized SIFs of a central crack in a isotropic FGM coating bonded to a homogeneous strip, namely $h_2/h_1 = 1.0$ and $\frac{(b_2-a_2)}{2} = 0$, with the results provided by Yong and Zhou [19]. Fig.2 illustrates the variation of normalized SIFs for bonded isotropic medium with βh_1 =ln0.1, ln1.0 and ln10.0. The numerical result is in good agreement with that presented by Yong and Zhou [19].

Consider a practical case that a FGM coating is bonded to a homogeneous substrate with internal cracks. Fig.3 compares the effect of material properties on the SIFs for internal cracks in different FGM coating-substrate structures. The isotropic strip has a very similar curve of SIFs to that of orthotropic I strip and the orthotropic II strip. It can be found that the material properties (i.e., isotropic, orthotropic I and orthotropic II) will influence the SIFs. The value of SIFs of the orthotropic I strip is maximal. Therefore, it is necessary to take the orthotropy into account when analyzing the mechanical behaviors of FGMs.

Figs. 4 and 5 show the influence of the crack length c_1/h_1 and c_2/h_2 on the SIFs for two internal cracks in an orthotropic I FGM coating-substrate structure. Obviously, the crack length will influence the SIFs, and this further shows the necessity of taking the internal cracks into account when analyzing the mechanical behaviors of FGMs.

Figs. 6 and 7 illustrate the variation of the normalized SIFs with different nonhomogeneous constant βh_1 and βh_2 in an orthotropic I FGM coating-substrate structure. It can be seen that $K_{III}(-a_1)/\tau_0 c_1^{0.5}$ increases and $K_{III}(-b_1)/\tau_0 c_1^{0.5}$ decreases with the increasing of βh_1. However, both $K_{III}(a_2)/\tau_0 c_2^{0.5}$ and $K_{III}(b_2)/\tau_0 c_2^{0.5}$ increase with the increasing of βh_2. With the increasing of c_2/h_1, both $K_{III}(a_2)/\tau_0 c_2^{0.5}$ and $K_{III}(b_2)/\tau_0 c_2^{0.5}$ generally increase. When β is positive, the material properties on the crack tip $-a_1$ and b_2 are stronger than ones on the crack tip $-b_1$ and a_2, and $K_{III}(-a_1)/\tau_0 c_1^{0.5}$ and $K_{III}(b_2)/\tau_0 c_2^{0.5}$ are greater than $K_{III}(-b_1)/\tau_0 c_1^{0.5}$ and $K_{III}(a_2)/\tau_0 c_2^{0.5}$. These results show where the stronger the material properties are, the larger the normalized stress intensity factors at the crack tip.

Figs. 8 and 9 show the influence of different thickness ratio h_2/h_1 on SIFs for an orthotropic I FGM coating-substrate structure when βh_1=ln0.5. It can be found that the SIFs of crack tip decrease with an increasing of h_2/h_1. For different h_2/h_1, $K_{III}(-b_1)/\tau_0 c_1^{0.5}$ are always greater than $K_{III}(-a_1)/\tau_0 c_1^{0.5}$. But $K_{III}(b_2)/\tau_0 c_2^{0.5}$ are greater than $K_{III}(a_2)/\tau_0 c_2^{0.5}$ when $c_1/h_1 < 0.2$, and $K_{III}(a_2)/\tau_0 c_2^{0.5}$ are greater than $K_{III}(b_2)/\tau_0 c_2^{0.5}$ when $c_1/h_1 > 0.2$.

Acknowledgements

This work is supported by the National Natural Science Foundation of China (51061015,11261045) and research fund for the doctoral program of higher education of China (20116401110002).

References

1. Ang, W.T., Clements, D.L., 1987. On some crack problems for inhomogeneous elastic materials. International Journal of Solid and Structures 23, 1089–1104.

2. Chen, Y.F., Erdogan, F., 1996. The interface crack problem for a nonhomogeneous coating bonded to a homogeneous substrate. Journal of The Mechanics and Physics of Solids 44, 771–787.
3. Li, C.Y., Weng, G.J., 2001. Dynamic stress intensity factors of a cylindrical interface crack with a functionally graded interlayer. Mechanics of Materials 33, 325–333.
4. Noda, N., Wang B.L., 2002. The collinear cracks in an inhomogeneous medium subjected to transient load. Acta Mechanica 153, 1–13.
5. Huang, G.Y., Wang, Y.S., Yu, S.W., 2004. Fracture analysis of a functionally graded interfacial zone under plane deformation. International Journal of Solid and Structures 41, 731–743.
6. Ding, S.H., Li, X., 2008a. An anti-plane shear crack in bonded functionally graded piezoelectric materials under electromechanical loading. Computational Materials Science 43, 337–344.
7. Ding, S.H., Li, X., 2008b. Periodic cracks in a functionally graded piezoelectric layer bonded to a piezoelectric half-plane. Theoretical and Applied Fracture Mechanics 49, 313–320.
8. Li, Y.D., Jia B., Zhang, N., Tang L.Q., Dai, Y., 2006. Dynamic stress intensity factor of the weak/micro-discontinuous interface crack of a FGM coating. International Journal of Solids and Structures 43, 4795–4809.
9. Li, Y.D., Kang Y.L., Yao D., 2008. Dynamic stress intensity factors of two collinear mode-III cracks perpendicular to and on the two sides of a bi-FGM weak-discontinuous interface. European Journal of Mechanics A/Solids 27 , 808–823.
10. Chen, Y.J., Chue, C.H., 2010. Mode III fracture problem of a cracked FGPM surface layer bonded to a cracked FGPM substrate. Archive of Applied Mechanics 80, 285–305.
11. Ding, S.H., Li, X., 2011. Mode-I crack problem for functionally graded layered structures. Internation Journal of Fracture 168, 209–226.
12. Sampath, S., Herman, H., Shimoda, N., Saito, T., 1995. Thermal spray processing of FGMs. Mrs Bulletin 20, 27–31.
13. Kaysser,W.A., Ilschner, B., 1995. FGM research activities in Europe. Mrs Bulletin 20, 22–26.
14. Chen, J., Liu, Z.X., 2005. Transient response of a mode III crack in an orthotropic functionally graded strip. European Journal of Mechanics A/Solids 24, 325–336.
15. Wang B.L., Mai Y.-W., 2006. Periodic antiplane cracks in graded coatings under static or transient loading. Journal of Applied Mechanics 73, 134–142.
16. Guo, L.C.,Wu, L.Z., Zeng, T.,Ma, L., 2004. Mode I crack problem for a functionally graded orthotropic strip. European Journal of Mechanics A-Solids 23, 219–234.
17. Guo, L.C.,Wu, L.Z., Zeng, T., 2005. The dynamic response of an edge crack in a functionally graded orthotropic strip. Mechanics Research Communications 32, 385–400.

18. Feng,W.J., Zhang, Z.G., Zou, Z.Z., 2003. Impact failure prediction of mode III crack in orthotropic functionally graded strip. Theoretical and Applied Fracture Mechanics 40, 97–104.
19. Yong, H.D., Zhou, Y.H, 2006. Analysis of a mode III crack problem in a functionally graded coating–substrate system with finite thickness. International Journal of Fracture 141, 459–467

A KIND OF RIEMANN BOUNDARY VALUE PROBLEM FOR TRIHARMONIC FUNCTIONS IN CLIFFORD ANALYSIS

LONGFEI GU

School of Mathematics and Statistics, Wuhan University,
430072/Wuhan Hubei, P. R. China
E-mail: gLfcooL@163.com
www.whu.edu.cn

In this article, we mainly deal with the boundary value problem for triharmonic function with value in universal Clifford algebra:

$$\begin{cases} \Delta^3[u](x) = 0 & x \in \mathcal{R}^n \backslash \partial\Omega \\ u^+(x) = u^-(x)G(x) + g(x), & x \in \partial\Omega \\ (D^j u)^+(x) = (D^j u)^-(x)A_j + f_j(x), & x \in \partial\Omega \\ u(\infty) = 0 \end{cases}$$

where $(j = 1, \ldots, 5)$ $\partial\Omega$ is a Liapunov surface in $\mathcal{R}^n$, the Dirac operator $D = \sum_{k=1}^{n} e_k \frac{\partial}{\partial x_k}$, $u(x) = \sum_A e_A u_A(x)$ are unknown functions with values in an universal Clifford algebra $Cl(V_{n,n})$. Under some hypotheses, it is proved that the boundary value problem has a unique solution.

Keywords: Symplectic schemes; elastodynamic; anti-plane shear; boundary integral method.

1. Introduction and preliminaries

Riemann boundary value problems theory in complex plane has been systematically developed in [7,8]. It is an interesting topic to generalize the classical Riemann boundary value problems theory to Clifford analysis. In [4,6,10,12], etc, many interesting results about boundary value problem and Riemann Hilbert problems for monogenic functions in Clifford analysis are presented. In [11], Green's function for the Dirichlet problem for polyharmonic equations were studied. In this article, with the aim to study the Riemann boundary value problem for triharmonic functions. At first, based on the higher order Cauchy integral representation formulas in [5, 9] and the Plemelj formula, we give some properties of triharmonic functions in

Clifford analysis, for example, the mean value theorem, the Painlevé theorem, etc. Furthermore, on the basis of the above results, we consider the following Riemann boundary value problems:

$$\begin{cases} \Delta^3[u](x) = 0 & x \in \mathcal{R}^n \backslash \partial\Omega \\ u^+(x) = u^-(x)A + g(x), & x \in \partial\Omega \\ (D^j u)^+(x) = (D^j u)^-(x)A_j + f_j(x), & x \in \partial\Omega \\ |u(\infty)| \leq M \end{cases} \tag{1}$$

and

$$\begin{cases} \Delta^3[u](x) = 0 & x \in \mathcal{R}^n \backslash \partial\Omega \\ u^+(x) = u^-(x)G(x) + g(x), & x \in \partial\Omega \\ (D^j u)^+(x) = (D^j u)^-(x)A_j + f_j(x), & x \in \partial\Omega \\ u(\infty) = 0 \end{cases} \tag{2}$$

where $(j = 1, \ldots, 5)$.

In (1) and (2), A and A_j are invertible constants, we denote the inverse elements as A^{-1} and A_j^{-1}. $u(x)$, $(D^j u)(x)$, $g(x)$, $f_j(x) \in H^\beta(\partial\Omega, Cl(V_{n,n}))$, $j = 1, \ldots, 5$, $0 < \beta \leq 1$. The explicit solutions for (1) is shown and the boundary value problem (2) is solvable under some hypotheses.

Let $V_{n,s}(0 \leq s \leq n)$ be an n-dimensional $(n \geq 1)$ real linear space with basis $\{e_1, e_2, \ldots, e_n\}$, $Cl(V_{n,s})$ be the universal Clifford algebra over $V_{n,s}$. For more information on $Cl(V_{n,s})(0 \leq s \leq n)$, we refer to [1, 2, 3].

Throughout this article, suppose Ω be an open, bounded non-empty subset of $\mathcal{R}^n$ with a Liapunov boundary $\partial\Omega$, denote $\Omega^+ = \Omega$, $\Omega^- = \mathcal{R}^n \backslash \overline{\Omega}$. In this article, for simplicity, we shall only consider the case of $s = n$, the operator D which is written as

$$D = \sum_{k=1}^{n} e_k \frac{\partial}{\partial x_k} : C^r(\Omega, Cl(V_{n,n})) \to C^{r-1}(\Omega, Cl(V_{n,n}))$$

Let u be a function with value in $Cl(V_{n,n})$ defined in Ω, the operator D acts on the function u from the left and from the right being governed by the rule

$$D[u] = \sum_{k=1}^{n} \sum_{A} e_k e_A \frac{\partial u_A}{\partial x_k}, \qquad [u]D = \sum_{k=1}^{n} \sum_{Ae_A} e_k \frac{\partial u_A}{\partial x_k}.$$

Definition 1.1. A compact surface Γ is called Liapunov surface with Hölder exponent α, if the following conditions are satisfied:

- At each point $x \in \Gamma$ there is a tangential space.

- There exists a number r, such that for any point $x \in \Gamma$ the set $\Gamma \bigcap B_r(x)$(Liapunov ball) is connected and parallel lines to the outer normal $\alpha(x)$ intersect at not more than one point.
- The normal $\alpha(x)$ is Hölder continuous on Γ, i.e. there are constants $C > 0$ and $0 < \alpha \leq 1$ such that for $x, y \in \Gamma$

$$|\alpha(x) - \alpha(y)| \leq C|x - y|^\alpha.$$

Let Ω be an open nonempty subset of $\mathcal{R}^n$ with a Liapunov boundary, $u(x) = \sum\limits_A e_A u_A(x)$, where $u_A(x)$ are real functions, $u(x)$ is called a Hölder continuous functions on $\overline{\Omega}$ if the following condition is satisfied,

$$|u(x_1) - u(x_2)| = [\sum_A |u_A(x_1) - u_A(x_2)|^2]^{1/2} \leq C|x_1 - x_2|^\alpha,$$

where for any $x_1, x_2 \in \overline{\Omega}$, $x_1 \neq x_2$, $0 < \alpha \leq 1$, C is a positive constant independent of x_1, x_2.

Let $H^\alpha(\partial\Omega, Cl(V_{n,n}))$ denote the set of Hölder continuous functions with values in $Cl(V_{n,n})$ on $\partial\Omega$ (the Hölder exponent is α, $0 < \alpha < 1$). We denote the norm in $H^\alpha(\partial\Omega, Cl(V_{n,n}))$ as

$$\|u\|_{(\alpha,\partial\Omega)} = \|u\|_\infty + \|u\|_\alpha \tag{3}$$

where $\|u\|_\infty := \sup_{x\in\partial\Omega} |u(x)|$, $\|u\|_\alpha := \sup\limits_{\substack{x_1,x_2\in\partial\Omega \\ x_1\neq x_2}} \frac{|u(x_1)-u(x_2)|}{|x_1-x_2|^\alpha}$.

Lemma 1.1. *The Hölder space $H^\alpha(\partial\Omega, Cl(V_{n,n}))$ is a Banach space with norm (3).*

Denote the fundamental solutions of D^j $(j = 1, 2, \ldots, 6)$ by

$$\begin{cases} H_1(x) = \frac{1}{\omega_n}\frac{\mathbf{x}}{\rho^n(x)} \\ H_2(x) = \frac{1}{2-n}\frac{1}{\omega_n}\frac{1}{\rho^{n-2}(x)} \\ H_3(x) = \frac{1}{2(2-n)}\frac{1}{\omega_n}\frac{\mathbf{x}}{\rho^{n-2}(x)} \\ H_4(x) = \frac{1}{2(2-n)(4-n)}\frac{1}{\omega_n}\frac{1}{\rho^{n-4}(x)} \\ H_5(x) = \frac{1}{8(2-n)(4-n)}\frac{1}{\omega_n}\frac{\mathbf{x}}{\rho^{n-4}(x)} \\ H_6(x) = \frac{1}{8(2-n)(4-n)(6-n)}\frac{1}{\omega_n}\frac{1}{\rho^{n-6}(x)} \end{cases} \tag{4}$$

where $\rho(x) = (\sum_{i=1}^n x_i^2)^{\frac{1}{2}}$ and ω_n denotes the area of the unit sphere in $\mathcal{R}^n(n \geq 3, n \neq 4, n \neq 6)$.

We shall introduce the following operators:

$$(F_{\partial\Omega}u)(x) = \int_{\partial\Omega} 2H_1(y - x)d\sigma_y u(y),\ x \in \mathcal{R}^n \setminus \partial\Omega, \tag{5}$$

$$(S_{\partial\Omega}u)(x) = \int_{\partial\Omega} 2H_1(y-x)d\sigma_y u(y),\ x \in \Omega \tag{6}$$

where $u \in H^\alpha(\partial\Omega, Cl(V_{n,n}))$.

Lemma 1.2. [10] *The integral operator $S_{\partial\Omega}$ be as in (6) is a bounded linear operator mapping from the functions space $H^\alpha(\partial\Omega, Cl(V_{n,n}))$ into itself, i.e. for any $u \in H^\alpha(\partial\Omega, Cl(V_{n,n}))$, there exists a positive constant M which doer not depend on u such that $\|S_{\partial\Omega}u\|_{(\alpha,\partial\Omega)} \leq M\|u\|_{(\alpha,\partial\Omega)}$.*

2. Some properties for triharmonic functions

Theorem 2.1. (***Gauss-mean value formula for triharmonic***)[12] *Suppose $\Delta^3[u] = 0$ in $\mathcal{R}^n$, then for $\forall x \in \mathcal{R}^n$,*

$$u(x) = \frac{1}{\omega_n R^{n-1}} \int_{\partial B(x,R)} u(y)dS - \frac{R^2}{2n}\Delta[u](x) - \frac{R^4}{8n(n+2)}\Delta^2[u](x) \tag{7}$$

or

$$u(x) = \frac{n}{\omega_n R^n} \int_{B(x,R)} u(y)dV - \frac{R^2}{2(n+2)}\Delta[u](x) - \frac{R^4}{8(n+2)(n+4)}\Delta^2[u](x) \tag{8}$$

where ω_n denotes the area of the unit sphere in $\mathcal{R}^n$.

Corollary 2.1. *Suppose $\Delta^3[u] = 0$ in $\mathcal{R}^n$ and $|u(x)| = O(|x|)$ $(|x| \to \infty)$, then $\Delta[u] = 0$ and $\Delta^2[u] = 0$ in $\mathcal{R}^n$.*

Corollary 2.2. *Suppose $\Delta^3[u] = 0$ in $\mathcal{R}^n$ and $u(x)$ is bounded in $\mathcal{R}^n$ then $u(x) \equiv C$.*

Theorem 2.2. *Suppose $\Delta^3[u] = 0$ in $\mathcal{R}^n \backslash \partial\Omega$, and for $x \in \partial\Omega$, $u \in C^6(\mathcal{R}^n \backslash \partial\Omega, Cl(V_{n,n}))$, $[u]^+(x) = [u]^-(x) \in C^5(\partial\Omega, Cl(V_{n,n}))$, moreover $D^j[u]^+(x) = D^j[u]^-(x) \in H^{\alpha_j}(\partial\Omega, Cl(V_{n,n})), 0 < \alpha_j \leq 1$, where $j = 0, 1, \ldots, 5$. then $\Delta^3[u] = 0$ in $\mathcal{R}^n$.*

Theorem 2.3. *Let $u \in C^6(\Omega^-, Cl(V_{n,n})) \bigcap C^5(\overline{\Omega^-}, Cl(V_{n,n})), \Delta^3[u] = 0$ in Ω^-, $D^j[u](x) \in H^{\alpha_j}(\partial\Omega, Cl(V_{n,n})), 0 < \alpha_j \leq 1$, $(j = 0, 1, \ldots, 5)$ and $|u(x)| = O(1)$ $(|x| \to \infty)$, then for $x \in \Omega^-$*

$$u(x) = \sum_{j=0}^{5} (-1)^{j+1} \int_{\partial\Omega} H_{j+1}(y-x)d\sigma_y D^j[u](y) + C \tag{9}$$

where $H_j(y-x)$ be as in (4), C is a constant.

Proof. For $y \in \partial\Omega$, denote $D^j[u](y) = -\Psi_j(y), (j = 0, 1, \ldots, 5)$. For $x \in \mathcal{R}^n \backslash \partial\Omega$, denoting

$$\Theta(x) = \sum_{j=0}^{5}(-1)^j \int_{\partial\Omega} H_{j+1}(y-x)d\sigma_y \Psi_j(y) \tag{10}$$

and

$$u^*(x) = \begin{cases} -\Theta(x), & x \in \Omega^+ \\ u(x) - \Theta(x), & x \in \Omega^- \end{cases} \tag{11}$$

then $\Delta^3[u^*] = 0$ in $\mathcal{R}^n \backslash \partial\Omega$. By using plemelj formula, combining with weak singularity of $H_j(x)$ $(j \geq 2)$. we obtain that

$$D^j[u^*]^+(x) = D^j[u^*]^-(x) \in H^{\alpha_j'}(\partial\Omega, Cl(V_{n,n})) \tag{12}$$

where $0 < \alpha_j' \leq 1, (j = 0, 1, \ldots, 5)$. By using Theorem 2.2, we have $\Delta^3[u^*] = 0$ in $\mathcal{R}^n$. It is clear that we have $|u^*(x)| = O(1)$ $(|x| \to \infty)$. In view of Corollary 2.2, then the results follows. □

Corollary 2.3. *Let* $u \in C^6(\Omega^-, Cl(V_{n,n})) \bigcap C^5(\overline{\Omega^-}, Cl(V_{n,n})), \Delta^3[u] = 0$ *in* Ω^-, $D^j[u](x) \in H^{\alpha_j}(\partial\Omega, Cl(V_{n,n})), 0 < \alpha_j \leq 1,$ $(j = 0, 1, \ldots, 5)$ *and* $|u(x)| = O(1)$ $(|x| \to \infty)$, *then*

$$D^j[u](\infty) = 0, \ j = 1, \ldots, 5. \tag{13}$$

Remark 2.1. When the condition $|u(x)| = O(1)$ $(|x| \to \infty)$ in Corollary 2.3 is replaced by $u(\infty) = 0$, then the results to Corollary 2.3 are still valid.

3. Riemann boundary value problem for triharmonic functions

In this section, we shall consider the Riemann boundary value problem (1), the explicit expression of the solution is given.

Theorem 3.1. *The Riemann boundary value problem* (1) *is solvable and the solution can be written as*

$$u(x) = \begin{cases} \sum_{i=1}^{6} \Psi_i(x) + C, & x \in \Omega, \\ \sum_{i=1}^{6} \Psi_i(x) + CA^{-1}, & x \in \Omega^-, \end{cases} \tag{14}$$

where

$$\Psi_1(x) = \begin{cases} -\int_{\partial\Omega} H_6(y-x)d\sigma_y f_5(y), & x \in \Omega^+ \\ -\int_{\partial\Omega} H_6(y-x)d\sigma_y f_5(y)A_5^{-1}, & x \in \Omega^- \end{cases} \tag{15}$$

$$\Psi_2(x) = \begin{cases} \int_{\partial\Omega} H_5(y-x)d\sigma_y \widetilde{f_4}(y), & x \in \Omega^+ \\ \int_{\partial\Omega} H_5(y-x)d\sigma_y \widetilde{f_4}(y)A_4^{-1}, & x \in \Omega^- \end{cases} \tag{16}$$

$$\Psi_3(x) = \begin{cases} -\int_{\partial\Omega} H_4(y-x)d\sigma_y \widetilde{f_3}(y), & x \in \Omega^+ \\ -\int_{\partial\Omega} H_4(y-x)d\sigma_y \widetilde{f_3}(y)A_3^{-1}, & x \in \Omega^- \end{cases} \tag{17}$$

$$\Psi_4(x) = \begin{cases} \int_{\partial\Omega} H_3(y-x)d\sigma_y \widetilde{f_2}(y), & x \in \Omega^+ \\ \int_{\partial\Omega} H_3(y-x)d\sigma_y \widetilde{f_2}(y)A_2^{-1}, & x \in \Omega^- \end{cases} \tag{18}$$

$$\Psi_5(x) = \begin{cases} -\int_{\partial\Omega} H_2(y-x)d\sigma_y \widetilde{f_1}(y), & x \in \Omega^+ \\ -\int_{\partial\Omega} H_2(y-x)d\sigma_y \widetilde{f_1}(y)A_1^{-1}, & x \in \Omega^- \end{cases} \tag{19}$$

$$\Psi_6(x) = \begin{cases} \int_{\partial\Omega} H_1(y-x)d\sigma_y \widetilde{g}(y), & x \in \Omega^+ \\ \int_{\partial\Omega} H_1(y-x)d\sigma_y \widetilde{g}(y)A^{-1}, & x \in \Omega^- \end{cases} \tag{20}$$

$$\widetilde{f_4}(x) = f_4(x) - \int_{\partial\Omega} H_2(y-x)d\sigma_y f_5(y)(-1 + A_5^{-1}A_4),\ x \in \partial\Omega \tag{21}$$

$$\begin{aligned} \widetilde{f_3}(x) = f_3(x) &+ \int_{\partial\Omega} H_3(y-x)d\sigma_y f_5(y)(-1 + A_5^{-1}A_3) \\ &- \int_{\partial\Omega} H_2(y-x)d\sigma_y \widetilde{f_4}(y)(-1 + A_4^{-1}A_3), \quad x \in \partial\Omega \end{aligned} \tag{22}$$

$$\begin{aligned} \widetilde{f_2}(x) = f_2(x) &- \int_{\partial\Omega} H_4(y-x)d\sigma_y f_5(y)(-1 + A_5^{-1}A_2) \\ &+ \int_{\partial\Omega} H_3(y-x)d\sigma_y \widetilde{f_4}(y)(-1 + A_4^{-1}A_2) \\ &- \int_{\partial\Omega} H_2(y-x)d\sigma_y \widetilde{f_3}(y)(-1 + A_3^{-1}A_2), \quad x \in \partial\Omega \end{aligned} \tag{23}$$

$$\begin{aligned}\widetilde{f_1}(x) = f_1(x) &+ \int_{\partial\Omega} H_5(y-x)d\sigma_y f_5(y)(-1+A_5^{-1}A_1)\\ &- \int_{\partial\Omega} H_4(y-x)d\sigma_y \widetilde{f_4}(y)(-1+A_4^{-1}A_1)\\ &+ \int_{\partial\Omega} H_3(y-x)d\sigma_y \widetilde{f_3}(y)(-1+A_3^{-1}A_1)\\ &- \int_{\partial\Omega} H_2(y-x)d\sigma_y \widetilde{f_2}(y)(-1+A_2^{-1}A_1), \quad x \in \partial\Omega \end{aligned} \tag{24}$$

$$\begin{aligned}\widetilde{g}(x) = g(x) &- \int_{\partial\Omega} H_6(y-x)d\sigma_y f_5(y)(-1+A_5^{-1}A)\\ &+ \int_{\partial\Omega} H_5(y-x)d\sigma_y \widetilde{f_4}(y)(-1+A_4^{-1}A)\\ &- \int_{\partial\Omega} H_4(y-x)d\sigma_y \widetilde{f_3}(y)(-1+A_3^{-1}A)\\ &+ \int_{\partial\Omega} H_3(y-x)d\sigma_y \widetilde{f_2}(y)(-1+A_2^{-1}A)\\ &- \int_{\partial\Omega} H_2(y-x)d\sigma_y \widetilde{f_1}(y)(-1+A_1^{-1}A), \quad x \in \partial\Omega \end{aligned} \tag{25}$$

4. Existence of solutions for Riemann boundary value problem for triharmonic functions

Theorem 4.1. *Suppose that $G(x) \in H^\alpha(\partial\Omega, Cl(V_{n,n}))$ $0 < \alpha < 1$ and $G(x)$ satisfies the following condition*

$$2^{n-2}\|1 - G(x)\|_{(\alpha,\partial\Omega)}(M+1) < 1, \tag{26}$$

where M is a positive constant mentioned in Lemma 1.2. Then (2) exists solution to Riemann boundary value problem.

Proof. Denoting $w(x) = D^5[u](x)$ then $w^+(x) = w^-(x)A_5 + f_5(x)$, $x \in \partial\Omega$. Moreover, by $D^5[u](\infty) = 0$. We get

$$\omega(x) = \begin{cases} \int_{\partial\Omega} H_1(y-x)d\sigma_y f_5(y), & x \in \Omega^+ \\ \int_{\partial\Omega} H_1(y-x)d\sigma_y f_5(y)A_5^{-1}, & x \in \Omega^- \end{cases} \tag{27}$$

With $\Psi_1(x)$ as in [15], it is easy to check that

$$D^5(u - \Psi_1)(x) = 0, \quad x \in \mathcal{R}^n \setminus \partial\Omega \tag{28}$$

Denoting $\Delta^2 u(x) - \Delta^2 \Psi_1(x) := \varphi_1(x)$, $x \in \mathcal{R}^n \setminus \partial\Omega$, and using $(\Delta^2 u)^+(x) = (\Delta^2 u)^+(x)A_4 + f_4(x)$, $x \in \partial\Omega$. We conclude

$$\varphi_1^+(x) = \varphi_1^-(x)A_4 + \widetilde{f}_4(x), \quad x \in \partial\Omega \tag{29}$$

where $\widetilde{f}_4(x) \in H^{\widetilde{\beta}}(\partial\Omega, Cl(V_{n,n}))$, $0 < \widetilde{\beta} \leq 1$ be as in (21). By Corollary 2.3, it is clear that $\varphi_1^+(\infty) = 0$, we then get the representation formula

$$\varphi_1(x) = \begin{cases} \int_{\partial\Omega} H_1(y - x) d\sigma_y \widetilde{f}_4(y), & x \in \Omega^+ \\ \int_{\partial\Omega} H_1(y - x) d\sigma_y \widetilde{f}_4(y) A_4^{-1}, & x \in \Omega^- \end{cases} \tag{30}$$

Analogously, we find with $\Psi_2(x)$ from (16), that $\Delta^2 u(x) - \Delta^2 \Psi_1(x) - \Delta^2 \Psi_2(x) = 0$, $x \in \mathcal{R}^n \setminus \partial\Omega$. Denoting

$$D^3 u(x) - D^3 \Psi_1(x) - D^3 \Psi_2(x) := \varphi_2(x), \quad x \in \mathcal{R}^n \setminus \partial\Omega$$

Using the condition $(D^3 u)^+(x) = (D^3 u)^-(x)A_3 + f_3(x)$, $x \in \partial\Omega$, we obtain that

$$\varphi_2^+(x) = \varphi_2^-(x)A_3 + \widetilde{f}_3(x), \quad x \in \partial\Omega \tag{31}$$

where $\widetilde{f}_3(x) \in H^{\widetilde{\beta}}(\partial\Omega, Cl(V_{n,n}))$, $0 < \widetilde{\beta} \leq 1$ be as in (22). By Corollary 2.3, it is clear that $\varphi_2^+(\infty) = 0$, we then get the representation formula

$$\varphi_2(x) = \begin{cases} \int_{\partial\Omega} H_1(y - x) d\sigma_y \widetilde{f}_3(y), & x \in \Omega^+ \\ \int_{\partial\Omega} H_1(y - x) d\sigma_y \widetilde{f}_3(y) A_3^{-1}, & x \in \Omega^- \end{cases} \tag{32}$$

Here $\Psi_3(x)$ be as in (17), we have $D^3 u(x) - D^3 \Psi_1(x) - D^3 \Psi_2(x) - D^3 \Psi_3(x) = 0$, $x \in \mathcal{R}^n \setminus \partial\Omega$. We denote

$$\Delta u(x) - \Delta \Psi_1(x) - \Delta \Psi_2(x) - \Delta \Psi_3(x) := \varphi_3(x) \tag{33}$$

and use the condition $(\Delta u)^+(x) = (\Delta u)^-(x)A_2 + f_2(x)$, $x \in \partial\Omega$. We conclude

$$\varphi_3^+(x) = \varphi_3^-(x)A_2 + \widetilde{f}_2(x), x \in \partial\Omega \tag{34}$$

where $\widetilde{f}_2(x) \in H^{\widetilde{\beta}}(\partial\Omega, Cl(V_{n,n}))$, $0 < \widetilde{\beta} \leq 1$ be as in (23). Corollary 2.3 ensure that $\varphi_3(\infty) = 0$ then

$$\varphi_3(x) = \begin{cases} \int_{\partial\Omega} H_1(y - x) d\sigma_y \widetilde{f}_2(y), & x \in \Omega^+ \\ \int_{\partial\Omega} H_1(y - x) d\sigma_y \widetilde{f}_2(y) A_2^{-1}, & x \in \Omega^- \end{cases} \tag{35}$$

Use the same way again, $\Psi_4(x)$ be as in (18) that $\Delta u(x) - \Delta\Psi_1(x) - \Delta\Psi_2(x) - \Delta\Psi_3(x) - \Delta\Psi_4(x) = 0$, $x \in \mathcal{R}^n \setminus \partial\Omega$. Denoting

$$Du(x) - \sum_{i=1}^{4} D\Psi_i(x) := \varphi_4(x),\ x \in \mathcal{R}^n \setminus \partial\Omega \tag{36}$$

and using $(Du)^+(x) = (Du)^-(x)A_1 + f_1(x)$, $x \in \partial\Omega$, we get

$$\varphi_4^+(x) = \varphi_4^-(x)A_1 + \widetilde{f_1}(x),\ x \in \partial\Omega \tag{37}$$

where $\widetilde{f_1}(x) \in H^{\widetilde{\beta}}(\partial\Omega, Cl(V_{n,n}))$, $0 < \widetilde{\beta} \leq 1$ is taken from (24). By Corollary 2.3, it is clear that $\varphi_4^+(\infty) = 0$, we then get the representation formula

$$\varphi_4(x) = \begin{cases} \int_{\partial\Omega} H_1(y-x)d\sigma_y \widetilde{f_1}(y), & x \in \Omega^+ \\ \int_{\partial\Omega} H_1(y-x)d\sigma_y \widetilde{f_1}(y)A_1^{-1}, & x \in \Omega^- \end{cases} \tag{38}$$

Finally, we use $\Psi_5(x)$ as defined in (19) and get $Du(x) - \sum_{i=1}^{4} D\Psi_i(x) = 0$, $x \in \mathcal{R}^n \setminus \partial\Omega$. Defining

$$u(x) - \sum_{i=1}^{5} \Psi_i(x) := \varphi_5(x), \quad x \in \mathcal{R}^n \setminus \partial\Omega.$$

Working with the condition $u^+(x) = u^-(x)G(x) + g(x)$ we arrive at

$$\varphi_5^+(x) = \varphi_5^-(x)G(x) + \widetilde{g}(x),\ x \in \partial\Omega \tag{39}$$

where $\widetilde{g}(x) \in H^{\widetilde{\beta}}(\partial\Omega, Cl(V_{n,n}))$, $0 < \widetilde{\beta} \leq 1$ is taken from (25). It is clear that $\varphi_5(\infty) = 0$. We obtain that

$$\begin{cases} D\varphi_5 = 0, & \mathcal{R}^n \setminus \partial\Omega \\ \varphi_5^+(x) = \varphi_5^-(x)G(x) + \widetilde{g}(x), & x \in \partial\Omega \\ \varphi_5(\infty) = 0 \end{cases} \tag{40}$$

We only need to consider the existence of solutions about (40). The solution to this problem may be written in the form

$$\varphi_5(x) = \int_{\partial\Omega} H_1(y-x)d\sigma_y \varphi_6(y) \tag{41}$$

where $\varphi_6(y)$ is a Hölder continuous function to be determined on $\partial\Omega$. Then, by using Plemelj formula, the (40) can be reduced to an equivalent singular integral equation for φ_6,

$$\varphi_6(x) = [\frac{\varphi_6(x)}{2} - \int_{\partial\Omega} H_1(y-x)d\sigma_y \varphi_6(y)](1 - G(x)) + \widetilde{g}(x), \quad x \in \partial\Omega \tag{42}$$

Let T denote an integral operator defined by the right hand side of (42), we get

$$(T\varphi_6)(x) = [\varphi_6(x) - (S_{\partial\Omega}\varphi_6)(x)]\frac{(1-G(x))}{2} + \widetilde{g(x)} \tag{43}$$

For any $\omega_1, \omega_2 \in H^\alpha(\partial\Omega, Cl(V_{n,n}))$, we have

$$\|T\omega_1 - T\omega_2\|_{(\alpha,\partial\Omega)} \leq \frac{2^{n-1}}{2}\|\omega_1 - \omega_2\|_{(\alpha,\partial\Omega)}\|1-G\|_{(\alpha,\partial\Omega)}(1+M) \tag{44}$$

Under the condition (26), then the integral operator T is a contraction operator mapping the Banach space $H^\alpha(\partial\Omega, Cl(V_{n,n}))$ into itself, then has a unique fixed point for the operator T. Thus there exists a unique solution to (40). The proof is done. □

References

1. Delanghe R, On regular analytic functions with values in a Clifford algebra, *Math. Ann.* **185**, 91-111 (1970).
2. Delanghe R, On the singularities of functions with values in a Clifford algebra, *Math. Ann.* **196**, 293-319 (1972).
3. Brackx F, Delanghe R, Sommen F, *Clifford analysis*, Research Notes in Mathematics, vol.76, Pitman Books Ltd, London, 1982.
4. Yafang G, Jinyuan D, A kind of Riemann and Hilbert boundary value problem for left monogenic functions in $\mathbf{R}^m (m \geq 2)$, *Complex. Var.* **49**(5), 303-318 (2004).
5. Zhongxiang Z, On k-regular functions with values in a universal Clifford algebra, *J. Math. Anal. Appl.* **315**(2), 491-505 (2006).
6. Zhongxiang Z, Gürlebeck K, Some Riemann boundary value problems in Clifford analysis (1), *Complex. Var.* (2011). DOI 10.1080/17476933.2011.613119.
7. Jianke L. *Boundary Value Problems of Analytic Functions.* World Scientific Publisher: Singapore, 1993.
8. Muskhelishvilli NI. *Singular Integral Equations.* Nauka: Moscow, 1968.
9. Begehr H, Jinyuan D, Zhongxiang Z, On higher order Cauchy-Pompeiu formula in Clifford analysis and its applications, *Gen. Math.* (in Romania) **11**, 5-26 (2003).
10. Zhenyuan X, On linear and nonlinear Riemann-Hilbert problems for regular function with values in a Clifford algebra, *Chin. Ann. of Math* **11B**(3), 349-358 (1990).
11. Kalmenov T S, Koshanov B D, Representation for the Green's function of the Dirichlet problem for polyharmonic equations in a ball, *Siberian. Math. J.* **1**, 423-428 (2008).
12. Longfei Gu, Jinyuan Du, Zhang Zhongxiang, Riemann boundary value problems for triharmonic functions in Clifford analysis, *Adv.Appl. Cliford Algebras*, DOI 10.1007/s00006-012-03336-6.

A NEW DYNAMICAL SYSTEMS METHOD FOR NONLINEAR OPERATOR EQUATIONS *

XING-JUN LUO, FAN-CHUN LI and SU-HUA YANG

School of Mathematics and Computer Science,
Gannan Normal University, Ganzhou 341000, China
E-mail: lxj0298@126.com

A new dynamical systems method is studied for solving nonlinear operator equation $\mathcal{F}(u) := \mathcal{L}u + \mathcal{G}(u) = f$. This method consists of the construction of a nonlinear dynamical systems, that is, a Cauchy problem $\dot{u} = \Phi(t, u), u(0) = u_0$, where Φ is a suitable operator, which has the following properties: i) $\exists u(t) \quad \forall t > 0$, ii) $\exists u(\infty)$, and iii) $\mathcal{F}(u(\infty)) = f$. Conditions on $\mathcal{L}$ and $\mathcal{G}$ are given which allow one to chosen Φ such that i), ii), and iii) hold. This method yields also a convergent iterative method for finding true solution y. Stable approximation to a solution of the equation $\mathcal{F}(u) = f$ is constructed by this DSM when f is unknown but f_δ is known, when $\|f - f_\delta\| \leq \delta$.

Keywords: Dynamical systems method, nonlinear operator equation, iterative methods, inverse problems, ill-posed problems.
AMS No: 65J20, 65J10.

I. Introduction

Various inverse problems reduce to solving a nonlinear equation of the form

$$\mathcal{F}(u) = f, \quad \mathcal{F} : H \longrightarrow H, \tag{1}$$

in a real Hilbert Space H. The literature on methods for solving nonlinear operator equations is large([1-5]). Most of the results obtained so far are based on Newton-type methods and their modifications. The method used in this paper is a new version of the dynamical systems method. The idea of this method is described briefly below. The DSM for solving equation (1) consists of solving the problem:

$$\dot{u} = \Phi(t, u), \quad u(0) = u_0, \tag{2}$$

*This research is supported by NSFC(No. 11061001), NSF (No. 20114BAB201014) and SF (No. GJJ10586).

where $\dot{u} := \frac{du}{dt}$, and $\Phi(t,u)$ is a nonlinear operator chosen so that the following three conditions hold:

$$i)\exists u(t)\forall t > 0; \quad ii)\exists u(\infty) := \lim_{t\to\infty} u(t); \quad iii)\mathcal{F}(u(\infty)) = f. \tag{3}$$

In the well-posed case (the Fréchet derivative $\mathcal{F}'$ of the operator $\mathcal{F}$ is a bijection in a neighborhood of the solution of equation (1)) in order to solve equation (1) one can use the following continuous processes([4]):

• Choosing $\Phi(t,u) = -\big(\mathcal{F}(u(t)) - f\big)$ one gets the simple iteration method,

• Choosing $\Phi(t,u) = -[\mathcal{F}'(u(t))]^{-1}\big(\mathcal{F}(u(t)) - f\big)$, one gets a Newton's method,

• Choosing $\Phi(t,u) = -[\mathcal{F}'^*(u(t))\mathcal{F}'(u(t))]^{-1}\mathcal{F}'^*(u(t))\big(\mathcal{F}(u(t)) - f\big)$ yields the Gauss-Newton's method.

However if Fréchet derivative $\mathcal{F}'$ of the operator $\mathcal{F}$ is not continuously invertible (ill-posed case) one has to replace operator $\Phi(t,u)$ by the corresponding regularized operators:

• Choosing $\Phi(t,u) = -\big[\mathcal{F}(u(t)) - f + \varepsilon(t)(u(t) - u_0)\big]$ one gets the regularized simple iteration method,

• Choosing $\Phi(t,u) = -\big[\mathcal{F}'(u(t)) + \varepsilon(t)I\big]^{-1}\big[\mathcal{F}(u(t)) - f + \varepsilon(t)(u(t) - u_0)\big]$, one gets the regularized Newton's method,

• Choosing

$$\Phi(t,u) = -\big[\mathcal{F}'^*(u(t))\mathcal{F}'(u(t)) + \varepsilon(t)I\big]^{-1}\big[\mathcal{F}'^*(u(t))\big(\mathcal{F}(u(t)) - f\big) + \varepsilon(t)(u(t) - u_0)\big]$$

yields the regularized Gauss-Newton's method, here the function $\varepsilon(t)$ and point u_0 is a appropriate chosen.

The goal of this paper is to develop a new approach to such continuous methods and to establish fairly convergence theorem. This approach is based on the decomposition of the nonlinear operator $\mathcal{F}$.

This paper is organized as follows. In section 2 Theorem 2.2 and Theorem 2.4 about the convergence of a new regularized continuous process and a iterative process are formulated and proved respectively. In section 3, the situation is consider when the right hand of the nonlinear operator equation are not known in exact.

II. DSM for nonlinear equations

In this section we propose a class of new DSM, which is based on equation (4), to solve nonlinear operator equation (1). We show, that for linearization satisfying some conditions it is possible to choose the operator $\Phi(t,u)$ that it will guarantee convergence of solution of the problem (2). We consider a linearization of the operator $\mathcal{F}$ achieved by its splitting into a linear part $\mathcal{L}$ and nonlinear part $\mathcal{G}$, i.e.

$$\mathcal{F}(u) = \mathcal{L}u + \mathcal{G}(u). \tag{4}$$

To estimate the difference between the solution of the problem (2) and equation (1), we introduce the following assumption:

Assumption A Assume equation (1) has a solution y, possibly non-unique, for any $u, v \in B_r(y) := \{u \in H, \|u - y\| \leq r\}$, there is $\beta(u, v) \in H$ such that

$$\mathcal{G}(u) - \mathcal{G}(v) = \mathcal{L}\beta(u, v), \tag{5}$$

$$\|\beta(u, v)\| \leq \rho\|u - v\| \tag{6}$$

where $\rho > 0$ is some constant.

Remark 1 Conditions (5) and (6) was first presented by [6], and a example satisfying equation (4) can be founded in [6].

Lemma 2.1. *([2]) Let $f(t, w), g(t, u)$ be continuous on region $[0, T) \times D (D \subset R, T \leq \infty)$ and $f(t, w) \leq g(t, u)$ if $w \leq u, t \in (0, T), w, u \in D$. Assume that $g(t, u)$ is such that the Cauchy problem*

$$\dot{u} = g(t, u), \quad u(0) = u_0, \quad u_0 \in D$$

has a unique solution. If

$$\dot{w} \leq f(t, w), \quad w(0) = w_0 \leq u_0, \quad w_0 \in D,$$

then $u(t) \geq w(t)$ for all t for which $u(t)$ and $w(t)$ are defined.

We are especially interested in the case when the operator $\mathcal{L}$ does not have bounded inverse. Thus, We consider following new DSM problem

$$\dot{u} = -\Big[u + \mathcal{R}_\varepsilon(\mathcal{L}^*\mathcal{L})\mathcal{L}^*\Big(\mathcal{G}(u) - f\Big)\Big], \quad u(0) = u_0, \tag{7}$$

we assume that for the family $\{\mathcal{R}_\varepsilon\}$ there are positive constants $\gamma_1, \gamma_2, \gamma_3$ and ν such that for all $\lambda \in [0, \|\mathcal{L}\|^2]$ it holds that

$$|\lambda\mathcal{R}_\varepsilon(\lambda)| \leq \gamma_1, \tag{8}$$

$$|\sqrt{\lambda}\mathcal{R}_\varepsilon(\lambda)| \leq \frac{\gamma_2}{\sqrt{\varepsilon}}, \tag{9}$$

$$|1 - \lambda\mathcal{R}_\varepsilon(\lambda)|\lambda^\nu \leq \gamma_3\varepsilon^\nu, \tag{10}$$

let $\varepsilon(t) > 0$ be continuous, monotonically decaying function on $[0, \infty)$,

$$\varepsilon(t) > 0, \quad \lim_{t\to\infty} \varepsilon(t) = 0, \quad \lim_{t\to\infty} \dot{\varepsilon}\varepsilon^{-2} = 0. \tag{11}$$

Remark 2([2]) It follows from condition (11) that $\int_0^\infty \varepsilon(\tau)d\tau = \infty$.

Example. For the famous Tikhonov-Philips regularization the function $\mathcal{R}_\varepsilon(\lambda)$ has the following form

$$\mathcal{R}_\varepsilon(\lambda) = \frac{1}{\varepsilon + \lambda}.$$

Then (8)-(10) are satisfied with $\gamma_1 = 1, \gamma_2 = \frac{1}{2}, \gamma_3 = 1, \nu = 1$.

Theorem 2.2. *Let the Assumption A and condition (8), (10) and (11) hold. If $u_0 \in B_r(y)$ and $y = \big(\mathcal{L}^*\mathcal{L}\big)^\nu v, v \in H, \|v\| \le \frac{(1-\rho\gamma_1)r}{\gamma_3 \varepsilon_0^\nu}, \rho < \frac{1}{\gamma_1}$, then problem (7) has a unique local solution $u(t) \in B_r(y)$, the conclusions i), ii), and iii) hold for (7), and $u(\infty) = y$.*

Proof. Let $\Phi(t,u) = -\Big[u + \mathcal{R}_\varepsilon(\mathcal{L}^*\mathcal{L})\mathcal{L}^*\Big(\mathcal{G}(u) - f\Big)\Big]$, for any $u, v \in B_r(y), t > 0$, from (5), (6) and (8) we have

$$\begin{aligned}
\|\Phi(t,u) - \Phi(t,v)\| &\le \|u - v\| + \|\mathcal{R}_\varepsilon(\mathcal{L}^*\mathcal{L})\mathcal{L}^*\big(\mathcal{G}(u) - \mathcal{G}(v)\big)\| \\
&= \|u - v\| + \|\mathcal{R}_\varepsilon(\mathcal{L}^*\mathcal{L})\mathcal{L}^*\mathcal{L}\beta(u,v)\| \\
&\le (\rho\gamma_1 + 1)\|u - v\|,
\end{aligned}$$

hence $\Phi(t,u)$ is locally Lipschitz with respect to u, so the problem (7) has a unique local solution u(t). Let

$$u(t) - y := w(t), \quad \|w(t)\| := p(t),$$

then from the conditions (8) and (10) we obtain

$$\begin{aligned}
p\dot{p} &= (\dot{w}, w) = (\dot{u}, w) \\
&= -\Big(u + \mathcal{R}_\varepsilon(\mathcal{L}^*\mathcal{L})\mathcal{L}^*\big(\mathcal{G}(u) - f\big), w\Big) \\
&= -\Big(u - y + \mathcal{R}_\varepsilon(\mathcal{L}^*\mathcal{L})\mathcal{L}^*\big(\mathcal{G}(u) - \mathcal{G}(y) + \mathcal{G}(y) - f\big) + y, w\Big) \\
&= -p^2 - \Big(\mathcal{R}_\varepsilon(\mathcal{L}^*\mathcal{L})\mathcal{L}^*\mathcal{L}\beta(u,y), w\Big) + \Big(\mathcal{R}_\varepsilon(\mathcal{L}^*\mathcal{L})\mathcal{L}^*\mathcal{L}y - y, w\Big) \\
&\le -p^2 + \rho\gamma_1 p^2 - \Big(\big(I - \mathcal{R}_\varepsilon(\mathcal{L}^*\mathcal{L})\mathcal{L}^*\mathcal{L}\big)\big(\mathcal{L}^*\mathcal{L}\big)^\nu v, w\Big) \\
&\le -(1 - \rho\gamma_1)p^2 + \gamma_3\varepsilon^\nu(t)\|v\|p.
\end{aligned}$$

Because $p > 0$, it follows from Lemma 2.1 that

$$p(t) \le e^{-(1-\rho\gamma_1)t}\Big[p(0) + \gamma_3\int_0^t e^{(1-\rho\gamma_1)\tau}\varepsilon^\nu(\tau)\|v\|d\tau\Big], \quad p_0 = \|u_0 - y\|. \tag{12}$$

Using L'Hospital's rule we obtain

$$\lim_{t\to\infty} \frac{\int_0^t e^{(1-\rho\gamma_1)\tau}\varepsilon^\nu(\tau)\|v\|d\tau}{e^{(1-\rho\gamma_1)t}} = \lim_{t\to\infty} \varepsilon^\nu(t)\|v\| = 0, \tag{13}$$

it follows from (12) and (13) that

$$\lim_{t\to\infty} p(t) = \lim_{t\to\infty} \|u(t) - y\| = 0.$$

Next, let us prove $u(t) \in B_r(y), \forall t > 0$. It follows from (12) and condition on v that

$$\begin{aligned}
p(t) &\le e^{-(1-\rho\gamma_1)t}\Big[p(0) + \gamma_3\varepsilon^{\nu}(0)\int_0^t e^{(1-\rho\gamma_1)\tau}\|v\|d\tau\Big] \\
&= e^{-(1-\rho\gamma_1)t}\Big[p(0) + \frac{\gamma_3\|v\|\varepsilon^{\nu}(0)}{1-\rho\gamma_1}\Big(e^{(1-\rho\gamma_1)t} - 1\Big)\Big] \\
&\le e^{-(1-\rho\gamma_1)t}\Big[r + r\Big(e^{(1-\rho\gamma_1)t} - 1\Big)\Big] = r,
\end{aligned}$$

thus $u(t) \in B_r(y), \forall t > 0$.

Lemma 2.3. *([7]) Let the sequence of positive number ν_n satisfies the inequality*

$$\nu_{n+1} \le (1-\alpha_n)\nu_n + \theta_n,$$

where

$$0 < \alpha_n \le 1, \quad \sum_{n=0}^{\infty}\alpha_n = \infty, \quad \lim_{n\to\infty}\frac{\theta_n}{\alpha_n} = 0.$$

Then

$$\lim_{n\to\infty}\nu_n = 0.$$

Theorem 2.4. *Under the assumptions of Theorem 2.2, the iterative process*

$$u_{n+1} = u_n - h_n\Big[u_n + \mathcal{R}_{\varepsilon_n}(\mathcal{L}^*\mathcal{L})\mathcal{L}^*\big(\mathcal{G}(u_n) - f\big)\Big], \quad u_0 = u_0, \tag{14}$$

where $\varepsilon_n := \varepsilon(t_n), t_n = \sum_{i=0}^{n} h_n, 0 < h_n \le 1, \sum_{i=0}^{\infty} h_n = \infty$, generates the sequence u_n converging to y and $u_n \in B_r(y), \forall n \in N$.

Proof. Let $w_n := u_n - y, \quad \|w_n\| := p_n$. We rewrite (14) as

$$w_{n+1} = w_n - h_n\Big[u_n + \mathcal{R}_{\varepsilon_n}(\mathcal{L}^*\mathcal{L})\mathcal{L}^*\big(\mathcal{G}(u_n) - f\big)\Big], \quad w_0 = u_0 - y.$$

From the conditions (5) we get

$$\begin{aligned}
w_{n+1} &= w_n - h_n\Big[u_n + \mathcal{R}_{\varepsilon_n}(\mathcal{L}^*\mathcal{L})\mathcal{L}^*\big(\mathcal{G}(u_n) - f\big)\Big] \\
&= w_n - h_n\Big[w_n + y + \mathcal{R}_{\varepsilon_n}(\mathcal{L}^*\mathcal{L})\mathcal{L}^*\big(\mathcal{G}(u_n) - \mathcal{G}(y) + \mathcal{G}(y) - f\big)\Big] \\
&= (1-h_n)w_n - h_n\Big[\mathcal{R}_{\varepsilon_n}(\mathcal{L}^*\mathcal{L})\mathcal{L}^*\mathcal{L}\beta(u_n, y)\Big] + h_n\Big(\mathcal{R}_{\varepsilon_n}(\mathcal{L}^*\mathcal{L})\mathcal{L}^*\mathcal{L} - I\Big)y.
\end{aligned}$$

Then we obtain from above equation and (10) the following inequality:

$$p_{n+1} \leq \Big(1 - (1-\rho\gamma_1)h_n\Big)p_n + h_n\varepsilon_n^{\nu}\|v\|\gamma_3. \tag{15}$$

Denote $\alpha_n = (1-\rho\gamma_1)h_n, \quad \theta_n = h_n\varepsilon_n^{\nu}\|v\|\gamma_3$, then

$$0 < \alpha_n < 1, \quad \sum_{i=0}^{\infty}\alpha_n = \infty, \quad \lim_{n\to\infty}\frac{\theta_n}{\alpha_n} = \lim_{n\to\infty}\frac{\varepsilon_n^{\nu}\|v\|\gamma_3}{1-\rho\gamma_1} = 0,$$

it follows from Lemma 2.3 that

$$\lim_{n\to\infty} p_n = \lim_{n\to\infty}\|u_n - y\| = 0.$$

Next, let us proof $u_n \in B_r(y), \forall n \in N$. It follows from (15) and condition on v that

$$\begin{aligned} p_1 &\leq \Big(1-(1-\rho\gamma_1)h_0\Big)p_0 + h_0\gamma_3\varepsilon_0^{\nu}\|v\| \\ &\leq \Big(1-(1-\rho\gamma_1)h_0\Big)p_0 + h_0\gamma_3\varepsilon_0^{\nu}\frac{(1-\rho\gamma_1)r}{\gamma_3\varepsilon_0^{\nu}} \\ &\leq r. \end{aligned}$$

Assume that $p_n \leq r$, then

$$\begin{aligned} p_{n+1} &\leq \Big(1-(1-\rho\gamma_1)h_n\Big)p_n + h_n\gamma_3\varepsilon_n^{\nu}\|v\| \\ &\leq \Big(1-(1-\rho\gamma_1)h_n\Big)p_n + h_n\gamma_3\varepsilon_0^{\nu}\frac{(1-\rho\gamma_1)r}{\gamma_3\varepsilon_0^{\nu}} \\ &\leq r. \end{aligned}$$

Theorem 3.3 is proved.

Remark 3 If we set $h_n = 1$, then the iterative process (14) is just the iteration developed by [6].

III. Stability of the solution

Assume that $\mathcal{F}(y) = f$, where the exact date f are not known but the noisy data f_δ are given, $\|f_\delta - f\| \leq \delta$. Then the DSM yields a stable approximation of the solution y if the stopping time t_δ is properly chosen. Consider the DSM similar to (7):

$$\dot{u}_\delta = -\Big[u_\delta + \mathcal{R}_\varepsilon(\mathcal{L}^*\mathcal{L})\mathcal{L}^*\Big(\mathcal{G}(u_\delta) - f_\delta\Big)\Big], \quad u_\delta(0) = u_0. \tag{16}$$

Theorem 3.1. *Let the assumptions of Theorem 2.2 hold. Assume that* $u_\delta := u_\delta(t_\delta)$, *where* $u_\delta(t)$ *solves problems (16) and* t_δ *is chosen as (18). Then* $\lim_{\delta\to 0}\|u_\delta - y\| = 0$.

Proof. Let $w_\delta := u_\delta(t) - y$, $p_\delta(t) := \|w_\delta\|$. As in the proof of the Theorem 2.2 we derive the inequality:

$$\dot{p_\delta} \leq -(1-\rho\gamma_1)p_\delta + \gamma_3\varepsilon^\nu(t)\|v\| + \frac{\gamma_2\delta}{\sqrt{\varepsilon(t)}}, \tag{17}$$

if the stopping time t_δ is chosen from the following equation

$$\gamma_3\varepsilon^\nu(t)\|v\|\| = \frac{\gamma_2\delta}{\sqrt{\varepsilon(t)}}, \tag{18}$$

it is obvious that $\lim_{\delta\to 0} t_\delta = 0$, from (17) and (18) we obtain

$$\dot{p_\delta} \leq -(1-\rho\gamma_1)p_\delta + 2\gamma_3\varepsilon^\nu(t_\delta)\|v\|, \tag{19}$$

as in the proof of Theorem 2.2 we have

$$\lim_{\delta\to 0}\|u_\delta - y\| = 0$$

Consider the iterative process similar to (14):

$$v_{n+1} = v_n - h_n[v_n + \mathcal{R}_\varepsilon(\mathcal{L}^*\mathcal{L})\mathcal{L}^*\Big(\mathcal{G}(v_n) - f_\delta\Big)], \quad v_0 = u_0, \tag{20}$$

Theorem 3.2. *Suppose that the assumptions of the Theorem 2.2 are satisfied. Let $v_n := v_{n(\delta)}$, where v_n is defined by iterative process (20) and $n(\delta)$ is chosen as (22). Then $\lim_{\delta\to 0}\|v_{n(\delta)} - y\| = 0$.*

Proof. Let $w_n := v_n - y$, $\psi_n := \|w_n\|$, and chosen $h_n = h$ independent of $n, h \in (0,1)$, then the inequality similar to (15) is:

$$\psi_{n+1} \leq \Big(1 - (1-\rho\gamma_1)h\Big)\psi_n + h\gamma_3\varepsilon_n^\nu\|v\| + \frac{h\gamma_2\delta}{\sqrt{\varepsilon_n}}, \quad \psi_0 = \|u_0 - y\|, \tag{21}$$

we stop iterations in (20) when $n = n(\delta)$, where $n(\delta)$ is the largest integer for which the following inequality

$$\frac{\gamma_2\delta}{\sqrt{\varepsilon_n}} \leq \gamma_3\varepsilon_n^\nu\|v\|, \tag{22}$$

then (21) and (22) imply

$$\psi_{n+1} \leq \Big(1 - (1-\rho\gamma_1)h\Big)\psi_n + 2h\gamma_3\varepsilon_n^\nu\|v\|, \quad n < n(\delta), \tag{23}$$

where $n(\delta)$ is the largest integer for which the inequality (22) holds. We have

$$\lim_{\delta\to 0} n(\delta) = 0. \tag{24}$$

It follows from Lemma 2.3, (23) and (24) that

$$\lim_{\delta \to 0} \psi_n = \lim_{\delta \to 0} \|v_{n(\delta)} - y\| = 0$$

Remark 4 The iterative processes (14) and (20) are obtained by applying Euler's method to Cauchy problem (7) and (16) respectively, we can also get another iterative process by using other methods, such as the implicit Euler method, Runge-Kutta method.

References

[1] Airapetyan, R.G., Ramm, A.G., Smirnova, A.B., Continuous analog of Gauss-Newton method, *Math. Models Methods Appl. Sci.,* 1999, 9: 463-474

[2] Airapetyan, R.G., Ramm, A.G., Smirnova, A.B., Continuous methods for solving nonlinear ill-posed problems. Operator theorem and its applications, *Amer. Math. Soc.,* Providence RI, Fields Inst. Commun, 2000, 25: 111-137

[3] Airapetyan, R.G., Ramm, A.G., Dynamical systems and discrete methods for solving nonlinear ill-posed problems, *Appl. Math. Reviews*, vol. 1, world Sci. Publishers, 2000, 1: 491-536

[4] Ramm, A.G., Smirnova, A.B., Favini, A., Continuous modified Newton's method for nonlinear operator equations.*Annali di Matematica Pure Appl.,* 2002, 182 (1): 37-52

[5] Ramm, A.G., Dynamical systems method for solving nonlinear operator equations, *Int. J. Appl. Math. Sci.,* 2004, 1: 97-110

[6] Pereverzyev, S., Pinnau, R., Siedow, N., Regularized Fixed Point Iterations for Nonlinear Inverse Problems, *Inverse Problems*, 2006, 22: 1-22

[7] Ramm, A.G., Smirnova, A.B., On stable numerical differention, *Math. Comput.,* 2001, 70: 1131-1153

A CLASS OF INTEGRAL INEQUALITY AND APPLICATION *

WU-SHENG WANG

Department of Mathematics, Hechi University,
Yizhou, Guangxi, 546300, P. R. China
E-mail: wang4896@126.com

We discuss a class of generalized retarded nonlinear integral inequalities, which not only include nonlinear compound function of unknown function but also include retard items, and give upper bound estimation of the unknown function by integral inequality technique. This estimation can be used as tool in the study of differential equations with the initial conditions.

Keywords: Integral inequality; Gronwall-Bellman inequality; Integral inequality technique; Retarded Differential-integral equation; Estimation.
2000 MSC 26D10, 26D15, 26D20, 34A12, 34A40

1. Introduction

Gronwall-Bellman inequalities [Gronwall,1 and Bellman,2] and their various generalizations can be used important tools in the study of existence, uniqueness, boundedness, stability and other qualitative properties of solutions of differential equations, integral equations and integral-differential equations.

Lemma 1. [*Gronwall*, 1] *Let $u(t)$ be a continuous function defined on the interval $[a, a+h]$, a, h are nonnegative constants and*

$$0 \le u(t) \le \int_{\alpha}^{t} [bu(s) + a]ds, \quad t \in [a, a+h]. \tag{1}$$

Then, $0 \le u(t) \le ah\exp(bh), \forall t \in [a, a+h]$.

*This research was supported by National Natural Science Foundation of China(Project No. 11161018), Scientific Research Foundation of the Education Department of Guangxi Province of China (Project No. 201106LX599). The Key Discipline of Applied Mathematics of Hechi University of China(200725).

Lemma 2. [*Bellman*, 2] *Let $f, u \in C([0,h],[0,\infty))$, h, c are positive constants. If u satisfy the inequality*

$$u(t) \leq c + \int_{\alpha}^{t} f(s)u(s)ds, \quad t \in [0,h]. \tag{2}$$

Then, $u(t) \leq c\exp(\int_0^t f(s)ds)$, $t \in [0,h]$.

Lemma 3. [*Lipovan*, 3] *Let $u, f \in C([t_0,T),\mathbf{R}_+)$. Further, let $\alpha \in C^1([t_0,T),[t_0,T))$ be nondecreasing with $\alpha(t) \leq t$ on $[t_0,T)$, and let c be a nonnegative constant. Then the inequality*

$$u(t) \leq c + \int_{\alpha(t_0)}^{\alpha(t)} f(s)u(s)ds, \quad t \in [t_0,T) \tag{3}$$

implies that $u(t) \leq c\exp\left(\int_{\alpha(t_0)}^{\alpha(t)} f(s)ds\right)$, $t \in [t_0,T)$.

Lemma 4. [*Abdeldaim and Yakout*, 4] *We assume that $u(t)$ and $f(t)$ are nonnegative real-valued continuous functions defined on $I = [0,\infty)$ and satisfy the inequality*

$$u(t) \leq \left(u_0 + \int_0^t f(s)u(s)ds\right)^2 + \int_0^t f(s) \int_0^s g(\lambda)\left(u(\lambda) + \int_0^{\lambda} f(\tau)u(\tau)d\tau\right)d\lambda\Big)ds \tag{4}$$

for all $t \in I$, where u_0 be a positive constant. Then

$$u(t) \leq u_0^2 + \int_0^t f(s)B_1(s)ds, \ \forall t \in I, \tag{5}$$

where

$$B_1(t) = \exp\left(2\int_0^t f(s)B_2(s)ds\right)\left(u_0^3 + \int_0^t g(s)B_2(s)\exp\left(-2\int_0^s f(\tau)B_2(\tau)d\tau\right)ds\right) \tag{6}$$

for all $t \in I$,

$$B_2(t) = \frac{u_0(1+2u_0^2)\exp(\int_0^t (f(s)+g(s))ds)}{1 - 2u_0(1+2u_0^2)\int_0^t f(s)\exp(\int_0^s (f(\tau)+g(\tau))d\tau)ds}, \forall t \in I \tag{7}$$

such that $2u_0(1+2u_0^2)\int_0^t f(s)\exp(\int_0^s (f(\tau)+g(\tau))d\tau)ds < 1$.

There can be found a lot of generalizations of Gronwall-Bellman inequalities in various cases from literature (e.g., [3-13]).

In this paper, we discuss a class of retarded nonlinear integral inequality and give upper bound estimation of the unknown function by integral inequality technique.

2. Main result

In this section, we discuss a class of retarded nonlinear integral inequality of Gronwall-Bellman type. Throughout this paper, let $I = [0, \infty)$.

Theorem 1. *Let $\varphi, \varphi', \alpha \in C(I, I)$ be increasing functions with $\varphi'(t) \leq k, \alpha(t) \leq t, \alpha(0) = 0$, $\forall t \in I$; k, u_0 be positive constants with $2ku_0 \geq 1$; we assume that $u(t)$ and $f(t)$ are nonnegative real-valued continuous functions defined on I and satisfy the inequality.*

$$u(t) \leq \Big(u_0 + \int_0^{\alpha(t)} f(s)\varphi(u(s))ds\Big)^2 + \int_0^{\alpha(t)} f(s)\Big(\int_0^s g(\lambda)\Big(\varphi(u(\lambda)) + \int_0^\lambda f(\tau)\varphi(u(\tau))d\tau\Big)d\lambda\Big)ds, \quad (1)$$

for all $t \in I$, If $2(u_0 + \varphi(u_0^2)) \int_0^{\alpha(t)} kf(s) \exp\Big(\int_0^s \Big(g(s)/(2u_0) + f(s)\Big)d\tau\Big)ds < 1$, then

$$u(t) \leq u^2(0) + \int_0^t \alpha'(s)f(\alpha(s))\theta_1(s)ds, \quad \forall t \in I \quad (2)$$

where

$$\theta_1(t) = \exp\Big(2\int_0^t k\alpha'(s)f(\alpha(s))\theta_2(s)ds\Big)\Big(2\varphi(u_0^2)u_0 -$$

$$\int_0^t \alpha'(s)g(\alpha(s))\theta_2(s)\exp\Big(-2\int_0^s k\alpha'(\tau)f(\alpha(\tau))\theta_2(\tau)d\tau\Big)ds\Big), \quad (3)$$

$$\theta_2(t) = \frac{(u_0+\varphi(u_0^2))\exp\Big(\int_0^{\alpha(t)} (g(s)/(2u_0)+f(s))ds\Big)}{1-2(u_0+\varphi(u_0^2))\int_0^{\alpha(t)} kf(s)\exp\Big(\int_0^s \Big(g(s)/(2u_0)+f(s)\Big)d\tau\Big)ds} \quad (4)$$

Remark 1: If $\alpha(t) = t, \varphi(u(s)) = u(s)$, then Theorem 1 reduces Lemma 4.

Proof Let $z(t)$ denote the function on the right-hand side of (1), which is a nonnegative and nondecreasing function on I with $z(0) = u_0^2$. Then (1) is equivalent to

$$u(t) \leq z(t), u(\alpha(t)) \leq z(\alpha(t)), \quad \forall t \in I. \quad (5)$$

Differentiating $z(t)$ with respect to t, using (5) we have

$$\begin{aligned}
\frac{dz}{dt} &= 2\alpha'(t)f(\alpha(t))\varphi(u(\alpha(t)))\Big(u_0 + \int_0^{\alpha(t)} f(s)\varphi(u(s))ds\Big) \\
&\quad + \alpha'(t)f(\alpha(t))\int_0^{\alpha(t)} g(\lambda)\Big(\varphi(u(\lambda)) + \int_0^{\lambda} f(\tau)\varphi(u(\tau))d\tau\Big)d\lambda \\
&\le \alpha'(t)f(\alpha(t))\Big(2\varphi(z(t))\Big(u_0 + \int_0^{\alpha(t)} f(s)\varphi(z(s))ds\Big) \\
&\quad + \int_0^{\alpha(t)} g(\lambda)\Big(\varphi(z(\lambda)) + \int_0^{\lambda} f(\tau)\varphi(z(\tau))d\tau\Big)d\lambda\Big) \\
&= \alpha'(t)f(\alpha(t))Y_1(t), \ \forall t \in I, \qquad (6)
\end{aligned}$$

where

$$\begin{aligned}
Y_1(t) :&= 2\varphi(z(t))\Big(u_0 + \int_0^{\alpha(t)} f(s)\varphi(z(s))ds\Big) \\
&\quad + \int_0^{\alpha(t)} g(\lambda)\Big(\varphi(z(\lambda)) + \int_0^{\lambda} f(\tau)\varphi(z(\tau))d\tau\Big)d\lambda, \ \forall t \in I. \qquad (7)
\end{aligned}$$

Hence $Y_1(0) = 2\varphi(u_0^2)u_0$, and $\varphi(z(t)) \le Y_1(t)/(2u_0)$. Differentiating $Y_1(t)$ with respect to t, and using (6) and (7), we get

$$\begin{aligned}
\frac{dY_1(t)}{dt} &\le 2\varphi'(z(t))\alpha'(t)f(\alpha(t))Y_1(t)\Big(u_0 + \int_0^{\alpha(t)} f(s)\varphi(z(s))ds\Big) \\
&\quad +2\alpha'(t)\varphi(z(t))f(\alpha(t)) * \varphi(z(\alpha(t))) \\
&\quad +\alpha'(t)g(\alpha(t))\Big(\varphi(z(\alpha(t))) + \int_0^{\alpha(t)} f(\tau)\varphi(z(\tau))d\tau\Big) \\
&\le 2k\alpha'(t)f(\alpha(t))Y_1(t)\Big(u_0 + \int_0^{\alpha(t)} \frac{f(s)Y_1(s)}{2u_0}ds\Big) + \alpha'(t)f(\alpha(t))\frac{Y_1^2(t)}{2u_0^2} \\
&\quad +\alpha'(t)g(\alpha(t))\Big(\frac{Y_1(t)}{2u_0} + \int_0^{\alpha(t)} \frac{f(s)Y_1(s)}{2u_0}ds\Big), \\
&\le \Big(2k\alpha'(t)f(\alpha(t))Y_1(t) + \alpha'(t)g(\alpha(t))\Big) \\
&\quad \Big(u_0 + \frac{Y_1(t)}{2u_0} + \int_0^{\alpha(t)} \frac{f(s)Y_1(s)}{2u_0}ds\Big), \qquad (8)
\end{aligned}$$

for all $t \in I$, where we use the conditions $\varphi'(t) \le k, 2ku_0 \ge 1$. Let

$$Y_2(t) = u_0 + \frac{Y_1(t)}{2u_0} + \int_0^{\alpha(t)} \frac{f(s)Y_1(s)}{2u_0}ds,$$

then $Y_2(0) = u_0+\varphi(u_0^2), Y_1(t) < 2u_0Y_2(t)$. Differentiating $Y_2(t)$ with respect to t, and using (8), we have

$$\begin{aligned}\frac{dY_2(t)}{dt} &\le \frac{2k\alpha'(t)f(\alpha(t))Y_1(t)+\alpha'(t)g(\alpha(t))}{2u_0}Y_2(t)+\frac{\alpha'(t)f(\alpha(t))Y_1(\alpha(t))}{2u_0}\\ &\le 2k\alpha'(t)f(\alpha(t))Y_2^2(t)\\ &\quad+\Big(\alpha'(t)g(\alpha(t))/(2u_0)+\alpha'(t)f(\alpha(t))\Big)Y_2(t) \qquad (9)\end{aligned}$$

Since $Y_2(0)>0$, from (9) we have

$$\begin{aligned}Y_2^{-2}(t)\frac{dY_2(t)}{dt} &\le 2k\alpha'(t)f(\alpha(t))\\ &\quad+\Big(\alpha'(t)g(\alpha(t))/(2u_0)+\alpha'(t)f(\alpha(t))\Big)Y_2^{-1}(t) \qquad (10)\end{aligned}$$

Let $S_1(t)=Y_2^{-1}(t)$, then $S_1(0)=(u_0+\varphi(u_0^2))^{-1}$, from (10) we obtain

$$\frac{dS_1(t)}{dt}+\Big(\alpha'(t)g(\alpha(t))/(2u_0)+\alpha'(t)f(\alpha(t))\Big)S_1(t)\ge -2k\alpha'(t)f(\alpha(t)) \quad (11)$$

Consider the ordinary differential equation

$$\begin{cases}\frac{dS_2(t)}{dt}+(\alpha'(t)g(\alpha(t))/(2u_0)+\alpha'(t)f(\alpha(t)))S_2(t)=-2k\alpha'(t)f(\alpha(t)),\ \forall t\in I,\\ S_2(0)=(u_0+\varphi(u_0^2))^{-1}.\end{cases} \qquad (12)$$

The solution of Equation (12) is

$$\begin{aligned}S_2(t)=&(u_0+\varphi(u_0^2))^{-1}\exp\Big(-\int_0^{\alpha(t)}(g(s)/(2u_0)+f(s))ds\Big)\Big(1-2(u_0+\varphi(u_0^2))\\ &*\int_0^{\alpha(t)}kf(s)\exp\Big(\int_0^s\Big(g(s)/(2u_0)+f(s)\Big)d\tau\Big)ds\Big) \qquad (13)\end{aligned}$$

By (4), (11), (12) and (13), we obtain

$$Y_2(t)=\frac{1}{S_1(t)}\le\frac{1}{S_2(t)}=\theta_2(t),\ \forall t\in I, \qquad (14)$$

where $\theta_2(t)$ as defined in (4). From (8) and (14), we have

$$\frac{dY_1(t)}{dt}\le 2k\alpha'(t)f(\alpha(t))\theta_2(t)Y_1(t)+\alpha'(t)g(\alpha(t))\theta_2(t),\ \forall t\in I.$$

The above inequality implies the following estimation for $Y_1(t)$:

$$Y_1(t)\le\theta_1(t),\quad \forall t\in I, \qquad (15)$$

where $\theta_1(t)$ as defined in (3), thus from (6) and (15) we have

$$\frac{dz}{dt}\le\alpha'(t)f(\alpha(t))\theta_1(t),\quad \forall t\in I.$$

By taking $t = s$ in the above inequality and integrating from 0 to t, we get

$$u(t) \le z(t) \le u^2(0) + \int_0^{\alpha(t)} \alpha'(s) f(\alpha(s)) \theta_1(s) ds$$

The estimation (2) of unknown function in the inequality (1) is obtained.

3. Application

In this section we give an example for the application of our Theorem 1 to the following retarded differential-integral equation

$$\begin{cases} \dfrac{dx(t)}{dt} = H\big(t, x(t), K(\alpha(t), x(\alpha(t))), K(t, x(t))\big), \quad t \in I_0 = [t_0, \infty), \\ x(t_0) = x_0, \end{cases} \tag{16}$$

where $\alpha(t)$ satisfies $\alpha(t) \le t$, $0 < \alpha'(t) \le 1$ and $\alpha(t_0) = t_0$. We assume that $K \in C(I_0, \mathbb{R})$, $H \in C(I_0, \mathbb{R} \times \mathbb{R} \times \mathbb{R})$ and

$$\begin{aligned} &H\big(t, x(t), K(\alpha(t), x(\alpha(t))), K(t, x(t))\big) \\ &= K(\alpha(t), x(\alpha(t)))\Big(2u_0 + \int_{t_0}^{\alpha(t)} K(\tau, x(\tau)) d\tau\Big) \\ &\quad + f(\alpha(t))\Big(\int_{t_0}^{\alpha(t)} g(\lambda)\Big(\varphi(x(\lambda)) + \int_{t_0}^{\lambda} K(\tau, x(\tau)) d\tau\Big) d\lambda\Big), \end{aligned} \tag{17}$$

$$|K(t, x(t))| \le f(t)\varphi(x(t)), \tag{18}$$

where f, g are nonnegative real-valued continuous functions defined on I_0.

Corollary 1. *Consider nonlinear system (16) and suppose that $\varphi, \varphi', \in C^1(I, I)$ be increasing functions with $\varphi'(t) \le k, \forall t \in I$; and K, H satisfy the conditions (17) and (18), then the estimation of solution of Eq. (16) can be given by*

$$|x(t)| \le u_0^2 + \int_0^t \alpha'(s) f(\alpha(s)) \theta_1(s) ds, \quad \forall t \in I, \tag{19}$$

where $\theta_1(t)$ is given by (3) in Theorem 1, $u_0^2 = |x_0|$.

Proof Integrating both sides of the equation (16) from t_0 to t, we get

$$x(t) = x_0 + \int_{t_0}^{t} H\big(t, x(t), K(\alpha(t), x(\alpha(t))), K(t, x(t))\big)ds, \qquad t \in I_0. \quad (20)$$

Using the conditions (17) and (18), from (20) we obtain

$$\begin{aligned}
|x(t)| &\le u_0^2 + 2u_0 \int_{t_0}^{t} f(\alpha(s))\varphi(x(\alpha(s)))ds \\
&\quad + \int_{t_0}^{t} f(\alpha(s))\varphi(x(\alpha(s))) \int_{t_0}^{\alpha(s)} f(\tau)\varphi(x(\alpha(\tau)))d\tau ds \\
&\quad + \int_{t_0}^{t} f(\alpha(s)) \int_{t_0}^{\alpha(s)} g(\lambda)\Big(\varphi(x(\alpha(\lambda))) + \int_{t_0}^{\lambda} f(\tau)\varphi(x(\alpha(\tau)))d\tau\Big)d\lambda ds \\
&\le \Big(u_0 + \int_{0}^{\alpha(t)} f(s)\varphi(u(s))ds\Big)^2 \\
&\quad + \int_{0}^{\alpha(t)} f(s)\Big(\int_{0}^{s} g(\lambda)\Big(\varphi(u(\lambda)) + \int_{0}^{\lambda} f(\tau)\varphi(u(\tau))d\tau\Big)d\lambda\Big)ds. \quad (21)
\end{aligned}$$

Applying Theorem 1 to (21), we get the estimation (16). This completes the proof of Corollary 1. □

References

1. T.H. Gronwall, Note on the derivatives with respect to a parameter of the solutions of a system of differential equations, *Ann. of Math.*, **20**(1919), 292-296.
2. R. Bellman, The stability of solutions of linear differential equations, *Duke Math. J.*, **10**(1943), 643-647.
3. O. Lipovan, A retarded Gronwall-like inequality and its applications, *J. Math. Anal. Appl.*, **252**(2000), 389-401.
4. A. Abdeldaim, M. Yakout, On some new integral inequalities of Gronwall-Bellman-Pachpatte type, *Appl. Math. Comput.*, **217**(2011),7887-7899
5. R.P. Agarwal, S. Deng and W. Zhang, Generalization of a retarded Gronwall-like inequality and its applications, *Appl. Math. Comput.*, **165**(2005), 599-612.
6. I.A. Bihari, A generalization of a lemma of Bellman and its application to uniqueness problem of differential equation, *Acta Math. Acad. Sci. Hung.*, **7**(1956), 81-94.
7. B.G. Pachpatte, *Inequalities for Differential and Integral Equations*, Academic Press, London, 1998.
8. Y.H. Kim, On some new integral inequalities for functions in one and two variables, *Acta Math. Sinica,* **21**(2005), 423-434.
9. W.S. Cheung, Some new nonlinear inequalities and applications to boundary value problems, *Nonlinear Anal.*, **64**(2006), 2112-2128.

10. W. S. Wang, A generalized retarded Gronwall-like inequality in two variables and applications to BVP, *Appl. Math. Comput.,* **191**(1)(2007), 144-154.
11. W. S. Wang and C. Shen, On a generalized retarded integral inequality with two variables, *J. Inequ. Appl.*, **2008**(2008), Art. ID 518646, 9 pages.
12. W. S. Wang, Z. Li, Y. Li and Y. Huang, Nonlinear retarded integral inequalities with two variables and applications, *J. Inequ. Appl.*, **2010**(2010), Art. ID 240790, 8 pages
13. W. S. Wang, R. C. Luo and Z. Li, A new nonlinear retarded integral inequality and its application , *J. Inequ. Appl.*, **2010**(2010), Art. ID 462163, 8 pages

AN EFFICIENT SPECTRAL BOUNDARY INTEGRAL EQUATION METHOD FOR THE SIMULATION OF EARTHQUAKE RUPTURE PROBLEMS

WENSHUAI WANG*, BAOWEN ZHANG

School of Mathematics and Computer Science, Ningxia University, Yinchuan, China
** E-mail: wws@nxu.edu.cn*

In this paper, we study on elastodynamic fracture problems and present a symplectic scheme, combined with the spectral boundary integral method, for solving anti-plane spontaneous rupture problems via constructing a nonautonomous Hamiltonian system by introducing independent variables in extended phase space. Slip-weakening friction law is considered with the approach. Numerical experiments are employed to test the effectiveness of symplectic scheme, and the comparisons of numerical results show that the present method is of higher accuracy compared to the non-symplectic method. The results show that the method is a efficient tool for simulating quasi-static and dynamic earthquake processes.

Keywords: Symplectic schemes; elastodynamic; anti-plane shear; boundary integral method.

1. Introduction

The spontaneously propagating shear crack on a frictional interface has proven to be a useful idealization of a natural earthquake. However, it remains quite challenging due to a wide range of spatial and temporal scales involved in modeling long-term slip histories of faults. Simulating spontaneous earthquake sequence on a fault needs to incorporate and resolve slow tectonic loading during the interseismic periods, nucleation and propagation of rupture during earthquakes that involve rapid changes in stress and slip rate at the propagating dynamic rupture tips, and the subsequent postseismic deformation and aseismic afterslip. To simulate the earthquake sequence as truly as possible, several friction laws was proposed, such as purely velocity-dependent law [1], slip-dependent law [2], and the rate and state law [3]. The purely velocity-dependent or slip-dependent laws were introduced as plausible descriptions that allowed to simulate unstable slip. A nonlinear friction proposed by Abercrombie and Rice [4] is employed to describe the rupture process, which is initially linear slip-weakening law, then non-linear law of Abercrombie and Rice, supported from lab [5]. Based on the above friction laws, many numerical algorithms have been proposed to simulate the elastodynamic problems in the past years, such as finite differences (FD) [6,7], boundary integral (BI) [8,9], finite volume (FV) [10,11], finite element (FE) [12,13] or spectral element (SE) [14,15] and discontinuous Galerkin method (DGM) [16]. All these methods have its advantages and disadvantages[16]. Among the methods, The boundary integral equation method is a powerful tool for solving certain boundary value problems. A spectral scheme combined with BI method in its original formulation was developed for simulating spontaneous crack propagation in homogeneous, isotropic materials and has been extended to handle various problems. As you known, slow slip transients, lasting from days to years [17], involving slip rates as large as 100 cm/yr [18], have now been observed in northern California [19], in Japan [18], and in New Zealand [20]. If we simulate the kinds of long-term slip histories of faults, it demands that a method can deal with long-time tracing effectively. To author's knowledge, symplectic integrators have a more remarkable capacity for capturing the long-time tracing problem than non-symplectic integrators. Here, we develop a time-integration of symplectic scheme for non-autonomous Hamilton system to deduce the evolution of the displacement field from the velocity values computed at discrete time intervals by using the boundary integral method in spatial discretization, and give an efficient numerical procedure for elastodynamic analysis of earthquake sequences on slowly loaded faults.

2. Formulation and numerical implementation

We follow Perrin et al. [21] in adopting a spectral representation of the relation between the traction and resulting displacement discontinuities, and the conversion between spectral and real domains is again performed efficiently through the FFT method. A detailed description of the steps leading to the spectral formulation for the problems can be found in Ref. [22]. We employ the spectral formulation of boundary integral method for planar interfaces pioneered by Perrin et al. [22] for two-dimensional anti-plane problems and extended by Geubelle and Rice [22], because the method have several advantages over purely space-time formulation. Firstly, the stress transfer functionals are written as integrals on both space and time, the resulting is that the value of the stress transfer functional for each cell would be determined by histories of displacement discontinuities for all relevant cell. Another advantage is having the transient elastodynamic response separated into Fourier modes and the convolution kernels are oscillating with decaying amplitude and hence at large enough times the convolutions can be truncated. In addition, the arguments of the kernels contain the magnitude of the wave vector, which is larger for higher modes. Hence the convolution for higher modes can be truncated sooner than for the lower modes, and such mode-dependent truncation can save a lot of computational time, as discussed by Lapusta et al. [3] for a 2-D case.

An outline of the method is presented hereafter for completeness purpose, followed by description of numerical implementation in time step. Let us consider an unbounded, homogeneous, linear elastic body of shear modulus μ and shear wave speed c_s and define a cartesian co-ordinate system xyz such that the fault plane coincides with the $x - y$ plane. The only nonzero displacement u_z is independent of the z-coordinate and satisfies the scalar wave equation.

$$c_s^2(u_{z,xx} + u_{z,yy}) = u_{z,tt}$$

We use the spectral representation of Perrin et al.[22] to write as

$$u_z(x, y, t) = e^{ikx}\Omega(y, t; k) \tag{1}$$

and taking a Laplace transform in time, giving

$$\hat{\Omega}(y, p; k)_{,yy} = k^2\alpha_s^2\hat{\Omega}(y, p; k) \tag{2}$$

with $\alpha_s = \sqrt{1 + p^2/k^2c_s^2}$.

Considering equation (1), a bounded solution of (2) for the upper plane $y > 0$ must have the form

$$\hat{u}_z(x, y, p) = e^{ikx}\hat{\Omega}_0(p; k)e^{-|k|\alpha_s y} \tag{3}$$

Defining the displacement Fourier coefficient $U_z(t; k)$ and the traction coefficient $T(t; k)$ as

$$u_z(x, 0^+, t) = U_z(t; k)e^{ikx}$$
$$\tau(x, t) = \tau_{yz}(x, 0^{\pm}, t) = T(t; k)e^{ikx}$$

Combining equation (3) and the linear elastic stress-strain relations, the traction and displacement Fourier coefficient can be represented by

$$\hat{T}(p; k) = -\mu\,|k|\,\alpha_s\hat{U}_z(p; k) \tag{4}$$

We define slip $\delta(x,t)$ on the fault as the displacement discontinuity and the relation of its Fourier coefficient $D(t;k)$ gives

$$\delta(x,t) = u_z(x,0^+,t) - u_z(x,0^-,t) = D(t;k)e^{ikx} \tag{5}$$

and

$$\hat{T}(p;k) = -\frac{\mu}{2}|k|\,\alpha_s\hat{D}(p;k) \tag{6}$$

We modify equation (6) by extracting a term corresponding to the instantaneous response

$$\hat{T}(p;k) = -\frac{\mu}{2c_s}p\hat{D}(p;k) - \frac{\mu}{2}\{|k|\,\alpha_s - \frac{p}{c_s}\}\hat{D}(p;k) \tag{7}$$

Taking inverse Laplace transform back to the time domain

$$T(t) = -\frac{\mu}{2c_s}\dot{D}(t;k) + F(t)$$

where the upper dot denotes differentiation with respect to time, and the $F(t)$ reads

$$F(t) = -\mu\,|k|\int_{-\infty}^{t} C_{III}(|k|\,c_s(t-t'))D(t')\,|k|\,c_s dt' \tag{8}$$

where $C_{III} = \frac{J_1(T)}{T}$, and $J_1(T)$ is the Bessel function of the first kind.

Follow the work of Cochard & Madariaga [8] and Perrin et al.[22], the elastodynamic relations between the stress and the slip history give

$$\tau(x,t) = \tau^0(x,t) + f(x,t) - \frac{\mu}{2c_s}V(x,t) \tag{9}$$

where $V(x,t) = \dot{\delta}(x,t)$ is the slip rate, $\tau^0(x,t)$ is external loading stress, and $f(x,t)$, incorporating stress transfers, is a linear functional of the history of the displacement discontinuities up to the present time.

An analogous 'velocity' representation can be integrated by parts and rewrite (8) as

$$F(t,k) = -\frac{1}{2}\mu\,|k|\,D(t) + \frac{1}{2}\mu\,|k|\int_{-\infty}^{t} W(|k|\,c_s(t-t'))\dot{D}(t';k)dt' \tag{10}$$

where $W(p) = \int_p^\infty J_1(\xi)/\xi d\xi$ with $W(0) = 1$.

3. Friction laws and spatial discretization

Crack-like rupture is one of styles in which Earthquake ruptures are thought to propagate. During crack-like rupture the frictional strength of the fault suffers irreversible reduction, the fault slides simultaneously over the entire ruptured area and slip continues until arrest fronts arrive from the terminal edges of the rupture. The local duration of slip, also known as rise time, scales with the shortest dimension of the final rupture area. In fact, the frictional strength evolves according to some constitutive functional which may in principle depend upon the history of the velocity discontinuity, and any number of the other mechanical or thermal quantities, but is here simplified to slip-weakening form proposed by Abercrombie & Rice [4]. The friction law is described as following:

$$\mu_f(l) = \begin{cases} \mu_s - (\mu_s - \mu_d)l/D_c, & l \leq D_c \\ \mu_d, & l > D_c \end{cases} \tag{11}$$

where μ_s and μ_d are coefficients of static and dynamic friction respectively, and D_c is the critical slip distance for the strength drop [23].

With respect to spatial discretization, as described in Ref. [3], the spectral formulation can be implemented with the discretization of a period X of the fracture plane by N uniformly spaced elements of size $\Delta x = \frac{X}{N}$, defining the grid points at which δ and f will be evaluated in a Fourier series

$$[\delta(x,t), f(x,t)] = \sum_{k=-N/2}^{N/2} [D_k(t), F_k(t)] e^{(2\pi i k x/X)} \tag{12}$$

The replication distance X has to be chosen larger enough to ensure that there is negligible influence of waves arriving from the periodic replicates of the rupture process. The Fourier coefficients are performed efficiently through the FFT algorithm with N sample points.

4. Symplectic schemes of evolution time step

From above discussion, we obtain the evolution of each Fourier mode as

$$\dot{\delta}_i(x,t) = V_i(x,t), V_i(x,t) = \phi_i(x,t) \tag{13}$$

Here $\phi_i(x,t)$ can be derived from (9) and (11) as

$$\phi_i(x,t) = \frac{2c_s}{\mu}(\tau_i^0(x,t) + f_i(x,t) - \tau_i(x,t)) \tag{14}$$

Considering the elastodynamic relations (5),(9) and (10), with the mode number k in (5) replaced by $\frac{2\pi k}{X}$ as suggested by equation (12), completed by a time-stepping scheme to update the slip from the slip rate (or velocity) distribution derived from (9).

In the previous studies, different updating schemes have been used, such as, the first-order accurate schemes used by Rice & Ben-Zion [24] and Ben-Zion & Rice [25]; So-called a semi-implicit velocity formulation was originally developed by Morrissey & Deubelle [26] for the case of constant evolution time steps and constitutive laws without state variables and implemented with delay, discretized kernel and convolutions by trapezoid rule, which is a second-order schemes; Another semi-implicit one used by Lapusta [3], given as a total variation diminishing Runge-Kutta schemes of second-order (TVDRK2), which performs better than both of the above mentioned methods. It does not use the delay, but captures more accurately the high slip velocities at the rupture tips, and has essentially the same stability performance. In this work, considering the differentiation of $V(x,t)$ with respect to time t, equation (9) can be rewritten as

$$\dot{V}(x,t) = \dot{\phi}(x,t) = \frac{2c_s}{\mu}(\dot{\tau}^0(x,t) + \dot{f}(x,t) - \dot{\tau}(x,t)) \tag{15}$$

where

$$\dot{f}(x,t) = \sum_{n=-N/2}^{N/2} \dot{F}_n(t) e^{ik_n x}$$

and

$$\dot{F}_n(t) = -\frac{\mu\,|k_n|}{2}\frac{J_1(|k_n|\,ct)}{t}D_n(0) - \frac{\mu\,|k_n|}{2}\int_0^t \frac{\partial}{\partial t}(\frac{J_1(|k_n|\,ct')}{t'}D_n(t-t'))dt'$$

The first equation (13) and equation (15) can be seen as a Hamilton systems by introducing variable $p_1 = \delta(x,t), q_1 = V(x,t)$. For the system of canonical equation given as

$$\frac{dp_1}{dt} = -\frac{\partial H}{\partial q_1} = V(x,t), \frac{dq_1}{dt} = \frac{\partial H}{\partial p_1} = \phi(x,t) \tag{16}$$

with Hamiltonian function $H(p_1, q_1, t)$. This is a nonautonomous Hamiltonian system. As given by Qin [27], considering the time t as an additional dependent variable, and choosing a parameter η as a new independent variable. We coordinate $t = q_2$, the corresponding phase space must comprise 4 dimensions, $z = (p_1, h, q_1, q_2)$, here t and h being merely alternative notations for q_2 and p_2. The new momentum h associated with the time t is, as its physical interpretation, the negative of the total energy. The new space is called the extended phase space, in which equation (16) becomes

$$\frac{dp_1}{d\eta} = -\frac{\partial H}{\partial q_1}, \frac{dq_1}{d\eta} = \frac{\partial H}{\partial p_1}, \tag{17a}$$

$$\frac{dp_2}{d\eta} = -\frac{\partial H}{\partial q_2}, \tag{17b}$$

$$\frac{dq_2}{d\eta} = 1. \tag{17c}$$

The equation (17a) is the original canonical equation. Equation (17b) shows that mormalized parameter now becomes equal to q_2 which is the time t. The last equation (17c) gives the law according to which the negative of the total energy, changes with the time.

We notice that after adding the time t to the mechanical variables, the system becomes conservative in the extended phase space. Ref. [27] also gave a symplectic scheme of second order(SS2), for a sufficiently small $s > 0$ and every x, as the time-step k, the symplectic scheme for equation (17) can be written as

$$\begin{aligned} p_1^{k+1} &= p_1^k + \tfrac{s}{2}(q_1^k + q_1^{k+1}), \\ q_1^{k+1} &= q_1^k + \tfrac{s}{2}(\dot{\phi}(x, q_2^{k+1}) + \dot{\phi}(x, q_2^k)), \\ q_2^{k+1} &= q_2^k + s. \end{aligned} \tag{18}$$

The time step selection criterion is given by Lapusta[3], here used the same criterion.

5. Numerical experiments

Our first numerical test is to solve the spontaneous rupture problem of anti-plane shear fault under the nonlinear slip-dependent. The model is a homogeneous and infinite linearly elastic medium subjected to anti-plane shearing. The assumed translational invariance along the z axis makes the problem 2D. The displacement is parallel to the z direction. The gray surface is the displacement of an iso-slice as induced by an external shear load. Non uniform slip can take place on a pre-existent fault plane Γ at $y = 0$. We employ the TVDRK3 method combined with BI method (TVDRK3-BI) to solve the equation (13) or (18). The model parameters are specified in Table 1. Fig. 1 gives the max slip rate changing with time on fault surface(upper panel) and the max slip on fault surface in whole simulating time(down panel). Fig. 2 gives numerical solution of the rupture speed at rupture time on fault surface generated by the three methods (upper panel), and the

Table 1. The model parameters

parameters	description	value
τ_0,MPa	Initial shear stress	70
$-\sigma_n$, MPa	Initial normal stress	120
f_s	Static friction coefficient	0.667
f_d	Dynamic friction coefficient	0.525
D_c	Critical slip distance	0.4
c_s, m s^{-1}	Shear wave speed	3464
μ	Shear modulus	3.2038e10
L	Fault length, km	100e3
ρ, kg m^{-3}	Density	2670
Δx, m	Space element size	97.656
T_{max}, s	Simulation time	43.303
Δt, s	Time increment	0.012318

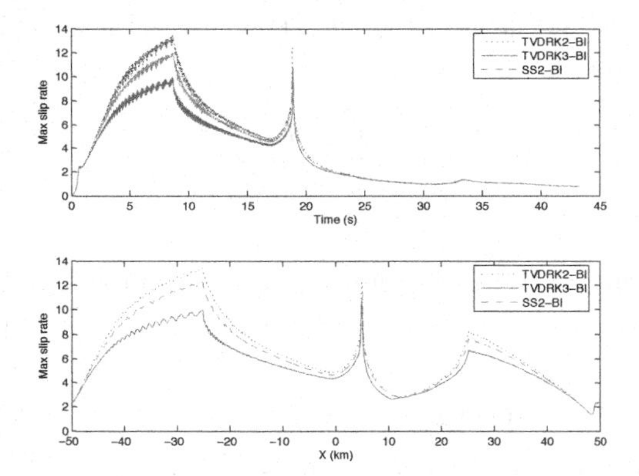

Fig. 1. The max slip rate generated by the three methods. Upper panel is the max slip rate on fault surface changing with time, down panel is the max slip in whole computing time at every spaced points of fault surface.

error generated by SS2-BI and TVDRK3-BI compared with the numerical solution of TVDRK3-BI as used a more accurate solution.

From Figs. 1 and 2 we can obtain that the present symplectic method is effective and perform better performance than the other published methods, it is also reveal that the symplectic method has a long-time tracing ability to simulate the long-term fracture process.

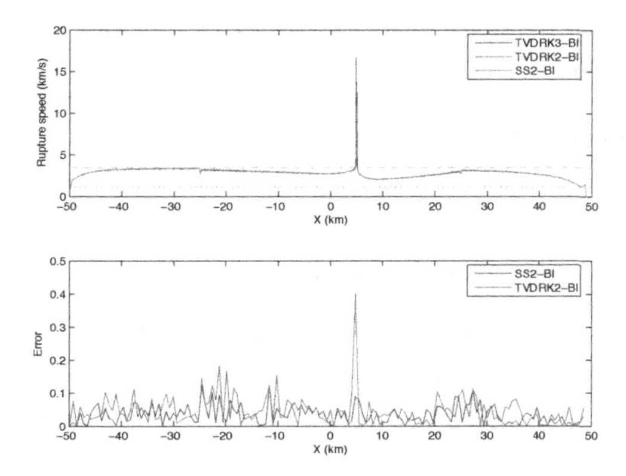

Fig. 2. The numerical solution of rupture speed at rupture time on fault surface, which generated by the three methods (upper panel), and their error solution (down panel).

6. Conclusions

In this paper, an alternative numerical method of modeling elastodynamic fracture problems of long-term histories of faults with nonlinear slip-weakening friction cases has been presented based on spatial discretization of spectral boundary integral method. Although the approach presented is only applied the problems of long-time simulation with nonlinear slip-weakening cases, it can be readily extended to others cases, such as exponential slip-weakening and rate- and state-dependent constitutive laws. From the simulation results in this paper, it has been shown that the present method is of higher accuracy and effective.

Acknowledgements

This study was supported by NNSF(Grant No. 41204041) of P.R. of China.

References

1. Dieterich, J. H. (1978) Time-dependent friction and the mechanism of stick-slip, Pure and Appl. Geophys., 116, 790-806.
2. Palmer A. C., and Rice J. R. (1973), The growth of slip surfaces in the progressive failure of overconsolidated clay slopes, Proc. R. Soc. Lond., A332, 537.
3. Lapusta N., Rice J., Ben-Zion Y., and Zheng G. (2000), Elastodynamic analysis for slow tectonic loading with spontaneous rupture episodes on faults with rate- and state-dependent friction, J. Geophys. Res., 105(23),765-789.

4. Rachel E. Abercrombie and James R. Rice. (2005), Can observations of earthquake scaling constrain slip weakening, Geophys. J. Int. 162, 406-424.
5. Chambon G., Schmittbuhl J., and Corfdir Alain. (2006), Frictional response of a thick gouge sample: 1. Mechanical measurements and microstructures. J. Geophys. Res., 111, B09308.
6. Day S., Dalguer L., Lapusta N., and Liu Y. (2005), Comparison of finite difference and boundary integral solutions to three dimensional spontaneous rupture, J. Geophys. Res., 110, B12307.
7. Moczo P., Kristek J., Galis M., Pazak P., and Balazovjech M. (2007), The finite-difference and finite-element modeling of seismic wave propagation and earthquake motion, Acta physica slovaca, 57(2), 177-406.
8. Cochard A., and Madariaga R. (1994), Dynamic faulting under rate-dependent friction, Pure and Applied Geophysics, 105(25),891- 907.
9. Geubelle P., and Rice J. (1995), A spectral method for three dimensional elastodynamic fracture problems, J.Mech. Phys. Solids, 43, 1791-1824.
10. Benjemaa M., Glinsky-Olivier N., Cruz-Atienza V., Virieux J., and Piperno S. (2007), Dynamic non-planar crack rupture by a finite volume method, Geophys. J. Int., 171, 271-285.
11. Benjemaa M., Glinsky-Olivier N., Cruz-Atienza V., and Virieux J. (2009), 3-D dynamic rupture simulations by a finite volume method, Geophys. J. Int., 178, 541-560.
12. Oglesby D., Archuleta R., and Nielsen S. (1998), The three dimensional dynamics of dipping faults, Bull. Seism. Soc. Am., 90, 616-628.
13. Oglesby D., Archuleta R., and Nielsen S. (2000), Earthquakes on dipping faults: the effects of broken symmetry, Science, 280, 1055-1059.
14. Kaneko Y., Lapusta N., and Ampuero J.-P. (2008), Spectral element modeling of spontaneous earthquake rupture on rate and state faults: Effect of velocity-strengthening friction at shallow depths, J. Geophys. Res., 113, B09317.
15. Galvez P., Ampuero J.-P., Dalguer L., and Nissen-Meyer T. (2011), Dynamic rupture modelling of the 2011 M9 Tohoku earthquake with unstructured 3D spectral element method, U51B-0043.
16. Dumbser M, Käser M. (2006), An arbitrary high order discontinuous Galerkin method for elastic waves on unstructured meshes II: The three-dimensional isotropic case[J]. Geophys. J. Int. 167 : 319-336.
17. Lowry A., Larson K., Kostoglodov V., and Bilham R. (2011), Transient fault slip in Guerrero, southern Mexico, Geophys. Res. Lett., 28, 3753-3756.
18. Miyazaki S., Segall P., McGuire J., Kato T., and Hatanaka Y. (2006), Spatial and temporal evolution of stress and slip rate during the 2000 Tokai slow earthquake, J. Geophys. Res., 111,B03409.
19. Szeliga W., Melbourne T., Miller M., and Santillan V. (2004), Southern Cascadia episodic slow earthquakes, Geophys. Res. Lett., 31, L16602.
20. Douglas A., Beavan J., Wallace L., and Townend J. (2005), Slow slip on the northern Hikurangi subduction interface, New Zealand, Geophys. Res. Lett., 32.

21. Perrin G., and Rice J. R. (1994), Disordering of a dynamic planar crack front in a model elastic medium of randomly variable toughness, J. Mech. Phys. Solids 42(6), 1047-1064.
22. Perrin G., Rice J. R., and Zheng G. (1995), Self-healing slip pulse on a frictional surface, J. Mech. Phys. Solids, 43, 1461-1495.
23. Dalguer L. A., Irikura K., and Riera J. D. (2003), Simulation of tensile crack generation by three-dimensional dynamic shear rupture propagation during an earthquake, J. Geophys. Res., 108(B3), 2144
24. Rice J. R., and Ben-Zion Y. (1996), Slip complexity in earthquake fault models, Proc. Natl. Acad. Sci. U.S.A., 93, 3811-3818.
25. Ben-Zion Y., and Rice J. R. (1997), Dynamic simulations of slip on a smooth fault in an elastic solid, J. Geophys. Res., 102(17),771-784.
26. Morrissey J. W., and Geubelle P. H. (1997), A numerical scheme for mode III dynamic fracture problems, Int. J. Numer. Meth. Eng., 40, 1181-1196.
27. Qin Mengzhao. (1996), Symplectic schemes for nonautonomous Hamiltonian system, Acta Mathematicae Applicatae Sinica, 12(3),284-288.

HIGH-FREQUENCY ASYMPTOTICS FOR THE MODIFIED HELMHOLTZ EQUATION IN A HALF-PLANE

MIN-HAI HUANG

College of Mathematics and Information Sciences, Zhaoqing University, Zhaoqing, GuangDong 526061, China
E-mail:hmh9520@sina.com

Base on the integral representations of the solution being derived via Fokas' transform method, the high-frequency asymptotics for the solution of the modified Helmholtz equation, in a half-plane and subject to the Dirichlet condition is discussed. For the case of piecewise constant boundary data, full asymptotic expansions of the solution are obtained by using Watson's lemma and the method of steepest descents for definite integrals.

Keywords: High-frequency asymptotics; Fokas' transform method; method of steepest descents; modified Helmholtz equation; Dirichlet condition

1. Introduction

There is huge mathematical and engineering interest in acoustic and electromagnetic wave scattering problems, driven by many applications such as modelling radar, sonar, acoustic noise barriers, atmospheric particle scattering, ultrasound and VLSI.[3] Many problems of scattering of time-harmonic acoustic or electromagnetic waves can be formulated as the Helmholtz and modified Helmholtz equations, supplemented with appropriate boundary conditions. Much effort has been put into the development of efficient numerical schemes and approximate methods to deal with the problems of high wave-numbers (i.e., high-frequencies).[1,2,4,8,9] It is noted in [4,7], that a question yet to be fully resolved, is to obtain accurate approximations of the solutions with a reasonable computational cost in the high-frequency case. Therefore it seems desirable, and of difficulty, to consider the high-frequency asymptotics of the equation and its modified version. This is the main motive of the present investigation. Applying the theory of asymptotic analysis,[5,10,12] one may achieve a high degree of accuracy with only a few leading terms in the asymptotic expansions of the solution being involved.

The objective of the present paper is to consider the following Dirichlet boundary value problem of the modified Helmholtz equation in the the upper half-plane Ω:

$$\triangle q(z,\bar{z}) - 4\beta^2 q(z,\bar{z}) = 0, \qquad z \in \Omega, \tag{1}$$

$$q = d(z), \qquad z \in \Gamma \tag{2}$$

where $\mathbf{n}$ is the outer normal vector, $\triangle = \frac{\partial^2}{\partial x^2} + \frac{\partial^2}{\partial y^2} = 4\frac{\partial^2}{\partial z \partial \bar{z}}$ is the usual Laplace operator, $z = x + iy$ and $\bar{z} = x - iy$, $\Gamma = \partial\Omega$ is the closed real line, d decay sufficiently at infinity (e.g. $d \in L^1(\mathbb{R}^+) \bigcap L^2(\mathbb{R}^+)$).

As a first step, we given the integral representation for the solution of the Dirichlet boundary value problem (1)-(2) for general $d(z)$ derived by Fokas' transform method.

And then, by using Watson's lemma and the method of steepest descents for definite integrals, we focus on the high-frequency asymptotics with respect to a specific Dirichlet data, namely,

$$q = d(z) := \begin{cases} D, & z \in [a,b], \\ 0, & z \in \Gamma \backslash [a,b], \end{cases} \tag{3}$$

where $[a,b]$ refers to a finite interval and D is an arbitrary constant.

2. The integral representation for the solution

For the Dirichlet boundary value problem (1)-(2), by using Fokas' transform method, we have the following lemma.

Lemma 2.1. *Assume that the function $q(x,y)$ solves the modified Helmholtz equation (1) in the upper half z-plane Ω, and satisfies the Dirichlet boundary conditions (2), then the integral representation is valid:*

$$q = \frac{\beta}{2\pi} \int_0^\infty e^{i\beta\left(kz - \frac{\bar{z}}{k}\right)} (k + \frac{1}{k}) D(k) \frac{dk}{k}, \tag{4}$$

where

$$D(k) = \int_{-\infty}^{\infty} e^{-i\beta(k-1/k)s} d(s) ds. \tag{5}$$

The interested reader is referred to [6, Ch. 11] and [11] for derived in detail. Accordingly, when specify (2) to the piecewise constant Dirichlet data (3), we have the following lemma.

Lemma 2.2. *Assume that the function $q(x,y)$ solves the modified Helmholtz equation (1) in the upper half z-plane Ω, and satisfies the Dirichlet boundary conditions (3), then the integral representation is valid:*

$$q(x,y) = \frac{D}{2\pi i}\int_l e^{i\beta\left(kz-\frac{\bar{z}}{k}\right)}\left[e^{i\beta a\left(\frac{1}{k}-k\right)} - e^{i\beta b\left(\frac{1}{k}-k\right)}\right]\frac{k^2+1}{k(k^2-1)}dk, \tag{6}$$

where l is an arbitrary ray in the first quadrant of the complex k-plane, oriented from the origin to infinity, with an angle between l and the positive real axis less than $\pi - \arg(z-b)$

3. Asymptotic approximations

Our goal in this section is to study the asymptotical behavior of the solution to the Dirichlet problem of the modified Helmholtz equation, as the frequency, or, equivalently, wavenumbers, approaches to infinity. This large-β asymptotic analysis will be based on the integral representation (6), and will be carried out by using the method of steepest descents.

First we note that (6) can be expressed as the sum of two integrals of the form

$$I(\beta) = \frac{D}{2\pi i}\int_l e^{i\beta\left[k(z+A)-\frac{1}{k}\overline{(z+A)}\right]}\frac{k^2+1}{k(k^2-1)}dk, \tag{7}$$

where $A = -a, -b$, and the path l, the same as in (6), is initially chosen as a ray in the first quadrant, emanating from the k-origin, with an open angle with the positive real line not exceeding $\pi - \arg(z-b)$. We denote

$$\varphi(k) = i\left[k(z+A) - \frac{1}{k}\overline{(z+A)}\right], \quad k = u+iv, \quad z = x+iy, \quad z+A = re^{i\theta}. \tag{8}$$

The phase function $\varphi(k)$ has a pair of saddle points $k_{\pm} = \pm ie^{-i\theta} = \pm e^{i(\pi/2-\theta)}$ determined by $\varphi'(k) = 0$, lying symmetrically on the unit circle. For large positive β, the steepest descent path passing through the saddle point k_+ is simply the ray Γ starts from the origin and passes through k_+. We note that the steepest descent path is defined by requiring $\operatorname{Im}\varphi(k) = \operatorname{Im}\varphi(k_+)$ and $\operatorname{Re}\varphi(k)$ decreasing as k goes away from the saddle; the definition and for basic background of the method of steepest descents can be found in [5,10,12]. Also the steepest ascent path through k_+ is the unit circle, ends at the other saddle k_-; compare Figure 1 for the paths.

Using Cauchy's integral theorem, the integration path l can be deformed to the steepest descent path Γ. Recalling that the argument θ may ranges over $(0,\pi)$ and that the integrand in (7) has a simple pole at $k = 1$, we

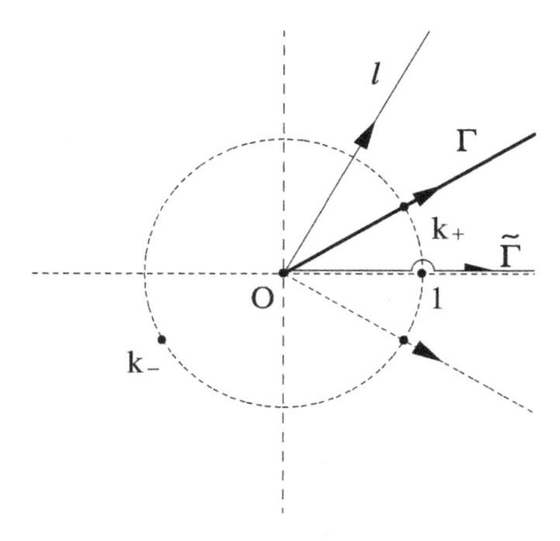

Fig. 1. The steepest descent path Γ.

divide our discussion into three cases, namely (i) $0 < \theta < \pi/2$, (ii) $\theta = \pi/2$, and (iii) $\pi/2 < \theta < \pi$. We deform the paths case by case:

Case (i). When $0 < \theta < \pi/2$, the steepest descent path passing through k_+ lies in the first quadrant of the complex k-plane. We simply deform l into Γ, and have

$$I(\beta) = \frac{D}{2\pi i} \int_\Gamma e^{i\beta\left[k(z+A) - \frac{1}{k}\overline{(z+A)}\,\right]} \frac{k^2+1}{k(k^2-1)} dk. \tag{9}$$

Case (ii). When $\theta = \pi/2$, the steepest descent path passing through $k_+ = 1$ coincides with the positive half real axis of the complex k-plane. Also $k = 1$ is a simple pole of the integrand. The seemly complicated situation turns out to be easily handled, since $I(\beta)$ can be explicitly given in this case:

$$I(\beta) = \frac{D}{2\pi i} \int_{\widetilde{\Gamma}} e^{-\beta y\left(k+\frac{1}{k}\right)} \frac{k^2+1}{k(k^2-1)} dk = -\frac{D}{2} e^{-2y\beta}, \tag{10}$$

where the integration path $\widetilde{\Gamma}$ consists of two segments $[0, 1-\varepsilon]$ and $[1+\varepsilon, \infty)$ and a upper half circle joining them, as illustrated in Figure 1. The last equality is obtained by using the symmetry of the integral under the transformation $k \leftrightarrow 1/k$, picking up half of the residue of the integrand at $k = 1$, and taking the limit as $\varepsilon \to 0$.

Case (iii). When $\pi/2 < \theta < \pi$, the steepest descent path Γ emanating from the saddle point k_+ is in the fourth quadrant of the complex k-plane.

To deform the original path l to Γ, one has to pick up the residue from the simple pole at $k = 1$. Accordingly, we have

$$I(\beta) = \frac{D}{2}e^{-2y\beta} + \frac{D}{2\pi i}\int_\Gamma e^{i\beta\left[k(z+A)-\frac{1}{k}\overline{(z+A)}\,\right]}\frac{k^2+1}{k(k^2-1)}dk. \tag{11}$$

We are now in a position to derive the asymptotic approximation for $q(x, y)$ in (7) for fixed $z = x + iy \in \Omega$. We denote

$$z - a = r_1 e^{i\theta_1}, \ z - b = r_2 e^{i\theta_2}. \tag{12}$$

Obviously we have $0 < \theta_1 < \theta_2 < \pi$. We divide the half z-plane Ω into three regions, namely, I, II and III, defined by $\mathrm{Re}\, z \in (-\infty, a)$, $\mathrm{Re}\, z \in (a, b)$ and $\mathrm{Re}\, z \in (b, \infty)$, respectively; cf. Figure 2. Then it is clear that when $z \in III$, $\theta_j \in (0, \pi/2)$ for $j = 1, 2$; for $z \in II$, $\theta_2 \in (\pi/2, \pi)$ and $\theta_1 \in (0, \pi/2)$; and for $z \in I$, $\theta_1, \theta_2 \in (\pi/2, \pi)$. On the boundaries, when $\mathrm{Re}\, z = b$, $\theta_2 = \pi/2$, while $\theta_1 = \pi/2$ as $\mathrm{Re}\, z = a$.

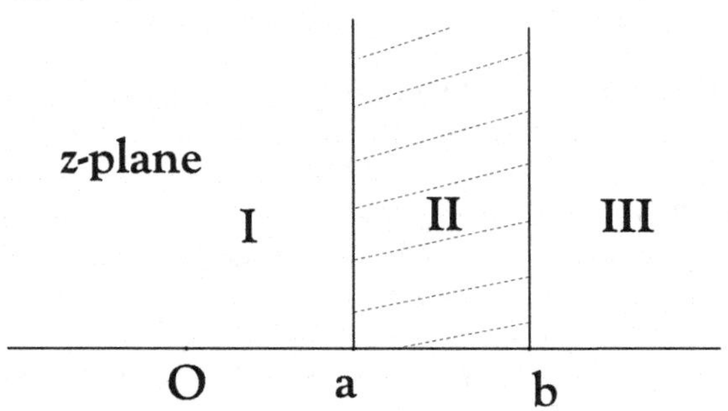

Fig. 2. The regions I, II and III in the complex half z-plane.

Now what remains is just a straightforward application of the method of steepest descents (cf. [5,10,12]). To be precise, we employ the standard argument of Watson's lemma. Firstly, we introduce

$$\tau := \varphi(k_+) - \varphi(k) = -2r - ir\left(ke^{i\theta} - \frac{1}{k}e^{-i\theta}\right) = r\left(|k| - 2 + \frac{1}{|k|}\right), \tag{13}$$

where $k = |k|ie^{-i\theta}$ for $k \in \Gamma$. Obviously $\tau \in [0, \infty)$ on the steepest descent path, and for each $\tau \in (0, \infty)$ there is a pair of points on Γ, say, k_1 and k_2, satisfying (13). Indeed, we may specify

$$k_1 = ie^{-i\theta}\left[1+\frac{\tau}{2r}-\sqrt{\left(1+\frac{\tau}{2r}\right)^2-1}\,\right], k_2 = ie^{-i\theta}\left[1+\frac{\tau}{2r}+\sqrt{\left(1+\frac{\tau}{2r}\right)^2-1}\,\right]. \tag{14}$$

Accordingly one obtains

$$\frac{D}{2\pi i}\int_{\Gamma} e^{\beta\varphi(k)}\frac{k^2+1}{k(k^2-1)}dk = \frac{D}{2\pi i}e^{-2r\beta}\int_0^{\infty}\Phi(\tau)e^{-\beta\tau}d\tau, \tag{15}$$

where

$$\Phi(\tau) = \frac{k_2^2+1}{k_2(k_2^2-1)}\frac{dk_2}{d\tau} - \frac{k_1^2+1}{k_1(k_1^2-1)}\frac{dk_1}{d\tau}$$

which can be expanded into convergent series for small τ in the form

$$\Phi(\tau) = \tau^{-1/2}\sum_{l=0}^{\infty} c_l\tau^l, \quad \tau \in (0, \delta r), \tag{16}$$

where δ is a positive constant, and the coefficients $c_l = c_l(r, \theta)$ can be evaluated in view of (14). Waston's lemma implies that all contribution to the asymptotic behavior of the integrals in (15) comes from the saddle $k = k_+$. Substituting (16) into (15) yields the full asymptotic expansion

$$\frac{D}{2\pi i}\int_{\Gamma} e^{\beta\varphi(k)}\frac{k^2+1}{k(k^2-1)}dk \sim \frac{D}{2\pi i}e^{-2r\beta}\sum_{l=0}^{\infty}\frac{c_l\Gamma(l+1/2)}{\beta^{l+1/2}}, \quad \beta \to +\infty, \tag{17}$$

where $\Gamma(t)$ is the usual Gamma function. We summarize our discussion as follows:

Theorem 3.1. *For large positive β (i.e., high frequency and equivalently, large wavenumbers), the asymptotic expansions of the solution $q(x, y)$ to the boundary value problem (1), (3) of the modified Helmholtz equation hold.*

$$\begin{aligned} q(x,y) \sim {} & \frac{D}{2\pi i}\sum_{j=1}^{2}(-1)^{j+1}e^{-2r_j\beta}\sum_{l=0}^{\infty}\frac{c_l(r_j,\theta_j)\Gamma(l+1/2)}{\beta^{l+1/2}} \\ & + \frac{D}{2}e^{-2y\beta}\cdot\begin{cases} 0, & z \in I \cup III \\ 1, & z \in II \end{cases} \end{aligned} \tag{18}$$

as $\beta \to +\infty$, where r_j and θ_j are defined in (12). Two exceptional cases are

$$q(x,y) \sim -\frac{D}{2\pi i}e^{-2r_2\beta}\sum_{l=0}^{\infty}\frac{c_l(r_2,\theta_2)\Gamma(l+1/2)}{\beta^{l+1/2}} + \frac{D}{4}e^{-2y\beta}, \tag{19}$$

as $\beta \to +\infty$ and $\operatorname{Re} z = a$, *and*

$$q(x,y) \sim \frac{D}{2\pi i}e^{-2r_1\beta}\sum_{l=0}^{\infty}\frac{c_l(r_j,\theta_1)\Gamma(l+1/2)}{\beta^{l+1/2}} + \frac{D}{4}e^{-2y\beta}, \tag{20}$$

as $\beta \to +\infty$ and $\operatorname{Re} z = b$.

Acknowledgement

This research was supported in part by the National Natural Science Foundation of China under grant numbers 10871212.

References

1. S. Arden, S.N. Chandler-Wilde and S. Langdon, A collocation method for high frequency scattering by convex polygons, *J. Comp. Appl. Math.*, **204**(2007), 334-343.
2. G. Bao, G.-W. Wei and S. Zhao, Numerical solution of the Helmholtz equation with high wavenumbers, *Int. J. Numer. Methods. Eng.*, **59**(2004), 389-408.
3. S.N. Chandler-Wilde and I.G. Graham, Boundary integral methods in high-frequency scattering, in "Highly Oscillatory Problems", B. Engquist, T. Fokas, E. Hairer, A. Iserles, editors, Cambridge University Press (2009), p.154-193.
4. S.N. Chandler-Wilde and S. Langdon, A Galerkin boundary element method for high frequency scattering by convex polygons, *SIAM J. Numer. Anal.*, **45**(2007), 610-640.
5. E.T. Copson, *Asymptotic expansions*, Cambridge tracts in Math. and Math. Phys. No.55, Cambridge University press, London, 1965.
6. A.S. Fokas, *A unified approach to boundary value problems*, SIAM, Philadelphia, 2008.
7. J.B. Keller, Progress and prospects in the theory of linear wave propagation, *SIAM Review*, **21**(1979), 229.
8. S. Kim, C.-S. Shin and J.B. Keller, High-frequency asymptotics for the numerical solution of the Helmholtz equation, *Appl. Math. Lett.*, **18**(2005), 797-804.
9. S. Langdon and S.N. Chandler-Wilde, A wavenumber independent boundary element method for acoustic scattering problem, *SIAM J. Numer. Anal*, **43**(2006), 2450-2477.
10. F.W.J. Olver, *Asymptotics and Special Functions*, Academic Press, New York, 1974.
11. E.A. Spence, Boundary Value Problems for Linear Elliptic PDEs, PhD Thesis, University of Cambridge, 2010.
12. R. Wong, *Asymptotic Approximations of Integrals*, Academic Press, Boston, 1989.

AN INVERSE BOUNDARY VALUE PROBLEM INVOLVING FILTRATION FOR ELLIPTIC SYSTEMS OF EQUATIONS*

ZUOLIANG XU   LI YAN †

School of Information, Renmin Unversity of China
Beijing 100872, P. R. China

In this paper, a inverse boundary value problem involving filtration for elliptic systems of equations is discussed. The problem can be reduced to a discontinuous Riemann-Hilbert boundary value problem for elliptic complex equations. By using the theory of complex equations and Newton imbedding method, we prove the existence and uniqueness of the inverse boundary value problem.

Keywords: Inverse boundary value problem; Elliptic complex equation; Filtration.

1. Introduction

It is known that there are many problems in mechanics and physics, where part or the whole boundary of a region is unknown, which are called inverse boundary value problems or free boundary problems. As a rule, a lot of problems involving filtration of a fluid are free boundary problems. This kinds of problems have received a great deal of attention recently([1]-[5]). It was seen that the inverse boundary value problem in planar fluid dynamics and gas dynamics might be transformed into a mixed boundary value problems for linear and nonlinear elliptic complex equations. The plannar question of the depression curve as it goes to infinity is important in many problems involving filtration of a fluid, which is a inverse boundary value problem. In this paper, we consider this kinds of inverse boundary value problems in nonhomogeneous anisotropic media by using complex boundary value problems([1],[6]-[9]). It is assumed that the filtration flow is the steady-state, finite at infinity and the impermeable surface goes out to a horizontal asymptote. The fluid filtration scheme is shown in Figure

---

*Supported by the National Natural Science Foundation of China(Grant No. 10971224).
†Corresponding author. E-mail: xuzl@ruc.edu.cn, yanli@ruc.edu.cn

1, where D_z is the filtration domain, AB is the permeable boundary, BC is the unknown curve which is a free boundary, AC is the impermeable surface. In the case of filtration of a fluid in nonhomogeneous anisotropic media the piezometric head $h(x,y) = p/\rho g + y$ satisfies the equation

$$\vec{q} = -k \cdot grad\, h \tag{1.1}$$

where $\vec{q} = (u,v)$ is the filtration velocity. $k = \{k_{ij}(x,y,h)\}$ is the symmetric filtration tensor. Introducing the generalized potential $\phi = -k_0 h$ (k_0 is constant), and the stream function $\psi(x,y)$ by setting $\psi_x = -v, \psi_y = u$, then we have the following system of equations for $\phi(x,y), \psi(x,y)$

$$\begin{cases} \frac{\partial\psi}{\partial y} = k_{11}^0 \frac{\partial\phi}{\partial x} + k_{12}^0 \frac{\partial\phi}{\partial y}, \\ -\frac{\partial\psi}{\partial x} = k_{21}^0 \frac{\partial\phi}{\partial x} + k_{22}^0 \frac{\partial\phi}{\partial y}, \end{cases} \tag{1.2}$$

where $k_{ij}^0 = k_{ij}/k_0$, and $k_{ij}^0 (i = 1,2, j = 1,2, k_{12}^0 = k_{21}^0)$ for $(x,y) \in D_z, h$ is continuous, and if k_{ij}^0 satisfies

$$|k_{ij}^0| \leq M_0 < \infty, k_{11}^0 k_{22}^0 - {k_{12}^0}^2 \geq \delta_0 > 0 \tag{1.3}$$

then the above system is uniformly elliptic(see [1]).

We consider the boundary conditions on the boundaries of the filtration domain D_z. For the permeable boundary AB, the piezometric head $h(x,y)$ is constant, therefore ϕ=const, and suppose that AB is known, denote it by $y = f(x)$, and $H = f(0)$ is given. On the free boundary BC, which is a streamline, it is usually assumed the pressure is a constant, and, consequently, $\phi + k_0 y$ =constant. However, the free boundary goes to infinity in such a filtration scheme the piezometric head $h = h(x,y)$ tends to a constant value, so that it is impossible to ignore the velocity head, and the influence of the velocity head begins to take effect. The boundary condition for the $h = h(x,y)$ is replaced by Bernoulli's equation for the filtration speed $q = q(x)$ beginning with some value $x = x_0$ ensures that the effect of the velocity head at infinity is taken into account, we suppose that the speed distribution $q = q(x)$ of filtration flow is given on $EC(x \geq x_0)$. Moreover, Bernoulli's equation gives us for the fluid that,

$$h(x,y) + q^2/2g = C_1 \tag{1.4}$$

Where C_1 is a known constant. Hence,$h = h(x)$ is a given function, and $\phi = g(x)$ is known on EC.

For the impermeable surface AC, it is also a streamline, which is a horizontal asymptote. Thus the boundary conditions can be written as follows

$$\begin{cases} AB: y = f(x),\ \ \phi = \phi_0, \\ BE: \phi + k_0 y = c,\ \ \psi = \psi_1, \\ EC: \phi = g(x),\ \ \psi = \psi_1, \\ AC: y = 0,\ \ \psi = \psi_0. \end{cases} \tag{1.5}$$

Without loss of generality, we may assume that $\psi_0 = 0$. Our problem is to find a continuous solution of the equation (1.2) in D_z satisfying the boundary condition (1.5), and determine the unknown boundary $BC(BE$ and $EC)$. This problem will be called Problem A, which is an inverse boundary value problem, is also called free boundary problem.

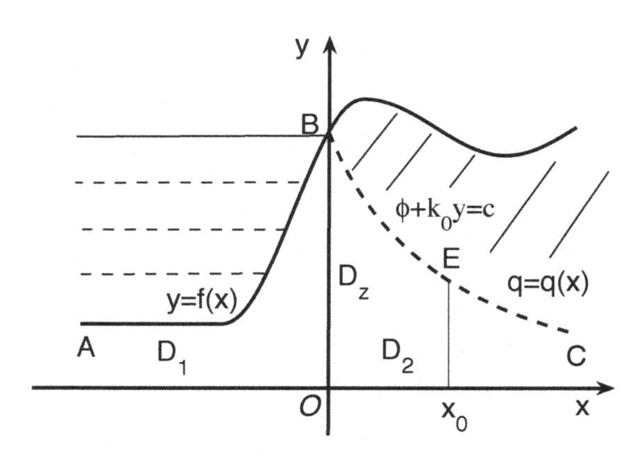

Figure 1

2. Problem transformation

We introduce the generalized complex filtration potential $w = \phi + i\psi$, noticing that the boundary condition (1.5), it is obvious to see that the domain D_w in the plane of the complex potential corresponding to the flow region D_z is a known fixed domain (see Figure 2).

And we can arrive the complex respective equations for the function $w(z) = \phi + i\psi$ and its inverse $z = x + iy = z(w)$:

$$w_{\overline{z}} + \mu_2(w, z)w_z + \mu_1(w, z)\overline{w}_{\overline{z}} = 0, \tag{2.1}$$

$$z_{\overline{w}} - \mu_1(w, z)z_w - \mu_2(w, z)\overline{z}_{\overline{w}} = 0, \tag{2.2}$$

where

$$\mu_1 = \frac{k_{22}^0 - k_{11}^0 - 2ik_{12}^0}{1 + k_{11}^0 + k_{22}^0 + k_{12}^0(k_{22}^0 - k_{11}^0)},$$

$$\mu_2 = \frac{1 - k_{22}^0 k_{11}^0 - 2ik_{12}^0}{1 + k_{11}^0 + k_{22}^0 + k_{12}^0(k_{22}^0 - k_{11}^0)}.$$

Because of the assumptions (1.3) made above about the filtration tensor, equations (2.1) and (2.2) are uniformly elliptic, i.e.,

$$|\mu_1(w, z)| + |\mu_2(w, z)| \leqslant \mu_0(\delta, M) < 1. \tag{2.3}$$

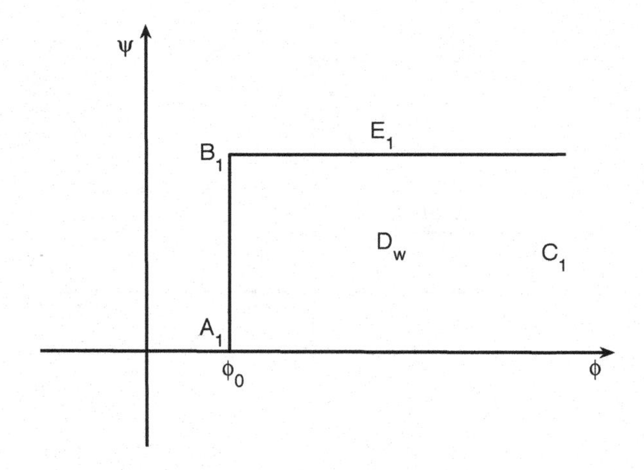

Figure 2

On the z-plane, let us define the transformation in the respective domains D_1 and D_2:

$$Z = f(x) - y + iy, \ (D_1) \ ; Z = x - y + H + iy, \ (D_2). \tag{2.4}$$

Then we arrive at the following boundary value problem for determining the function $Z(w) = Z[z(w)]$:

$$ReZ = 0, \quad on\ AB; \quad ImZ = 0, \quad on\ AC;$$

$$ImZ = \frac{c-\phi}{k_0}, \quad on\ BE; \quad ReZ = g^{-1}(\phi) - ImZ + H, \quad on\ EC.$$

Let $w = w(\zeta)$ be a conformal mapping of the disk $D_\zeta : |\zeta| < 1$ onto D_w, and $w(t_1) = \phi_0$, $w(t_2) = \phi_0 + i\psi_1$, $w(t_3) = \phi_1 + i\psi_1$, $w(t_4) = \infty$, $t_j \in \Gamma : |\zeta| = 1, j = 1,2,3,4$. This transformation takes (2.2) to the form

$$z_{\overline{\zeta}} - \hat{\mu_1}(\zeta, z)z_\zeta - \hat{\mu_2}(\zeta, z)\overline{z}_{\overline{\zeta}} = 0, \tag{2.5}$$

where

$$\hat{\mu_1} = \mu_1[w(\zeta), \zeta]exp\{-2iarg\frac{dw}{d\zeta}\}, \quad \hat{\mu_2} = \mu_2[w(\zeta), \zeta].$$

Assume that $f(x) \in C^1$, and

$$|f'(x)| \geq \varepsilon_0 \quad x \in (-\infty, 0]. \tag{2.6}$$

Since the Jacobian of this transformation has the form

$$\frac{DZ}{Dz} = |Z_z|^2 - |Z_{\overline{z}}|^2 = yf'(x)/f^2(x)$$

the transformation (2.4) is one-to-one, non-degenerate. For the line BO, when $(x, y) \to (0, y_0)$ in D_1 and D_2, we have $Z \to H - y_0 + iy_0$, thus $Z = Z(z)$ is continuous in D_z. Differentiation of the equality $Z = Z[z(\zeta)] \equiv Z(\zeta)$ gives us that

$$Z_\zeta = Z_{\overline{z}}\overline{z}_\zeta + Z_z z_\zeta, \quad Z_{\overline{\zeta}} = Z_{\overline{z}}\overline{z}_{\overline{\zeta}} + Z_z z_{\overline{\zeta}},$$

which, after determining z_ζ and $z_{\overline{\zeta}}$ and substituting them in (2.5), leads to the following equation for the unknown function $Z = Z(\zeta)$:

$$Z_{\overline{\zeta}} = q_1(\zeta, Z)Z_\zeta + q_2(\zeta, Z)\overline{Z}_{\overline{\zeta}}, \tag{2.7}$$

where

$$q_1 = [(Z_z)^2 - (Z_{\overline{z}})^2 - 2iZ_z Z_{\overline{z}} Im\hat{\mu_2}]Q(\zeta),$$

$$q_2 = \{\hat{\mu_2}(|Z_z|^2 + |Z_{\overline{z}}|^2 - \hat{\mu_2}Z_z Z_{\overline{z}}) + \overline{Z}_{\overline{z}} Z_{\overline{z}}(|\hat{\mu_1}| - \frac{Z_z\overline{Z}_z}{\overline{Z}_{\overline{z}}Z_{\overline{z}}})\}Q(\zeta),$$

$$\frac{1}{Q(\zeta)} = |Z_z - \mu_2 Z_{\overline{z}}|^2 - |\hat{\mu_1} Z_{\overline{z}}|^2.$$

Moreover, it is easy to see that $|q_1| + |q_2| \le q_0 < 1$ i.e., the equation (2.7) is also uniformly elliptic.(see [1])

Next, we consider the boundary condition (1.5), which can be reduced to

$$\begin{cases} ReZ = 0, \quad \zeta \in \widetilde{t_1 t_2}, \\ ImZ = \frac{1}{k_0}(c - \phi(\zeta)), \quad \zeta \in \widetilde{t_2 t_3}, \\ ReZ + ImZ = g^{-1}(\phi(\zeta)) + H, \ \zeta \in \widetilde{t_3 t_4}, \\ ImZ = 0, \quad \zeta \in \widetilde{t_4 t_1}. \end{cases} \tag{2.8}$$

Thus, we come to the discontinuous Riemann-Hilbert boundary value problem for equation (2.7), i.e. to find a continuous solution $Z(\zeta)$ of the complex equation (2.7) in D_ζ satisfying the boundary condition

$$\mathrm{Re}[\overline{\lambda(\zeta)} Z(\zeta)] = r(\zeta), \quad \zeta \in \Gamma : |\zeta| = 1, \tag{2.9}$$

where

$$\lambda(\zeta) = \begin{cases} 1, & \zeta \in \widetilde{t_1 t_2}, \\ i, & \zeta \in \widetilde{t_2 t_3}, \\ \frac{1}{\sqrt{2}} + \frac{i}{\sqrt{2}}, & \zeta \in \widetilde{t_3 t_4}, \\ i, & \zeta \in \widetilde{t_4 t_1}, \end{cases} \tag{2.10}$$

$$r(\zeta) = \begin{cases} 0, & \zeta \in \widetilde{t_1 t_2}, \\ \frac{1}{k_0}(c - \phi(\zeta)), & \zeta \in \widetilde{t_2 t_3}, \\ \frac{1}{\sqrt{2}}[g^{-1}(\phi(\zeta)) + H], & \zeta \in \widetilde{t_3 t_4}, \\ 0, & \zeta \in \widetilde{t_4 t_1}. \end{cases} \tag{2.11}$$

This problem will be simply called Problem B. Therefore, Problem A is transformed into Problem B for the nonlinear elliptic complex equation (2.7). In next section, we will prove that Problem B has a bounded solution $Z(\zeta)$ in D_ζ. Moreover, if $F(\zeta, Z, Z_\zeta)$ satisfies the condition

$$|F(\zeta, Z_1, V) - F(\zeta, Z_2, V)| \le R(\zeta)|Z_1 - Z_2|, \tag{2.12}$$

where $F(\zeta, Z, Z_\zeta) = q_1(\zeta, Z)Z_\zeta + q_2(\zeta, Z)\overline{Z_{\overline{\zeta}}}$, for $\zeta \in D_\zeta, Z_1, Z_2, V \in C$, $R(\zeta) \in L_p(\overline{D_\zeta})$, $p > 2$, then we can prove that the solution of Problem B is unique.

Setting $Z(w) = Z[\zeta(w)], Z(w) = Z(\phi + i\psi) = g(x) - y + iy(D_1), Z(w) = x - y + H + iy(D_2)$, from $y = y(\phi, \psi)$, we can find out $x = x(\phi + i\psi)$. Thus, we obtain $z = z(w), z = x + iy$,and $w = w(z), \phi = \phi(x, y), \psi = \psi(x, y)$, which is just a solution of Problem A. When we find out the complex potential $w(z)$, the quantity ψ_1 of filtration and the equation $\phi = -k_0 y = c,\ \psi = \psi_1$ and $\phi = g(x), \psi = \psi_1$ of unknown boundary BE and EC are obtained.

3. The existence and uniqueness of solution for problem B

If conditions (1.3), (2.6) and (2.12) hold, we can prove that the solution of Problem B is unique. Thus, the uniqueness theorem is obtained.

Theorem 3.1. Under assumption (1.3), conditions (2.6) and (2.12), the solution of Problem B for complex equation (2.7) is unique

Proof. Let $Z_1(\zeta), Z_2(\zeta)$ be the two solutions of Problem B for (2.7), and $Z^*(\zeta) = Z_1(\zeta) - Z_2(\zeta)$, then $Z^*(\zeta)$ is the solution of the following complex equation

$$Z_{\overline{\zeta}} = Q^*(\zeta)Z_\zeta + A^*(\zeta)Z, \quad \zeta \in D_\zeta, \tag{3.1}$$

where

$$Q^*(\zeta) = \begin{cases} \frac{F(\zeta,Z_1,Z_{1\zeta})-F(\zeta,Z_1,Z_{2\zeta})}{Z_{1\zeta}-Z_{2\zeta}}, & Z_{1\zeta} \neq Z_{2\zeta}, \zeta \in D_\zeta, \\ 0, & Z_{1\zeta} = Z_{2\zeta} \text{ or } \zeta \notin D_\zeta. \end{cases}$$

$$A^*(\zeta) = \begin{cases} \frac{F(\zeta,Z_1,Z_{2\zeta})-F(\zeta,Z_2,Z_{2\zeta})}{Z_1-Z_2}, & Z_1 \neq Z_2, \zeta \in D_\zeta, \\ 0, & Z_1 = Z_2 \text{ or } \zeta \notin D_\zeta \end{cases}$$

and we can get $|Q^*(\zeta)| \le Q_0^* < 1, A^*(\zeta) \in L_p(\overline{D_\zeta}), Q_0^*$ is a positive constant. According to the theorem of solution express [7], we have

$$Z^*(\zeta) = \Phi(\sigma(\zeta))e^{\varphi(\zeta)} \tag{3.2}$$

where $\sigma(\zeta),\ \varphi(\zeta) \in C_\alpha(\overline{D_\zeta}),\ 0 < \alpha < 1, \sigma(\zeta)$ is a homeomorphism which quasiconformally maps D_ζ onto the same type domain D_σ,and inverse function $\zeta(\sigma) \in C_\alpha(\overline{D_\sigma})$. Thus, $Z^*(\zeta)$ satisfying the homogeneous boundary condition

$$\mathrm{Re}[\overline{\lambda(\zeta)}Z^*(\zeta)] = 0, \quad \zeta \in \Gamma, \tag{3.3}$$

can be reduced to the analytic function $\Phi(\sigma)$ satisfying boundary condition

$$\mathrm{Re}[\overline{\lambda(\zeta(\sigma))}e^{\varphi[\zeta(\sigma)]}\Phi(\sigma)] = 0, \quad \zeta \in L := \sigma(\Gamma). \tag{3.4}$$

Denote by $\lambda(t_j - 0)$ and $\lambda(t_j + 0)$ the left limit and right limit of $\lambda(\zeta)$ as $\zeta \to t_j (j = 1, 2, 3, 4)$ on Γ, and noting that

$$e^{i\varphi_j} = \frac{\lambda(t_j - 0)}{\lambda(t_j + 0}, \quad \gamma_j = \frac{1}{\pi i} ln \frac{\lambda(t_j - 0)}{\lambda(t_j + 0} = \frac{\varphi_j}{\pi} - K_j. \tag{3.5}$$

By (2.10), we can get $\varphi_1 = \frac{\pi}{2}$, $\varphi_2 = -\frac{\pi}{2}$, $\varphi_3 = \frac{\pi}{4}$, $\varphi_4 = \frac{3\pi}{4}$. In order to seek the bounded solutions , it is required that $0 < \gamma_j < 1$, in which $K_j (j = 1, 2, 3, 4)$ are integers, hence $K_1 = K_3 = K_4 = 0, K_2 = -1$. This shows that the index

$$K = \frac{1}{2}(K_1 + K_2 + K_3 + K_4) = -\frac{1}{2}. \tag{3.6}$$

By using the results of Chapter 1 in [9], we can obtain the solution of Problem B for analytic functions is unique. It is easy to see that $\Phi(\sigma) = 0$, thus $Z^*(\zeta) = 0$, i.e. $Z_1(\zeta) = Z_2(\zeta)$. This completes the proof.

In the following, we prove the existence of Problem B for (2.7) by using Newton imbedding method. In order to seek the approximate solution of Problem B, we consider the complex equation with a parameter $\theta \in [0, 1]$:

$$Z_{\overline{\zeta}} - \theta F(\zeta, Z, Z_\zeta) = B(\zeta),\ B(\zeta) = (1 - \theta)F(\zeta, 0, 0), \tag{3.7}$$

and first suppose that $F(\zeta, Z, Z_\zeta) = 0$ in the $\varepsilon_m = 1/m$ (m is a positive integer) neighborhood U_m of points $t_j (j = 1, 2, 3, 4)$, it sufficies to multiply $F(\zeta, Z, Z_\zeta)$ by the function

$$\eta_m(\zeta) = \begin{cases} 0, & \zeta \in U_m = \cup_{j=1}^4 \{|\zeta - t_j| < 1/m\}, \\ 1, & \zeta \in D_m = \overline{D_\zeta} \backslash U_m, \end{cases} \tag{3.8}$$

we know that when $\theta = 0$, Problem B for the complex equation (2.7) has a unique bounded solution $Z(\zeta)$, which possesses the form

$$Z(\zeta) = \Phi(\zeta) + \Psi(\zeta), \Psi(\zeta) = TB = -\frac{1}{\pi}\int_{D_\zeta} \frac{B(\tau)}{\tau - \zeta} d\sigma_\tau, \tag{3.9}$$

where $\Phi(\zeta)$ is a bounded analytic function in D_ζ satisfying the boundary condition

$$\text{Re}[\overline{\lambda(\zeta)}\Phi(\zeta)] = r(\zeta) - \text{Re}[\overline{\lambda(\zeta)}\Psi(\zeta)], \zeta \in \Gamma^* = \Gamma \backslash \{t_1, t_2, t_3, t_4\}. \tag{3.10}$$

Introduce a function $g(\zeta) = \prod_{j=1}^4 (\zeta - t_j)^\eta$, it can be derived $g(\zeta)Z(\zeta) \in C_\beta(\overline{D_\zeta}) \cap W^1_{p_0}(D_\zeta)$, herein $p_0 (2 < p_0 < p)$, $\beta = 1 - \frac{2}{p_0}$, $0 < \eta < 1$.

Suppose that when $\theta = \theta_0 (0 < \theta_0 < 1)$, Problem B for (2.7) has a bounded solution, we shall prove that there exists a neighborhood of θ_0 : $E = \{|\theta - \theta_0| \leq \delta, 0 \leq \theta \leq 1, \delta > 0\}$, such that for every $\theta \in E$ and any function $B(\zeta) \in L_{p_0}(\overline{D_\zeta})$, Problem B for (3.7) is solvable. Practically, the complex equation (3.7) can be written in the form

$$Z_{\overline{\zeta}} - \theta_0 F(\zeta, Z, Z_\zeta) = (\theta - \theta_0)F(\zeta, Z, Z_\zeta) + B(\zeta). \tag{3.11}$$

We are free to choose a function $Z_0(\zeta)$ so that $g(\zeta)Z_0(\zeta) \in C_\beta(\overline{D_\zeta}) \cap W^1_{p_0}(D_\zeta)$, in particular $Z_0(\zeta) = 0$ on $\overline{D_\zeta}$. Let $Z_0(\zeta)$ in the right hand side of (3.11). From the condition mentioned in the Section 2, it is obvious that

$$B_0(\zeta) = (\theta - \theta_0)F(\zeta, Z_0, Z_{0\zeta}) + B(\zeta) \in L_{p_0}(\overline{D_m}). \tag{3.12}$$

Noting the assumption as before, Problem B for the complex equation (3.11) has a solution $Z_1(\zeta)$, so that $g(\zeta)Z_1(\zeta) \in C_\beta(\overline{D_\zeta}) \cap W^1_{p_0}(D_\zeta)$. By using the successive iteration, we can find out a sequence of functions: $Z_n(\zeta)$, $g(\zeta)Z_n(\zeta) \in C_\beta(\overline{D_\zeta}) \cap W^1_{p_0}(D_\zeta)$, which satisfy the complex equations

$$Z_{n+1\overline{\zeta}} - \theta_0 F(\zeta, Z_{n+1}, w_{n+1\zeta}) = (\theta - \theta_0)F(\zeta, Z_n, Z_{n\zeta}) + B(\zeta), n = 1, 2, \cdots. \tag{3.13}$$

The difference of the above equations for $n+1$ and n is as follows

$$\begin{aligned} &(Z_{n+1} - Z_n)_{\overline{\zeta}} - \theta_0[F(\zeta, Z_{n+1}, Z_{n+1\zeta}) - F(\zeta, Z_n, Z_{n\zeta})] \\ &= (\theta - \theta_0)[F(\zeta, Z_n, Z_{n\zeta}) - F(\zeta, Z_{n-1}, Z_{n-1\zeta})], n = 1, 2, \cdots. \end{aligned} \tag{3.14}$$

From the condition (1.3) and (2.12), it can be seen that

$$\begin{aligned} &F(\zeta, Z_{n+1}, Z_{n+1\zeta}) - F(\zeta, Z_n, Z_{n\zeta}) = [F(\zeta, Z_{n+1}, Z_{n+1\zeta}) \\ &-F(\zeta, Z_{n+1}, Z_{n\zeta})] + [F(\zeta, Z_{n+1}, Z_{n\zeta}) - F(\zeta, Z_n, Z_{n\zeta})] \\ &= \widetilde{Q}_{n+1}(\zeta)(Z_{n+1} - Z_n)_\zeta + \widetilde{B}_{n+1}(\zeta)(Z_{n+1} - Z_n), \\ &|\widetilde{Q}_{n+1}(\zeta)| \le Q_0 < 1,\ L_{p_0}[\widetilde{B}_{n+1}(\zeta), D_\zeta] \le K_0, n = 1, 2, \cdots, \end{aligned} \tag{3.15}$$

K_0 is a constant, and

$$\begin{aligned} &L_{p_0}[F(\zeta, Z_n, Z_{n\zeta}) - F(\zeta, Z_{n-1}, Z_{n-1\zeta}), D_m] \\ &\le Q_0 L_{p_0}[(Z_n - Z_{n-1})_\zeta, D_m] + K_0 C[Z_n - Z_{n-1}, D_m] \\ &\le (Q_0 + K_0)\{C_\beta[g(Z_n - Z_{n-1}), \overline{D_\zeta}] + L_{p_0}[|(g(Z_n - Z_{n-1}))_{\overline{\zeta}}| \\ &+|(g(Z_n - Z_{n-1}))_\zeta|, D_\zeta]\} = (Q_0 + K_0)S_n, \end{aligned} \tag{3.16}$$

in which $S_n = C_\beta[g(Z_n - Z_{n-1}), \overline{D_\zeta}] + L_{p_0}[|(g(Z_n - Z_{n-1}))_{\overline{\zeta}}| + |(g(Z_n - Z_{n-1}))_\zeta|, D_\zeta]$. Moreover, $Z_{n+1}(\zeta) - Z_n(\zeta)$ satisfies the homogeneous boundary condition

$$\mathrm{Re}[\overline{\lambda(\zeta)}(Z_{n+1}(\zeta) - Z_n(\zeta))] = 0, \zeta \in \Gamma^*. \tag{3.17}$$

On the basis of [7], we have

$$\begin{aligned} S_{n+1} &= C_\beta[g(Z_n - Z_{n-1}), \overline{D_\zeta}] + L_{p_0}[|(g(Z_n - Z_{n-1}))_{\overline{\zeta}}| \\ &+|(g(Z_n - Z_{n-1}))_\zeta|, \overline{D_\zeta}] \le M_1|\theta - \theta_0|(Q_0 + K_0)S_n, \end{aligned} \tag{3.18}$$

where M_1 is a positive integer. Provided that $\delta(> 0)$ is small enough, so that $\eta = \delta M_1(Q_0 + K_0) < 1$, it can be obtained that

$$S_{n+1} \le \eta^n S_1 = \eta^n[C_\beta[gZ_1, \overline{D_\zeta}] + L_{p_0}[|(gZ_1)_{\overline{\zeta}}| + |(gZ_1)_\zeta|, \overline{D_\zeta}] \tag{3.19}$$

for every $\theta \in E$. Thus

$$\begin{aligned} S_{nl} = & C_\beta[g(Z_n - Z_l), \overline{D_\zeta}] + L_{p_0}[|(g(Z_n - Z_l))_{\overline{\zeta}}| + |(g(Z_n - Z_l))_\zeta|, \overline{D_\zeta}] \\ & \leq S_n + S_{n-1} + \cdots + S_{l+1} \leq (\eta^{n-1} + \eta^{n-2} + \cdots + \eta^l) S_1 \\ & = \eta^l (1 + \eta + \cdots + \eta^{n-l-1}) S_1 \leq \eta^{N+1} \frac{1-\eta^{n-l}}{1-\eta} S_1 \leq \frac{\eta^{N+1}}{1-\eta} S_1 \end{aligned} \qquad (3.20)$$

for $n \geq l > N$, where N is a positive integer. This shows that $S(Z_n - Z_l) \to 0$ as $n, l \to \infty$. Following the completeness of the Banach space $B = C_\beta(\overline{D_\zeta}) \cap W^1_{p_0}(D_m)$, there is a function $Z_*(\zeta) g(\zeta) \in B$, such that when $n \to \infty$,

$$\begin{aligned} S(Z - Z_*) = C_\beta[g(Z_n - Z_*), \overline{D_\zeta}] + L_{p_0}[|(g(Z_n - Z_*))_{\overline{\zeta}}| \\ + |(g(Z_n - Z_*))_\zeta|, \overline{D_\zeta}] \to 0, \end{aligned}$$

from (3.12) it follows that $Z_*(\zeta)$ is a solution of Problem B for (3.11), i.e. (3.7) for $\theta \in E$. It is easy to see that the positive constant δ is independent of $\theta_0 (0 \leq \theta_0 < 1)$. Hence from Problem B for the complex equation (3.7) with $\theta = \theta_0 = 0$ is solvable, we can derive that when $\theta = \delta, 2\delta, \ldots, [1/\delta]\delta, 1$, Problem B for (3.7) are solvable, especially Problem B for (3.7) with $\theta = 1$ and $B(\zeta) = 0$, namely (2.7) has a bounded solution. Thus, we have

Theorem 3.2. Let the complex equation (2.7) satisfy the condition(1.30, (2.6) and (2.12), using Newton Imbedding Method, we can derive the approximate solution $Z_n(\zeta)$ of Problem B of complex equation (2.7), and Problem B has a bounded solution $Z(\zeta)$, such that $g(\zeta)Z(\zeta) \in B = C_\beta(\overline{D_\zeta}) \cap W^1_{p_0}(D_\zeta), \beta = 1 - 2/p_0, 2 < p_0 \leq p, g(\zeta) = \Pi^4_{j=1}(\zeta - t_j)^\eta, 0 < \eta < 1$.

We summarize the above discussion as

Theorem 3.3. Problem B for elliptic complex equation (2.7) has a unique solution, i.e. there exists a unique solution for the inverse boundary value problem A under the same conditions of Theorem 3.1.

References

1. V. N. Monakhov, Boundary value problem with free boundaries for elliptic systems of equations, Providence, RI, Amer. Math. Soc.,1983.
2. R. P. Gilbert, G. C. Wen, Free boundary problems occurring in plannar fluid dynamics, Nonlinear Analysis, 13(1989), 285-303.
3. Z. L. Xu, A free boundary problem involving filtration of nonhomogeneous anisotripic dam, J. Liaoning Inst. of Tech., Vol. 17,No.4 (1997), 57-60.(Chinese)
4. G. C. Wen, A free boundary problem in axisymmetric filtration with two fluids, Acta Math. App. Sinica, 21(1998), 139-143.(Chinese) Appl. Math., 15 (1962) 289-347.

5. Z. L. Xu, G. Q. Zhang, An inverse boundary value problem for cavity flows in gas dynamics, Nonlinear Analysis, 50(2002), 717-725.
6. I. N. Vekua, Generalized analytic functions, Pergamon Press, Oxford, 1962.
7. G. C. Wen, H. Begehr, Boundary value problems for elliptic equations and systems, Longman Scientific and Technical, Harlow, 1990.
8. C. K. Lu, Boundary value problems for analytic functions, World Scientific, Singapore,1993.
9. G. C. Wen, D. C. Chen, Z. L. Xu, Nonlinear complex analysis and its applications, Science Press, Beijing, 2008.

FIXED POINT THEOREMS OF CONTRACTIVE MAPPING IN EXTENDED CONE METRIC SPACES *

HUAPING HUANG[1†], XING LI[2‡]

1. School of Mathematics and Statistics, Hubei Normal University, Huangshi, 435002, China
2. School of Mathematics and Computer Science, Ningxia University, Yinchuan, 750021, China

In this paper we introduce extended cone metric spaces, prove some fixed point theorems of contractive mappings in such spaces. The results directly improve and generalize some fixed point results in cone metric spaces. In addition, we use our conclusions to obtain the existence and uniqueness of solution for an ordinary differential equation with periodic boundary condition.

Keywords: Extended cone metric space, Fixed point, periodic boundary condition

1. Introduction

Fixed point theory plays a basic role in applications of many branches of mathematics. Finding the fixed point of contractive mappings becomes the the center of strong research activity. There are many works about fixed point of contractive maps (see, for example, [1-2]). In recent investigations, the fixed point in non-convex analysis, especially in ordered normed space, occupies a prominent place in many areas. The authors define an ordering by using a cone, which naturally induces a partial ordering in Banach spaces. In [3], Huang and Zhang introduced cone metric spaces, as a generalization of metric spaces. Moreover, they proved some fixed point theorems for contractive mappings that generalized certain results of fixed point in metric spaces. Ever since [3], focusing on cone metric spaces for fixed points, papers have been emerged in large numbers (see [4-9]). The purpose of this

*The research is supported by Graduate initial fund of Hubei Normal University (2008D36) and National Science Foundation of China (No. 10962008)
†Corresponding author: Huaping Huang. E-mail: hhpa2000@163. com (H. Huang)
‡E-mail: li_x@nxu.edu.cn

work is to introduce extended cone metric spaces, as a generalization of cone metric spaces, which offer a strong tool for studying the positive fixed points of numerous nonlinear operators and the positive solutions of many operator equations. In addition, we obtain some fixed point theorems of contractive mappings in such spaces. Furthermore, we support our results by an example. The results greatly generalize the well-known work of [3-5]. As some applications, we show the existence and uniqueness of solution for a second-order ordinary differential equation with a periodic boundary condition.

For the convenience of the readers, we recall the following definitions and results, consistent with [3], as follows:

Let E be a real Banach space and θ the zero element of E. A nonempty closed subset P of E is called a cone if $P \neq \{\theta\}$, $P \cap (-P) = \{\theta\}$, $P + P \subset P$ and $\lambda P \subset P$ for all $\lambda \geq 0$. By int$P$ we denote the interior of P. On this basis, we define a partial ordering $\leq$ with respect to P by $x \leq y$ if and only if $y - x \in P$. We shall write $x < y$ to indicate that $x \leq y$ but $x \neq y$, while $x \ll y$ will stand for $y - x \in \text{int}P$. Write $\|\cdot\|$ as the norm on E. The cone P is called normal if there is a number $K > 0$ such that for all $x, y \in E$, $\theta \leq x \leq y$ implies $\|x\| \leq K\|y\|$. The least positive number satisfying above is called the normal constant of P. It is well known that $K \geq 1$.

In the following we always suppose that E is a Banach space, P is a cone in E with int$P \neq \emptyset$ and $\leq$ is a partial ordering with respect to P.

Definition 1.1([3]). Let X be a nonempty set. Suppose that the mapping $d : X \times X \to E$ satisfies:

(d1) $\theta < d(x, y)$ for all $x, y \in X$ with $x \neq y$ and $d(x, y) = \theta$ if and only if $x = y$;

(d2) $d(x, y) = d(y, x)$ for all $x, y \in X$;

(d3) $d(x, y) \leq d(x, z) + d(z, y)$ for all $x, y, z \in X$.

Then d is called a cone metric on X, and (X, d) is called a cone metric space.

Definition 1.2. Let X be a nonempty set and $s \geq 1$ a given real number. A mapping $d : X \times X \to E$ is said to be extended cone metric if and only if for all $x, y, z \in X$, the following conditions are satisfied:

(i) $\theta < d(x, y)$ with $x \neq y$ and $d(x, y) = \theta$ if and only if $x = y$;

(ii) $d(x,y) = d(y,x)$;
(iii) $d(x,y) \leq s[d(x,z) + d(z,y)]$.
A pair (X,d) is called an extended cone metric space.

Remark 1.3. The class of extended cone metric space is larger than the one of cone metric space, since any cone metric space must be an extended cone metric space. Therefore, it is obvious that extended cone metric spaces generalize cone metric spaces.

Example 1.4. Let $E = \mathbb{R}^2$, $P = \{(x,y) \in E : x, y \geq 0\}$, $X = \mathbb{R}$ and $d : X \times X \to E$ such that $d(x,y) = (|x-y|^p, \alpha|x-y|^p)$, where $\alpha \geq 0$ and $p > 1$ are two constants. Then (X,d) is an extended cone metric space, but not a cone metric space.

In fact, we only need to prove (iii) in Definition 1.2 as follows:

Let $x, y, z \in X$. Set $u = x - z$, $v = z - y$, so $x - y = u + v$. From the inequality
$(a+b)^p \leq (2\max\{a,b\})^p \leq 2^p(a^p + b^p)$, for all $a, b \geq 0$,
we have

$$|x-y|^p = |u+v|^p \leq (|u| + |v|)^p \leq 2^p(|u|^p + |v|^p) = 2^p(|x-z|^p + |z-y|^p),$$

which implies that $d(x,y) \leq s[d(x,z) + d(z,y)]$ with $s = 2^p > 1$. But

$$|x-y|^p \leq |x-z|^p + |z-y|^p$$

is impossible for all $x > z > y$.

Actually, Taking account of the inequality
$(a+b)^p > a^p + b^p$, for all $a, b > 0$,
we arrive at
$|x-y|^p = |u+v|^p = (u+v)^p > u^p + v^p = (x-z)^p + (z-y)^p = |x-z|^p + |z-y|^p$,
for all $x > z > y$. Thus, (d3) in Definition 1.1 is not satisfied, i.e. (X,d) is not a cone metric space.

Example 1.5. Let $X = \{1,2,3,4\}$, $E = \mathbb{R}^2$, $P = \{(x,y) \in E : x \geq 0,\ y \geq 0\}$. Define $d : X \times X \to E$ by

$$d(x,y) = \begin{cases} (|x-y|^{-1}, |x-y|^{-1}), & \text{if } x \neq y, \\ \theta, & \text{if } x = y. \end{cases}$$

Then (X,d) is an extended cone metric space with the coefficient $s = \dfrac{6}{5}$. But it is not a cone metric space since the triangle inequality is not satisfied.

Indeed,

$$d(1,2) > d(1,4) + d(4,2), \qquad d(3,4) > d(3,1) + d(1,4).$$

Example 1.6. The set $X = l^p$ with $0 < p < 1$, where $l^p = \{\{x_n\} \subset \mathbb{R} : \sum_{n=1}^{\infty} |x_n|^p < \infty\}$, together with the function $d : X \times X \to \mathbb{R}_+$,

$$d(x,y) = \left(\sum_{n=1}^{\infty} |x_n - y_n|^p \right)^{\frac{1}{p}},$$

where $x = \{x_n\}$, $y = \{y_n\} \in l^p$. Put $E = l^1$, $P = \{\{x_n\} \in E : x_n \geq 0,$ for all $n \geq 1\}$. Letting the mapping $\tilde{d} : X \times X \to E$ be defined by $\tilde{d}(x,y) = \{\frac{d(x,y)}{2^n}\}_{n\geq 1}$, we conclude that $(X, \tilde{d})$ is an extended cone metric space with the coefficient $s = 2^{\frac{1}{p}} > 1$.

Definition 1.7. Let (X, d) be an extended cone metric space, $x \in X$ and $\{x_n\}$ a sequence in X. Then

(i) $\{x_n\}$ converges to x whenever for every $c \in E$ with $\theta \ll c$ there is a natural number N such that $d(x_n, x) \ll c$ for all $n \geq N$. We denote this by $\lim_{n\to\infty} x_n = x$ or $x_n \to x (n \to \infty)$.

(ii) $\{x_n\}$ is a Cauchy sequence whenever for every $c \in E$ with $\theta \ll c$ there is a natural number N such that $d(x_n, x_m) \ll c$ for all $n, m \geq N$.

(iii) (X, d) is a complete extended cone metric space if every Cauchy sequence is convergent.

A same method as in [8], the following assertions will be used in the sequel(in particular when dealing with extended cone metric spaces in which the cone need not be normal).

(p1) If $u \leq v$ and $v \ll w$, then $u \ll w$.

(p2) If $\theta \leq u \ll c$ for each $c \in \text{int}P$, then $u = \theta$.

(p3) If $a \leq b + c$, for each $c \in \text{int}P$, then $a \leq b$.

(p4) If $\theta \leq d(x_n, x) \leq b_n \to \theta$ (as $n \to \infty$) and $c \in \text{int}P$, then there exists N such that, for all $n > N$, $d(x_n, x) \ll c$, where x_n, x are, respectively, a sequence and a given point in X.

(p5) If $a \leq \lambda a$, where $a \in P$ and $0 \leq \lambda < 1$, then $a = \theta$.

(p6) If $c \in \text{int}P$ and $\theta \leq a_n \to \theta$ (as $n \to \infty$), then there exists N such that, for all $n > N$, we have $a_n \ll c$.

2. Main results

Theorem 2.1. Let (X, d) be a complete extended cone metric space with the coefficient $s \geq 1$. Suppose the mapping $T : X \to X$ satisfies the contractive condition

$$d(Tx, Ty) \leq \lambda_1 d(x, y) + \lambda_2 d(x, Tx) + \lambda_3 d(y, Ty) + \lambda_4 d(x, Ty) + \lambda_5 d(y, Tx),$$

for all $x, y \in X$, where $\lambda_i \geq 0$ $(i = 1, \cdots, 5)$ and $\lambda_2 + \lambda_3 + s(\lambda_4 + \lambda_5) < \min\{1 - \lambda_1, \frac{2}{s}\}$. Then T has a unique fixed point in X. Furthermore, the iterative sequence $\{T^n x\}$ converges to the fixed point.

Proof. Choose $x_0 \in X$. We construct the iterative sequence $\{x_n\}$, where $x_n = Tx_{n-1}$, $n \geq 1$, i.e. $x_{n+1} = Tx_n = T^{n+1}x_0$. Firstly,

$$\begin{aligned} d(x_{n+1}, x_n) &= d(Tx_n, Tx_{n-1}) \\ &\leq \lambda_1 d(x_n, x_{n-1}) + \lambda_2 d(x_n, Tx_n) + \lambda_3 d(x_{n-1}, Tx_{n-1}) \\ &\quad + \lambda_4 d(x_n, Tx_{n-1}) + \lambda_5 d(x_{n-1}, Tx_n) \\ &= \lambda_1 d(x_n, x_{n-1}) + \lambda_2 d(x_n, x_{n+1}) + \lambda_3 d(x_{n-1}, x_n) + \lambda_5 d(x_{n-1}, x_{n+1}) \\ &\leq (\lambda_1 + \lambda_3 + s\lambda_5) d(x_n, x_{n-1}) + (\lambda_2 + s\lambda_5) d(x_{n+1}, x_n). \end{aligned}$$

It follows that

$$(1 - \lambda_2 - s\lambda_5) d(x_{n+1}, x_n) \leq (\lambda_1 + \lambda_3 + s\lambda_5) d(x_n, x_{n-1}). \tag{2.1}$$

Secondly,

$$\begin{aligned} d(x_{n+1}, x_n) &= d(Tx_n, Tx_{n-1}) = d(Tx_{n-1}, Tx_n) \\ &\leq \lambda_1 d(x_{n-1}, x_n) + \lambda_2 d(x_{n-1}, Tx_{n-1}) + \lambda_3 d(x_n, Tx_n) \\ &\quad + \lambda_4 d(x_{n-1}, Tx_n) + \lambda_5 d(x_n, Tx_{n-1}) \\ &= \lambda_1 d(x_n, x_{n-1}) + \lambda_2 d(x_n, x_{n-1}) + \lambda_3 d(x_{n+1}, x_n) + \lambda_4 d(x_{n+1}, x_{n-1}) \\ &\leq (\lambda_1 + \lambda_2 + s\lambda_4) d(x_n, x_{n-1}) + (\lambda_3 + s\lambda_4) d(x_{n+1}, x_n). \end{aligned}$$

It means that

$$(1 - \lambda_3 - s\lambda_4) d(x_{n+1}, x_n) \leq (\lambda_1 + \lambda_2 + s\lambda_4) d(x_n, x_{n-1}). \tag{2.2}$$

Adding up (2.1) and (2.2) yields that

$$d(x_{n+1}, x_n) \leq \frac{2\lambda_1 + \lambda_2 + \lambda_3 + s\lambda_4 + s\lambda_5}{2 - \lambda_2 - \lambda_3 - s\lambda_4 - s\lambda_5} d(x_n, x_{n-1}).$$

Put $\lambda = \frac{2\lambda_1 + \lambda_2 + \lambda_3 + s\lambda_4 + s\lambda_5}{2 - \lambda_2 - \lambda_3 - s\lambda_4 - s\lambda_5}$, then $0 \leq \lambda < 1$,

$$d(x_{n+1}, x_n) \leq \lambda d(x_n, x_{n-1}) \leq \cdots \leq \lambda^n d(x_1, x_0).$$

It follows that for any $m \geq 1,\ p \geq 1$, we arrive at

$$\begin{aligned}
d(x_{m+p}, x_m) &\leq s[d(x_{m+p}, x_{m+p-1}) + d(x_{m+p-1}, x_m)] \\
&= sd(x_{m+p}, x_{m+p-1}) + sd(x_{m+p-1}, x_m) \\
&\leq sd(x_{m+p}, x_{m+p-1}) + s^2[d(x_{m+p-1}, x_{m+p-2}) + d(x_{m+p-2}, x_m)] \\
&= sd(x_{m+p}, x_{m+p-1}) + s^2 d(x_{m+p-1}, x_{m+p-2}) + s^2 d(x_{m+p-2}, x_m) \\
&\leq sd(x_{m+p}, x_{m+p-1}) + s^2 d(x_{m+p-1}, x_{m+p-2}) \\
&+ s^3 d(x_{m+p-2}, x_{m+p-3}) + \cdots \\
&\quad + s^{p-1} d(x_{m+2}, x_{m+1}) + s^{p-1} d(x_{m+1}, x_m) \\
&\leq s\lambda^{m+p-1} d(x_1, x_0) + s^2\lambda^{m+p-2} d(x_1, x_0) \\
&+ s^3\lambda^{m+p-3} d(x_1, x_0) + \cdots \\
&\quad + s^{p-1}\lambda^{m+1} d(x_1, x_0) + s^{p-1}\lambda^m d(x_1, x_0) \\
&= (s\lambda^{m+p-1} + s^2\lambda^{m+p-2} + s^3\lambda^{m+p-3} + \cdots + s^{p-1}\lambda^{m+1}) d(x_1, x_0) \\
&\quad + s^{p-1}\lambda^m d(x_1, x_0) \\
&= \frac{s\lambda^{m+p}[(s\lambda^{-1})^{p-1} - 1]}{s - \lambda} d(x_1, x_0) + s^{p-1}\lambda^m d(x_1, x_0) \\
&\leq \frac{s^p\lambda^{m+1}}{s - \lambda} d(x_1, x_0) + s^{p-1}\lambda^m d(x_1, x_0) \\
&= \frac{s^p\lambda^{m+1}}{s - \lambda} d(x_1, x_0) + s^{p-1}\lambda^m d(x_1, x_0).
\end{aligned}$$

Let $\theta \ll c$ be given. Noticing that $\frac{s^p\lambda^{m+1}}{s - \lambda} d(x_1, x_0) + s^{p-1}\lambda^m d(x_1, x_0) \to \theta$ as $m \to \infty$ for any $p \geq 1$, we find $m_0 \in \mathbb{N}$ such that

$$\frac{s^p\lambda^{m+1}}{s - \lambda} d(x_1, x_0) + s^{p-1}\lambda^m d(x_1, x_0) \ll c,$$

for each $m > m_0$. Thus,

$$d(x_{m+p}, x_m) \leq \frac{s^p\lambda^{m+1}}{s - \lambda} d(x_1, x_0) + s^{p-1}\lambda^m d(x_1, x_0) \ll c,$$

for all $m \geq 1$, $p \geq 1$. So $\{x_n\}$ is a Cauchy sequence in (X, d). Since (X, d) is a complete extended cone metric space, there exists $x^* \in X$ such that $x_n \to x^*$. So for each $c \gg \theta$, there exists $n_0 \in \mathbb{N}$ such that

$$d(x_n, x^*) \ll \frac{2 - s\lambda_2 - s\lambda_3 - s^2\lambda_4 - s^2\lambda_5}{4s + 2s^2} c,$$

for all $n > n_0$. Now, on one hand,

$$\begin{aligned} d(x^*, Tx^*) &\leq s[d(x^*, Tx_n) + d(Tx_n, Tx^*)] \\ &\leq sd(x^*, x_{n+1}) + s[\lambda_1 d(x_n, x^*) + \lambda_2 d(x_n, Tx_n) \\ &\quad + \lambda_3 d(x^*, Tx^*) + \lambda_4 d(x_n, Tx^*) + \lambda_5 d(x^*, Tx_n)] \\ &= sd(x^*, x_{n+1}) + s\lambda_1 d(x_n, x^*) + s\lambda_2 d(x_n, x_{n+1}) \\ &\quad + s\lambda_3 d(x^*, Tx^*) + s\lambda_4 d(x_n, Tx^*) + s\lambda_5 d(x^*, x_{n+1}) \\ &\leq (s + s^2\lambda_2 + s\lambda_5) d(x_{n+1}, x^*) + (s\lambda_3 + s^2\lambda_4) d(x^*, Tx^*) \\ &\quad + (s\lambda_1 + s^2\lambda_2 + s^2\lambda_4) d(x_n, x^*). \end{aligned}$$

This implies that

$$\begin{aligned} &(1 - s\lambda_3 - s^2\lambda_4) d(x^*, Tx^*) \\ &\leq (s + s^2\lambda_2 + s\lambda_5) d(x_{n+1}, x^*) + (s\lambda_1 + s^2\lambda_2 + s^2\lambda_4) d(x_n, x^*). \end{aligned} \tag{2.3}$$

On the other hand,

$$\begin{aligned} d(Tx^*, x^*) &\leq s[d(Tx^*, Tx_n) + d(Tx_n, x^*)] \\ &\leq s[\lambda_1 d(x^*, x_n) + \lambda_2 d(x^*, Tx^*) + \lambda_3 d(x_n, Tx_n) \\ &\quad + \lambda_4 d(x^*, Tx_n) + \lambda_5 d(x_n, Tx^*)] + sd(x_{n+1}, x^*) \\ &= s\lambda_1 d(x^*, x_n) + s\lambda_2 d(x^*, Tx^*) + s\lambda_3 d(x_n, x_{n+1}) \\ &\quad + s\lambda_4 d(x^*, x_{n+1}) + s\lambda_5 d(x_n, Tx^*) + sd(x_{n+1}, x^*) \\ &\leq (s + s^2\lambda_3 + s\lambda_4) d(x_{n+1}, x^*) + (s\lambda_2 + s^2\lambda_5) d(x^*, Tx^*) \\ &\quad + (s\lambda_1 + s^2\lambda_3 + s^2\lambda_5) d(x_n, x^*). \end{aligned}$$

This indicates that

$$\begin{aligned} &(1 - s\lambda_2 - s^2\lambda_5) d(x^*, Tx^*) \\ &\leq (s + s^2\lambda_3 + s\lambda_4) d(x_{n+1}, x^*) + (s\lambda_1 + s^2\lambda_3 + s^2\lambda_5) d(x_n, x^*). \end{aligned} \tag{2.4}$$

Combing (2.3) and (2.4) concludes that

$$\begin{aligned} d(x^*, Tx^*) &\leq \frac{2s + s^2\lambda_2 + s^2\lambda_3 + s\lambda_4 + s\lambda_5}{2 - s\lambda_2 - s\lambda_3 - s^2\lambda_4 - s^2\lambda_5} d(x_{n+1}, x^*) \\ &\quad + \frac{2s\lambda_1 + s^2\lambda_2 + s^2\lambda_3 + s^2\lambda_4 + s^2\lambda_5}{2 - s\lambda_2 - s\lambda_3 - s^2\lambda_4 - s^2\lambda_5} d(x_n, x^*) \\ &\leq \frac{2s + s^2}{2 - s\lambda_2 - s\lambda_3 - s^2\lambda_4 - s^2\lambda_5} [d(x_{n+1}, x^*) + d(x_n, x^*)] \\ &\ll c. \end{aligned}$$

Hence, $Tx^* = x^*$. Finally, we prove the fixed point is unique. If there exists another fixed point y^*, then

$$\begin{aligned} d(x^*, y^*) &= d(Tx^*, Ty^*) \\ &\leq \lambda_1 d(x^*, y^*) + \lambda_2 d(x^*, Tx^*) + \lambda_3 d(y^*, Ty^*) \\ &\quad + \lambda_4 d(x^*, Ty^*) + \lambda_5 d(y^*, Tx^*) \\ &\leq \lambda_1 d(x^*, y^*) + s\lambda_4 [d(x^*, y^*) + d(y^*, Ty^*)] \\ &\quad + s\lambda_5 [d(y^*, x^*) + d(x^*, Tx^*)] \\ &= (\lambda_1 + s\lambda_4 + s\lambda_5) d(x^*, y^*). \end{aligned}$$

Observing that $0 \leq \lambda_1 + s\lambda_4 + s\lambda_5 < 1$, we have $x^* = y^*$. Therefore the claim holds.

Corollary 2.2. Let (X, d) be a complete extended cone metric space with the coefficient $s \geq 1$. Suppose the mapping $T : X \to X$ satisfies the contractive condition

$d(Tx, Ty) \leq \lambda d(x, y)$, for $x, y \in X$,

where $\lambda \in [0, 1)$ is a constant. Then T has a unique fixed point in X. Furthermore, the iterative sequence $\{T^n x\}$ converges to the fixed point.

Corollary 2.3. Let (X, d) be a complete extended cone metric space with the coefficient $s \geq 1$. Suppose the mapping $T : X \to X$ satisfies the contractive condition

$d(Tx, Ty) \leq \lambda_1 d(x, Tx) + \lambda_2 d(y, Ty) + \lambda_3 d(x, Ty) + \lambda_4 d(y, Tx)$,

for $x, y \in X$, where the constant $\lambda_i \in [0, 1)$ and $\lambda_1 + \lambda_2 + s(\lambda_3 + \lambda_4) < \min\{1, \frac{2}{s}\}$, $i = 1, 2, 3, 4$. Then T has a unique fixed point in X. Moreover, the iterative sequence $\{T^n x\}$ converges to the fixed point.

Corollary 2.4. Let (X, d) be a complete cone metric space. Suppose the mapping $T : X \to X$ satisfies the contractive condition

$$d(Tx, Ty) \leq \lambda_1 d(x, y) + \lambda_2 d(x, Tx) + \lambda_3 d(y, Ty) + \lambda_4 d(x, Ty) + \lambda_5 d(y, Tx),$$

for all $x, y \in X$, where $\lambda_i \geq 0$ $(i = 1, \cdots, 5)$ and $\sum_{i=1}^{5} \lambda_i < 1$. Then T has a unique fixed point in X. Furthermore, the iterative sequence $\{T^n x\}$ converges to the fixed point.

Example 2.5. Let $X = [0, 1]$, $E = \mathbb{R}^2$, $p > 1$ be a constant. Take $P = \{(x, y) \in E : x, y \geq 0\}$. We define $d : X \times X \to E$ as

$d(x,y) = (|x-y|^p, |x-y|^p)$, for all $x, y \in X$.
Then (X, d) is a complete extended cone metric space. Let us define $T : X \to X$ as

$Tx = \frac{1}{2}x - \frac{1}{4}x^2$, for all $x \in X$.

Therefore,

$$\begin{aligned} d(Tx,Ty) &= (|Tx-Ty|^p, |Tx-Ty|^p) \\ &= \left(|\frac{1}{2}(x-y) - \frac{1}{4}(x-y)(x+y)|^p, |\frac{1}{2}(x-y) - \frac{1}{4}(x-y)(x+y)|^p\right) \\ &= \left(|x-y|^p \cdot |\frac{1}{2} - \frac{1}{4}(x+y)|^p, |x-y|^p \cdot |\frac{1}{2} - \frac{1}{4}(x+y)|^p\right) \\ &\leq \frac{1}{2^p}(|x-y|^p, |x-y|^p) = \frac{1}{2^p}d(x,y). \end{aligned}$$

Here $0 \in X$ is the unique fixed point of T.

Remark 2.6. Normality of the cone is not assumed in Theorem 2.1 and Corollary 2.2-2.4. Moreover, Corollary 2.2 extends Theorem 1 in [3], Corollary 2.3 and 2.4 generalize Theorem 2.6-2.8 in [4].

Remark 2.7. Any fixed point theorem in the setting of metric space or cone metric space cannot cope with Example 2.5. So, Example 2.5 shows that the fixed point theory of extended cone metric spaces offers independently a strong tool for studying the positive fixed points of some nonlinear operators and the positive solutions of some operator equations.

3. Applications

In this section we shall apply Theorem 2.1 to the second-order periodic boundary problem

$$\begin{cases} \frac{\mathrm{d}^2y}{\mathrm{d}x^2} + a_1(x)\frac{\mathrm{d}y}{\mathrm{d}x} + a_2(x)y = f(x), & x \in [0, x_0], \\ y(0) = C_0, \ y'(0) = C_1, \end{cases} \tag{3.1}$$

where $a_1(x), a_2(x), f(x) \in C([0, x_0])$ are given.

Theorem 3.1. Consider boundary problem (3.1), and set $M = \max_{0 \leq t, x \leq x_0} [a_2(x)(t-x) - a_1(x)]$. If $x_0^2 M^2 < 1$, then there exists a unique solution of (3.1).

Proof. Put $u(x) = \frac{\mathrm{d}^2 y}{\mathrm{d}x^2}$, $p(x) = \frac{\mathrm{d}y}{\mathrm{d}x}$, then $u(x), p(x) \in C([0, x_0])$. Considering the initial conditions, we see

$$\frac{\mathrm{d}y}{\mathrm{d}x} = \int_0^x u(t)\mathrm{d}t + C_1, \tag{3.2}$$

$$\begin{aligned} y &= \int_0^x p(s)\mathrm{d}s + C_0 = \int_0^x [\int_0^s u(t)\mathrm{d}t + C_1]\mathrm{d}s + C_0 \\ &= \int_0^x [\int_0^s u(t)\mathrm{d}t]\mathrm{d}s + C_1 x + C_0 = \int_0^x \mathrm{d}t \int_t^x u(t)\mathrm{d}s + C_1 x + C_0 \\ &= \int_0^x (x-t)u(t)\mathrm{d}t + C_1 x + C_0. \end{aligned} \tag{3.3}$$

Substituting (3.2) and (3.3) into (3.1), we speculate (3.1) is equivalent to the following Volterra integral equation

$$u(x) = \int_0^x K(x,t)u(t)\mathrm{d}t + F(x),$$

Where $K(x,t) = a_2(x)(t-x) - a_1(x)$, $F(x) = f(x) - C_1 a_1(x) - C_1 x a_2(x) - C_0 a_2(x)$.

Let $X = E = C([0, x_0])$ and $P = \{\varphi \in E : \varphi \geq 0\}$. Put $d : X \times X \to E$ as $d(u,v) = f(x) \max\limits_{0 \leq x \leq x_0} |u(x) - v(x)|^2$ with $f : [0, x_0] \to \mathbb{R}$ such that $f(x) = e^x$. It is clear that (X, d) is a complete extended cone metric space.

Define a mapping $T : C([0, x_0]) \to C([0, x_0])$ by

$$Tu(x) = \int_0^x K(x,t)u(t)\mathrm{d}t + F(x).$$

For any $u, v \in C([0, x_0])$, we have

$$\begin{aligned} d(Tu, Tv) &= f(x) \max_{0 \leq x \leq x_0} |\int_0^x K(x,t)u(t)\mathrm{d}t - \int_0^x K(x,t)v(t)\mathrm{d}t|^2 \\ &= f(x) \max_{0 \leq x \leq x_0} |\int_0^x K(x,t)[u(t) - v(t)]\mathrm{d}t|^2 \\ &\leq x_0^2 M^2 f(x) \max_{0 \leq t \leq x_0} |u(t) - v(t)|^2 \\ &= x_0^2 M^2 d(u,v), \end{aligned}$$

which establishes that

$$d(Tu, Tv) \leq \lambda_1 d(u,v) + \lambda_2 d(u, Tu) + \lambda_3 d(v, Tv) + \lambda_4 d(u, Tv) + \lambda_5 d(v, Tu),$$

where $\lambda_1 = x_0^2 M^2$, $\lambda_2 = \lambda_3 = \lambda_4 = \lambda_5 = 0$.

Owing to the above statement, all conditions of Theorem 2.1 are satisfied. Hence T has a unique fixed point $u(x) \in C([0, x_0])$. That is to say, there exists a unique solution of (3.1).

References

1. K. Deimling, Nonlinear Functional Analysis. Springer-Verlag, 1985.
2. R. P. Pant, Common fixed points of noncommuting mappings, J. Math. Anal. Appl. 188(1994) 436-440.
3. L. -G. Huang, X. Zhang, Cone metric spaces and fixed point theorems of contractive mappings, J. Math. Anal. Appl. 332(2)(2007) 1468-1476.
4. Sh. Rezapour, R. Hamlbarani, Some notes on the paper"Cone metric spaces and fixed point theorems of contractive mappings", J. Math. Anal. Appl. 345(2008) 719-724.
5. M. Abbas, B. E. Rhoades, Fixed and periodic point results in cone metric spaces, Applied mathematics Letters. 22(2009) 511-515.
6. B. S. Choudhury, N. Metiya, Fixed points of weak contractions in cone metric spaces. 72(2010) 1589-1591.
7. M. Alimohammady et al., Conditions of regularity in cone metric spaces, Applied Mathematics and Computation. 217(2011) 6359-6363.
8. S. Janković, Z. Kadelburg, S. Radebović, On cone metric spaces: A survey, Nonlinear Analysis. 74(2011) 2591-2601.
9. S. -H. Cho, J. -S. Bae, Common fixed point theorems for mappings satisfying property (E.A) on cone metric space, Mathematical and Computer Modeling. 53(2011) 945-951.

POSITIVE SOLUTIONS OF SINGULAR THIRD-ORDER THREE-POINT BOUNDARY VALUE PROBLEMS*

BAOQIANG YAN,XUAN LIU†

Department of Mathematics,Shandong Normal University, Jinan 250014,PR China

In this paper we investigate the existence of at least one positive solutions and the existence of multiple positive solutions for the nonlinear singular third-order three-point boundary value problem. Our nonlinearity $f(t, y)$ may be singular at $t = 0$, $t = 1$, $y = 0$.

Keywords: The nonlinear alternatives of Leray-Schauder theorem; Ascoli-Arzela theorem; fixed point index theroy.

1. Introduction

In this paper, we consider the third-order three-point boundary value problems

$$\begin{cases} y'''(t) - a(t)f(t, y(t)) = 0, & 0 < t < 1, \\ y(0) = y'(\eta) = y''(1) = 0. \end{cases} \tag{1.1}$$

(see also [2,3,7]).

In [6], using the fixed point theorem, Sun considered the boundary value problems

$$\begin{cases} y'''(t) - \lambda a(t)f(t, y(t)) = 0, & 0 < t < 1, \\ y(0) = y'(\eta) = y''(1) = 0, \end{cases}$$

and got the existence of single and multiple positive solutions, where $a(t)$ is singular on $(0, 1)$, $\lambda > 0$ and $\eta \in [\frac{1}{2}, 1)$, but $f(t, y) \geq 0$ has no singularities. In [7], using the Guo-Krasnosel'skii fixed point theorem of cone

*The project is supported by the fund of National Natural Science(10871120),fund of Shandong Education Committee(J07WH08) and the fund of Shandong Natural Science(Y2008A06)

†Corresponding author: yanbqcn@yahoo.com.cn

expansion-compression type, Yao considered the existence of at least one positive solution and multiple positive solutions for BVP(1.1) when $f(t,y)$ is singular at $t=0$, $t=1$, and $y=0$. We notice that Yao needed that the nonlinearity $f(t,y)$ has some sepecial properties on some sepecial bounded set.

Inspired by papers [1,6], we consider the BVP (1) when the nonlinear term $f(t,y)$ is singular at $t=0$, $t=1$, $y=0$. The diffrence from [7] is that $f(t,y)$ satisfies a more direct condition (1.2.4) instead of that $f(t,y)$ has some sepecial properties on some sepecial bounded set. The idea of this paper is from [1].

Thoughout this paper, suppose that $a(t)$ *may be singular at* $t=0$, *or* $t=1$, $a>0$ *on* $(0,1)$, $f:(0,1)\times(0,+\infty)\to(0,+\infty)$ *is continuous*, and the following conditions are satisfied:
(H_1) $0\le f(t,y)\le g(y)+h(y)$ *on* $(0,1)\times(0,+\infty)$ *with* $g>0$ *continuous and nonincreasing on* $(0,+\infty)$, $h\ge 0$ *continuous on* $[0,\infty)$ *and* $\frac{h}{g}$ *nondecreasing on* $(0,+\infty)$.
(H_2) *for each constant* $H>0$ *there exists a function* φ_H *continuous on* $[0,1]$ *and positive on* $(0,1)$ *such that* $f(t,u)\ge\varphi_H(t)$ *on* $(0,1)\times(0,H]$.
(H_3) $\exists\ r>0$ *with* $\dfrac{r}{\left(1+\dfrac{h(r)}{g(r)}\right)\displaystyle\int_0^1 g(rx(1-x))a(x)dx}>\dfrac{1}{2}$ *holds*,
where $\displaystyle\int_0^1 g(rx(1-x))a(x)dx<+\infty$.
Theorem 1. Let $(H_1),(H_2)and(H_3)$ hold, then $BVP(1.1)$ has a solution $y\in C[0,1]\cap C^2(0,1)\cap C^3(0,1)$ with $y>0$ on $(0,1)$ and $|y|_0<r$.
Theorem 2. *Further suppose the following condition is satified* :
(H_4) *there exist* $[\alpha,\beta]\subset(0,1)$ *such that*

$$\lim_{x\to+\infty}\frac{f(t,x)}{x}=+\infty,\quad \lim_{x\to 0+}f(t,x)=+\infty,$$

both uniformly with respect to $t\in[\alpha,\beta]$, *then* BVP (1.1) *has at least two positive solutions* $x(t)$*and* $y(t)$.

In section 2, we make some preliminaries. In section 3, we give the proofs of Theorem 1 and Theorem 2. In section 4, an example is worked out to demonstrate our main results.

2. Preliminaries

First we begin with the nonlinear alternative of Leray-Schauder type, i.e Lemma 2.1. Then we present existence principle for nonsingular boundary value problem, which is Lemma 2.2.

Lemma 2.1 (see [4]) Let E be a banach space, C a convex subset of E, U an open subset of C and $0 \in U$. Suppose $F : \bar{U} \to C$ (here $\bar{U}$ denotes the closure of U in C) is a continous, compact map.

Then either

$(A1)$. *F has a fixed point in $\bar{U}$; or*

$(A2)$. *there exists $u \in \partial U$ (the boundary of U in C) and $\lambda \in (0,1)$ with $u = \lambda F(u)$.*

Lemma 2.2 *Suppose the following two condtions are satisfied :*

$$\text{the map } y \mapsto f(t,y) \text{ is continuous for a.e } t \in [0,1] \tag{2.1}$$

and

$$\text{the map } t \mapsto f(t,y) \text{ is measurable for all } y \in R. \tag{2.2}$$

Assume

$$\begin{cases} \text{for each } r > 0 \text{ there exists } h_r \in L^1_{loc}(0,1) \text{ with } \int_0^1 t(1-t)h_r(t)dt < \infty \\ \text{such that } |y| \le r \text{ implies } |f(t,y)| \le h_r(t) \text{ for a.e } t \in (0,1) \end{cases} \tag{2.3}$$

holds. In addition suppose there is a constant $M > a$, independent of λ, with $|y|_0 = \sup_{t\in[0,1]} |y(t)| \neq M$ for any solution $y \in AC[0,1]$ (with $y' \in AC_{loc}(0,1)$) to

$$\begin{cases} y''' - \lambda f(t,y) = 0, \quad 0 < t < 1, \\ y(0) = a, \ y'(\eta) = y''(1) = 0, \end{cases} \tag{2.4}$$

for each $\lambda \in (0,1)$. Then (1) have a solution y with $|y|_0 \le M$.

Proof. We begin by showing that solving $(2.4)_\lambda$ is equivalent to finding a solution $y \in C[0,1]$ to

$$y(t) = a + \lambda \int_0^t (t\min\{\eta,s\} + \frac{1}{2}s^2 - ts)f(s,y(s))ds$$

$$+\lambda \int_t^1 (t\min\{\eta,s\} - \frac{1}{2}t^2)f(s,y(s))ds. \tag{2.5}$$

To see this notice if $y \in C[0,1]$ satisfies $(2.5)_\lambda$ then it is easy to see that $y' \in L^1[0,1]$. Thus $y \in AC[0,1]$, $y' \in AC_{loc}[0,1]$ and note

$$y'(t) = \lambda \int_0^1 (\min\{\eta, s\}) f(s, y(s))ds - \lambda \int_0^t sf(s, y(s))ds - \lambda t \int_t^1 f(s, y(s))ds,$$

$$y''(t) = -\lambda \int_t^1 f(s, y(s))ds.$$

Obviously $y'(\eta) = 0,\ y''(1) = 0$.

Integrate $y'(t)$ from 0 to x to get

$$\begin{aligned} y(x) - y(0) &= \int_0^x y'(t)dt \\ &= \lambda x \int_0^1 \min\{\eta, s\} f(s, y(s))ds - \lambda \int_0^x (x - \frac{1}{2}s) sf(s, y(s))ds \\ &-\frac{1}{2}\lambda x^2 \int_x^1 f(s, y(s))ds \\ &= y(x) - a, \end{aligned}$$

which implies $y(0) = a$. Thus if $y \in C[0,1]$ satisfies $(2.5)_\lambda$, y is a solution of $(2.4)_\lambda$.

Define the operator $N : C[0,1] \to C[0,1]$ by

$$Ny(t) = a + \int_0^t (t \min\{\eta, s\} + \frac{1}{2}s^2 - ts) f(s, y(s))ds$$

$$+ \int_t^1 (t \min\{\eta, s\} - \frac{1}{2}t^2) f(s, y(s))ds.$$

Then $(2.5)_\lambda$ is equivalent to the fixed point problem

$$y = (1 - \lambda)a + \lambda N y.$$

By (2.3), it is easy to see that $N : C[0,1] \to C[0,1]$ is continuous and completely continuous. Set

$$U = \{u \in C[0,1] : |u|_0 < M\}.$$

Now the nonlinear alternative of Leray-Schauder type guarantees that N has a fixed point i.e $(2.4)_1$ has a solution.

Lemma 2.3 (see [4]) *Let E be a Banach space, $R > 0$, $B_R = \{x \in E : \|x\| \leq R\}$, $F : B_R \to E$ be a completely continuous operator. If $x \neq \lambda F(x)$ for any $x \in E$ with $\|x\| = R$ and $0 < \lambda \leq 1$, F has a fixed point in B_R.*

Lemma 2.4 (see [4]) *Let E be a Banach space, Let P be a cone of E, Ω be a bounded open set of E, $\theta \in \Omega$, the map $F : P \cap \bar{\Omega} \to P$ be completely continuous. If $\inf_{x \in P \cap \partial\Omega} \|Fx\| > 0$ and $Fx \neq \mu x$ for $x \in P \cap \partial\Omega$ and $\mu \in (0, 1]$, then $i(F, P \cap \Omega, P) = 0$.*

3. Proofs of Theorems

In this section, we give the existence of positive solutions for the third-order three-point boundary value problem (1.1), in which $f(t, y)$ is singular at $t = 0, t = 1$ and $y = 0$.

Proof of Theorem 1. From (H_3), there exist an $r > 0$ such that

$$\frac{r}{\left(1 + \frac{h(r)}{g(r)}\right) \int_0^1 g(rx(1-x))a(x)dx} > \frac{1}{2}.$$

Choose $0 < \epsilon < r$ small enough such that

$$\frac{r - \epsilon}{\left(1 + \frac{h(r)}{g(r)}\right) \int_0^1 g(rx(1-x))a(x)dx} > \frac{1}{2}. \tag{3.1}$$

Let $n_0 \in \{1, 2 \ldots\}$ be chosen so that $\frac{1}{n_0} < \epsilon$ and let $N_0 = \{n_0, n_0 + 1, \ldots\}$. To show

$$\begin{cases} y''' - a(t)f(t, y(t)) = 0, & 0 < t < 1, \\ y(0) = \frac{1}{m}, \ y'(\eta) = y''(1) = 0, \end{cases} \tag{3.2}$$

has a solution, we examine

$$\begin{cases} y''' - a(t)F(t, y(t)) = 0, & 0 < t < 1, \\ y(0) = \frac{1}{m}, \ y'(\eta) = y''(1) = 0, & m \in N_0, \end{cases} \tag{3.3}$$

where

$$F(t, y) = \begin{cases} f(t, y), & y \geq \frac{1}{m}, \\ f(t, \frac{1}{m}), & y < \frac{1}{m}. \end{cases}$$

To show $(3.3)^m$ has a solution for each $m \in N_0$ we will apply Lemma 2.2. Consider the family of problems

$$\begin{cases} y''' - \lambda a(t)F(t, y(t)) = 0, & 0 < t < 1, \\ y(0) = \dfrac{1}{m}, \ y'(\eta) = y''(1) = 0, & m \in N_0, \end{cases} \tag{$(3.4)^m_\lambda$}$$

where $0 < \lambda < 1$. Let y be a solution of $(3.4)^m_\lambda$. Then $y''' \geq 0$, consequently $y'' < 0$ and y' is nonincreasing. If $t \in (0, \eta]$, we have $y'(t) > y'(\eta) = 0$. So, y is nondecreasing on $(0, \eta]$. Let $\alpha(t)$ be a function defined as follows,

$$\alpha(t) = \begin{cases} \dfrac{y(\eta)}{\eta} t, & t \in [0, \eta], \\ \dfrac{y(\eta)}{1-\eta}(1-t), & t \in [\eta, 1]. \end{cases} \tag{3.5}$$

Obviously, g is nonincreasing, we get $g(y(x)) \leq g(\alpha(x)) \leq g(y(\eta)x(1-x))$. Since $\dfrac{h}{g}$ is nondecreasing, $1 + \dfrac{h(y(x))}{g(y(x))} \leq 1 + \dfrac{h(y(\eta))}{g(y(\eta))}$ holds.
For $x \in (0, 1)$ we have

$$\begin{aligned} y'''(x) &= \lambda a(x)F(x, y(x)) \leq a(x)F(x, y(x)) \\ &\leq a(x)(g(y(x)) + h(y(x))) = a(x)g(y(x))\left(1 + \frac{h(y(x))}{g(y(x))}\right) \\ &\leq \left(1 + \frac{h(y(\eta))}{g(y(\eta))}\right) a(x)g(y(x)) \\ &\leq \left(1 + \frac{h(y(\eta))}{g(y(\eta))}\right) a(x)g(y(\eta)x(1-x)). \end{aligned} \tag{3.6}$$

Integrate from t to 1 to obtain

$$\int_t^1 y'''(x)dx \leq \int_t^1 \left(1 + \frac{h(y(\eta))}{g(y(\eta))}\right) a(x)g(y(\eta)x(1-x)),$$

and so

$$-y''(t) \leq \left(1 + \frac{h(y(\eta))}{g(y(\eta))}\right) \int_t^1 a(x)g(y(\eta)x(1-x))dx, \ t \in (0, 1).$$

For $t \in (0, \eta]$, we have $y'(t) \geq 0$. Then

$$-y''(t)y'(t) \leq \left(1 + \frac{h(y(\eta))}{g(y(\eta))}\right) \int_t^1 a(x)g(y(\eta)x(1-x))dx \cdot y'(t), \ t \in (0, \eta]. \tag{3.7}$$

Integrate from t to η to get

$$-\int_t^\eta y'(\tau)dy'(\tau) \leq \left(1+\frac{h(y(\eta))}{g(y(\eta))}\right) \int_t^\eta \int_\tau^1 a(x)g(y(\eta)x(1-x))dx \cdot y'(\tau)d\tau, \ t \in (0, \eta],$$

and so

$$\frac{1}{2}(y'(t))^2 \leq \left(1+\frac{h(y(\eta))}{g(y(\eta))}\right)\int_t^\eta \int_\tau^1 g(y(\eta)x(1-x))a(x)dx \cdot y'(\tau)d\tau, \ t \in (0,\eta].$$

Since

$$\begin{aligned}
&\int_t^\eta \int_\tau^1 g(y(\eta)x(1-x))a(x)dx \cdot y'(\tau)d\tau \\
&= \int_t^\eta \int_t^x g(y(\eta)x(1-x))a(x)y'(\tau)d\tau dx + \int_\eta^1 \int_t^\eta g(y(\eta)x(1-x))a(x)y'(\tau)d\tau dx \\
&= \int_t^\eta g(y(\eta)x(1-x))a(x)\int_t^x y'(\tau)d\tau dx + \int_\eta^1 g(y(\eta)x(1-x))a(x)\int_t^\eta y'(\tau)d\tau dx \\
&= \int_t^\eta g(y(\eta)x(1-x))a(x)y(x)dx + \int_\eta^1 g(y(\eta)x(1-x))a(x)y(\eta)dx \\
&- \textstyle\int_t^1 g(y(\eta)x(1-x))a(x)y(t)dx \\
&\leq \int_t^\eta g(y(\eta)x(1-x))a(x)y(\eta)dx + \int_\eta^1 g(y(\eta)x(1-x))a(x)y(\eta)dx \\
&- \textstyle\int_t^1 g(y(\eta)x(1-x))a(x)y(t)dx \\
&= \int_t^1 g(y(\eta)x(1-x))a(x)dx(y(\eta)-y(t)) \\
&\leq (y(\eta)-y(t))\int_0^1 g(y(\eta)x(1-x))a(x)dx,
\end{aligned}$$

$$\frac{1}{2}(y'(t))^2 \leq \left(1+\frac{h(y(\eta))}{g(y(\eta))}\right)(y(\eta)-y(t))\int_0^1 g(y(\eta)x(1-x))a(x)dx, \ t \in (0,\eta]$$

and so

$$y'(t) \leq \sqrt{2\left(1+\frac{h(y(\eta))}{g(y(\eta))}\right)(y(\eta)-y(t))\int_0^1 g(y(\eta)x(1-x))a(x)dx}, \ t \in (0,\eta]. \tag{3.8}$$

Integrate from 0 to η to obtain

$$\int_0^\eta \frac{y'(t)}{\sqrt{2\left(1+\frac{h(y(\eta))}{g(y(\eta))}\right)(y(\eta)-y(t))\int_0^1 g(y(\eta)x(1-x))a(x)dx}}dt \leq \eta \leq 1,$$

i.e

$$\frac{1}{\sqrt{2\left(1+\frac{h(y(\eta))}{g(y(\eta))}\right)\int_0^1 g(y(\eta)x(1-x))a(x)dx}}\int_0^\eta \frac{y'(t)}{\sqrt{y(\eta)-y(t)}}dt \leq 1.$$

It is easy to get

$$\frac{1}{\sqrt{2\left(1+\frac{h(y(\eta))}{g(y(\eta))}\right)\int_0^1 g(y(\eta)x(1-x))a(x)dx}}\int_{\frac{1}{m}}^{y(\eta)}\frac{1}{\sqrt{y(\eta)-u}}du \le 1,$$

i.e

$$\frac{2}{\sqrt{2\left(1+\frac{h(y(\eta))}{g(y(\eta))}\right)\int_0^1 g(y(\eta)x(1-x))a(x)dx}}\sqrt{y(\eta)-\frac{1}{m}} \le 1,$$

which yields

$$2(y(\eta)-\frac{1}{m}) \le \left(1+\frac{h(y(\eta))}{g(y(\eta))}\right)\int_0^1 g(y(\eta)x(1-x))a(x)dx.$$

Consequently

$$2(y(\eta)-\epsilon) \le \left(1+\frac{h(y(\eta))}{g(y(\eta))}\right)\int_0^1 g(y(\eta)x(1-x))a(x)dx, \tag{3.9}$$

this together with (3.1) implies $|y|_0 \neq r$. Then Lemma 2.2 implies that $(3.3)^m$ has a solution y_m with $|y_m|_0 \le r$.

Now we consdier $\{y_m\}$. Obviously,

$$\frac{1}{m} \le y_m(t) < r, \ \ for\ t \in [0,1].$$

Next we will show that there exists a constant $k > 0$, independent of m, with

$$y_m(t) \ge kt(1-t), \ \ for\ t \in [0,1]. \tag{3.10}$$

Notice (H_2) gurantees the existence of a function $\varphi_r(t)$ continuous on [0,1] and positive on (0,1) with $F(t,u) \ge \varphi_r(t)$ for $(t,u) \in (0,1) \times (0,r]$. Now, using the Green's function representation for the solution of $(3.3)^m$, we have

$$G(t,s) = \begin{cases} t\min\{\eta,s\}+\frac{1}{2}s^2-ts, & 0 \le s \le t \le 1, \\ t\min\{\eta,s\}-\frac{1}{2}t^2, & 0 \le t \le s \le 1, \end{cases}$$

and

$$y_m(t) = \frac{1}{m} + \int_0^1 G(t,s)a(s)F(s,y_m(s))ds$$

$$\geq \int_0^1 G(t,s)a(s)\varphi_r(s)ds \equiv \Phi_r(t),\ t\in[0,1],$$

and

$$\Phi_r'(t) = \int_0^1 \min\{\eta,s\}a(s)\varphi_r(s)ds - \int_0^t sa(s)\varphi_r(s)ds - t\int_t^1 a(s)\varphi_r(s)ds.$$

It is easy to check that $\Phi_r(0)=0$ and there exists k_1, independent of m such that $\Phi_r'(0) = \displaystyle\int_0^1 \min\{\eta,s\}a(s)\varphi_r(s)ds \geq k_1$, which implies that there is an $\epsilon_0 > 0$ with $\Phi_r(t) \geq k_1 t \geq k_1 t(1-t)$ for $t\in[0,\epsilon_0]$.

Since $\Phi_r(1) = \displaystyle\int_0^1 (\min\{\eta,s\} + \frac{1}{2}s^2 - s)a(s)\varphi_r(s)ds \equiv k_2$ and $\Phi_r'(1) = \Phi_r(1) - \frac{1}{2}\int_0^1 s_2 a(s)\varphi_r(s)ds \leq \Phi_r(1) \equiv k_2$, there exists a $\delta > 0$ such that $\Phi_r'(t) \leq k_2$ for $t\in[1-\delta,1]$.

Then, $\displaystyle\int_t^1 \Phi'(\tau)d\tau \leq k_2(1-t)$ i.e $\Phi_r(t) \geq k_2 t \geq k_2 t(1-t)$. Since $\Phi_r(t)$ is continuous, $\dfrac{\Phi_r(t)}{t(1-t)}$ is bounded on $[\epsilon_0, 1-\delta]$, there is a constant k, independent of m, with $\Phi_r(t) \geq kt(1-t)$ on $[0,1]$ i.e (3.10) is true.

Next we will show $\{y_m\}_{m\in N_0}$ is equicontinuous family on any closed interval of $(0,1)$. Returning to (1), with y replaced by y_m, we have

$$\begin{aligned} y_m'''(t) &= a(t)f(t,y_m(t)) \\ &\leq a(t)(g(y_m(t)) + h(y_m(t))) \\ &\leq a(t)g(kt(1-t))(1+\frac{h(r)}{g(r)}),\ t\in(0,1), \end{aligned} \tag{3.11}$$

for any $\varepsilon > 0$, which is sufficiently small. Consequently, $y_m''(t)$ is equicontinuous on $[\varepsilon, 1-\varepsilon]$.
Integrate (3.11) from t to 1, we have

$$0 < -y_m''(t) \leq (1+\frac{h(r)}{g(r)})\int_t^1 a(s)g(ks(1-s))ds,\ t\in(0,1),$$

so

$$|y_m''(t)| \leq d =^{def} \max_{t\in[\varepsilon,1-\varepsilon]}(1+\frac{h(r)}{g(r)})\int_t^1 a(s)g(ks(1-s))ds,$$

and

$$|y_m'(t)| = |y_m'(t) - y_m'(\eta)| = |y_m''(\xi)||t-\eta| \leq |y_m''(\xi)| \leq d,\ t\in[\varepsilon,1-\varepsilon].$$

Then $y'_m(t)$, $y_m(t)$ is equicontinuous on $[\varepsilon, 1-\varepsilon]$ and since ε is small and arbitrary, $y_m(t)$ is equicontinous on any interval $[c,d] \subseteq (0,1)$. The Arzela-Ascoli Theorem guarantees the existence of subsequence N of N_0 and a function $y \in C[0,1]$ with y_m converging uniformly on (0,1) to y as $m \to \infty$ through N. Fix $t \in (0,1)$ (without loss of generality assume $t \neq \frac{1}{2}$). Now $y_m, m \in N$ satisfies the integral equation

$$y_m(x) = y_m(\frac{1}{2}) + y'_m(\frac{1}{2})(x-\frac{1}{2}) + \frac{1}{2}y''_m(\frac{1}{2})(x-\frac{1}{2})^2 + \frac{1}{2}\int_{\frac{1}{2}}^{x}(x-s)^2 a(s) f(s, y_m(s))ds,$$

for $x \in (0,1)$. Notice (take $x = \frac{2}{3}$) that $\{y'_m(\frac{1}{2})\}, \{y''_m(\frac{1}{2})\}, m \in N$ are bounded sequences since $kt(1-t) \leq y_m(t) \leq r$ for $t \in [0,1]$. Thus they have convergent subsequences. For convennience let $\{y'_m(\frac{1}{2})\}, \{y''_m(\frac{1}{2})\}$ denote the subsequences and r_0, r_1 be their limits. Further we have

$$0 < f(t, y_m(t)) \leq g(y_m(t)) + h(y_m(t))$$

$$= g(y_m(t))\left(1 + \frac{h(y_m(t)}{g(y_m(t))}\right) \leq g(kt(1-t))\left(1 + \frac{h(r)}{g(r)}\right),$$

i.e

$$|f(t, y_m(t))| \leq g(kt(1-t))\left(1 + \frac{h(r)}{g(r)}\right).$$

Notice that $g(kt(1-t)) \in L^1(0,1)$, using Lebesgue dominated convengence theorem, for the above fixed t,

$$y_m(t) = y_m(\frac{1}{2}) + y'_m(\frac{1}{2})(t-\frac{1}{2}) + \frac{1}{2}y''_m(\frac{1}{2})(t-\frac{1}{2})^2 + \frac{1}{2}\int_{\frac{1}{2}}^{t}(t-s)^2 a(s) f(s, y_m(s))ds,$$

let $m \to \infty$ through N to obtain

$$y(t) = y(\frac{1}{2}) + r_0(t-\frac{1}{2}) + \frac{1}{2}r_1(t-\frac{1}{2})^2 + \frac{1}{2}\int_{\frac{1}{2}}^{t}(t-s)^2 a(s) f(s, y(s))ds.$$

We can do this argument for each $t \in (0,1)$. Then $y'''(t) = a(t)f(t,y)$ for $0 < t < 1$.

Next we will show $y(0) = y'(\eta) = y''(1) = 0$.

Similar to (3.6)-(3.8) and replace y to y_m, we have

$$\frac{y'_m(t)}{\sqrt{r - y_m(t)}} \leq \sqrt{2(1 + \frac{h(r)}{g(r)}) \int_0^1 a(s) g(ks(1-s))ds}. \tag{3.12}$$

Fixed $z \in (0,\eta)$, integrate (3.12) from 0 to z, we obtain easily

$$\int_{y_m(0)}^{y_m(z)} \frac{1}{\sqrt{r-s}} ds \leq \int_0^z \sqrt{2(1+\frac{h(r)}{g(r)})\int_0^1 a(s)g(ks(1-s))dsdt}, \quad z \in (0,\eta]. \tag{3.13}$$

Letting $m \to +\infty$ in (3.13) and noticing $y_m(0) = \frac{1}{m} \to 0$, we have

$$\int_0^{y(z)} \frac{1}{\sqrt{r-s}} ds \leq \int_0^z \sqrt{2(1+\frac{h(r)}{g(r)})\int_0^1 a(s)g(ks(1-s))dsdt}, \quad z \in (0,\eta]. \tag{3.14}$$

Equation (3.14) leads to $y(0) = \lim_{z\to 0^+} y(z) = 0$.

Fixed $z \in (\eta, 1)$, integrate (3.11) from z to 1, we obtain easily

$$\int_{y_m''(z)}^{y_m''(1)} ds \leq (1+\frac{h(r)}{g(r)})\int_z^1 a(s)g(ks(1-s))ds, \quad z \in [\eta, 1). \tag{3.15}$$

Letting $m \to +\infty$ in (3.15) and noticing $y_m''(1) = 0$, we have

$$\int_{y''(z)}^{0} ds \leq (1+\frac{h(r)}{g(r)})\int_z^1 a(s)g(ks(1-s))ds, \quad z \in [\eta, 1). \tag{3.16}$$

Equation (3.16) leads to $y''(1) = \lim_{z\to 1^-} y''(z) = 0$.

Also we know $y_m'(t)$ is equicontinous on any interval $[c,d] \subseteq (0,1)$ and $y_m'(\eta) = 0$, by Ascoli-Arzela theorem, we obtain $y'(\eta) = \lim_{m\to+\infty} y_m'(\eta) = 0$. This complete the proof. Finally it is easy to see that $|y|_0 < r$ (note if $|y|_0 = r$ then following essentially the argument from (3.7)-(3.8) will yield a contradiction). Theorem 1 follows.

Proof of Theorem 2. Let $\|x\| = \max_{t\in(0,1)} |x(t)|$ for any $x \in C[0,1]$. It is well known that $C[0,1]$ is a Banach space with the norm $\|\cdot\|$. Let $P = \{x | x \in C[0,1],\ x(t) \geq 0 \ for \ t \in [0,1]\}$ and $\bar{Q} = \{x \in P : x(t) \geq \|x\| t(1-t) \ for \ t \in [0,1]\}$. Clearly, P and $\bar{Q}$ are cones of $C[0,1]$.

From (H_3), there exists an $r > 0$ such that

$$\frac{r}{\left(1+\frac{h(r)}{g(r)}\right)\int_0^1 g(rx(1-x))a(x)dx} > \frac{1}{2}.$$

Choose $0 < \epsilon < r$ small enough such that

$$\frac{r}{\left(1+\frac{h(r+\epsilon)}{g(r+\epsilon)}\right)\int_0^1 g(rx(1-x))a(x)dx} > \frac{1}{2}. \tag{3.17}$$

Let $n_0 \in \{1, 2 \ldots\}$ be chosen so that $\dfrac{1}{n_0} < \epsilon$ and let $N = \{n_0, n_0 + 1, \ldots\}$.

Define the mapping

$$(T_n x)(t) = \int_0^1 G(t,s)a(s)f(s, x(s) + \frac{1}{n})ds, \ n \in N.$$

By (H_1) and (3.5), we can easily get $T_n : P \to \bar{Q}$ is continous and completely continuous.

Now we claim that $i(T_n, Q_r, \bar{Q}) = 1$, where $Q_r = \{x \in \bar{Q}, \|x\| < r\}$. If

$$x \neq \mu T_n x, \forall \mu \in [0, 1], x \in \partial Q_r, \tag{3.18}$$

is false, there exists $x_0 \in \partial Q_r$ and $\mu_0 \in [0, 1]$ such that

$$x_0(t) = \mu_0(T_n x_0)(t), \forall t \in [0, 1].$$

Noticing $x_0 \in \partial Q_r$ and by (3.5), we get

$$x_0(t) \geq \|x_0\| t(1 - t) = rt(1 - t), \forall t \in [0, 1].$$

On the other hand,

$$\begin{aligned} x_0(t) &= \mu_0(T_n x_0)(t) \\ &= \mu_0 \int_0^1 G(t,s)a(s)f(s, x_0(s) + \frac{1}{n})ds. \end{aligned}$$

Then we get

$$x_0'''(t) = \mu_0 a(t) f(t, x_0(t) + \frac{1}{n}) \leq a(t) f(t, x_0(t) + \frac{1}{n}).$$

Similar to (3.6)-(3.8), we can obtain

$$x_0'(t) \leq \sqrt{2\left(1 + \frac{h(r+\epsilon)}{g(r+\epsilon)}\right)(r - x_0(t)) \int_0^1 a(s)g(rs(1-s))ds}$$

i.e

$$\frac{x_0'(t)}{\sqrt{r - x_0(t)}} \leq \sqrt{2\left(1 + \frac{h(r+\epsilon)}{g(r+\epsilon)}\right) \int_0^1 a(s)g(rs(1-s))ds}.$$

Integrate from 0 to η to obtain

$$r \leq \frac{1}{2}\left(1 + \frac{h(r+\epsilon)}{g(r+\epsilon)}\right) \int_0^1 a(s)g(rs(1-s))ds,$$

which is contradicted with (3.17). Then (3.18) follows. By Lemma 2.3, we can get $i(T_n, Q_r, \bar{Q}) = 1$.

Now we claim that $i(T_n, \tilde{Q}_R, \bar{Q}) = 0$, where $\tilde{Q}_R = \{x \in \bar{Q} : \|x\| < R\}$. By (H_4), we know that there exists $L > r > 0$, such that

$$\frac{f(t,x)}{x} \geq 2\left[\min_{t\in[\alpha,\beta]} t(1-t)\int_\alpha^\beta G(t,s)a(s)ds\right]^{-1}, \forall x > L, t \in [\alpha,\beta]. \quad (3.19)$$

Set

$$R = \max\left\{r, \frac{L}{\alpha(1-\beta)}\right\}.$$

Then

$$T_n x \neq \mu x, \forall x \in \partial\tilde{Q}_R,\ \mu \in [0,1]. \quad (3.20)$$

If it is not ture, there exists $x_0 \in \partial\tilde{Q}_R$ and $\mu_0 \in [0,1]$ such that $T_n x_0 = \mu_0 x_0 \leq x_0$. Since $x_0 \in \bar{Q},\ x_0(t) \geq \|x_0\| t(1-t) = Rt(1-t) \geq \alpha(1-\beta) > L$. By (3.19), we have

$$\begin{aligned}
x_0(t) &\geq \int_0^1 G(t,s)a(s)f(s, x_0(s) + \frac{1}{n})ds \\
&\geq \int_\alpha^\beta G(t,s)a(s)2\left[\min_{t\in[\alpha,\beta]} t(1-t)\int_\alpha^\beta G(t,s)a(s)ds\right]^{-1}(x_0(s) + \frac{1}{n})ds \\
&\geq \int_\alpha^\beta G(t,s)a(s)2\left[\min_{t\in[\alpha,\beta]} t(1-t)\int_\alpha^\beta G(t,s)a(s)ds\right]^{-1} x_0(s)ds \\
&\geq \int_\alpha^\beta G(t,s)a(s)2\left[\min_{t\in[\alpha,\beta]} t(1-t)\int_\alpha^\beta G(t,s)a(s)ds\right]^{-1} Rs(1-s)ds \\
&\geq 2R > R,
\end{aligned}$$

which is in contradiction with $x_0 \in \partial\tilde{Q}_R$. (3.20) is ture. By Lemma 2.4, we have $i(T_n, \tilde{Q}_R, \bar{Q}) = 0$.
Now we prove $i(T_n, Q_{r'}, \bar{Q}) = 0$, where $Q_{r'} = \{x \in \bar{Q} : \|x\| < r'\}$.

By (H_4), we can get for $\forall\ L' > 0$, there exists $\delta > 0$ such that

$$f(t,x) > L' \ for\ x \in (0,\delta)\ and\ t \in [\alpha,\beta].$$

Choose $r' \in (0.\delta)$ and n sufficiently large such that

$$r' < L'[\min_{t\in[\alpha,\beta]} \int_\alpha^\beta G(t,s)a(s)ds]^{-1},\ and\ r' + \frac{1}{n} < \delta.$$

Next, we claim

$$T_n x \neq \mu x, \forall x \in \partial Q_{r'},\ \mu \in (0,1].$$

Suppose this is false, then there exists $x_0 \in \partial Q_{r'}$ and $\mu_0 \in (0,1]$ such that $T_n x_0 = \mu_0 x_0$. Since $x_0 \in \partial Q_{r'}, x_0(t) \le \|x_0(t)\| = r'$, $x_0(t)+\frac{1}{n} \le r'+\frac{1}{n} < \delta$. Therefore,

$$\begin{aligned} x_0(t) \ge (T_n x_0)(t) &= \int_0^1 G(t,s)a(s)f(s,x_0(s)+\frac{1}{n})ds \\ &> L' \int_\alpha^\beta G(t,s)a(s)ds \ge L' . \min_{t\in[\alpha,\beta]} \int_\alpha^\beta G(t,s)a(s)ds > r', \\ &\forall t \in [\alpha,\beta]. \end{aligned}$$

This is in contradiction with $x_0 \in \partial Q_{r'}$ and immediately by Lemma 2.4 our result follows.

From the above instructions, we get

$$i(T_n, Q_r, \bar{Q}) = 1,$$

$$i(T_n, Q_r \setminus \overline{Q_{r'}}, \bar{Q}) = 1 - 0 = 1,$$

$$i(T_n, \tilde{Q}_R \setminus \overline{Q_r}, \bar{Q}) = 0 - 1 = -1.$$

By fixed index theroy, T_n has two fixed point x_n and y_n satisfing that $r' < \|x_n\| < r < \|y_n\| < R$.

Similar to the proof of $\{y_m\}$, we can get $x'''(t) = a(t)f(t,x(t))$, $t \in (0,1)$ and $x(0) = x'(\eta) = x''(1) = 0$, where $x_n(t) \to x(t)$, $n \to +\infty$.

This means $x(t)$ is a positive solution of BVP(1.1). In the same way, we can get $y(t)$ is another positive solution of BVP(1.1), where $y_n(t) \to y(t)$, $n \to +\infty$. We can also get $x(t)$ and $y(t)$ satisfying that $r' < \|x\| \le r \le \|y\| \le R$. Then Theorem 2 follows.

4. Example

Consider the following singular boundary value problem:

$$\begin{cases} y'''(t) - \dfrac{t^{\frac{7}{3}}}{(1+t)^{\frac{1}{2}}(1-t)^{\frac{1}{6}}}(y^{-\frac{1}{3}} + y^2) = 0, \\ y(0) = y'(\eta) = y''(1) = 0. \end{cases} \tag{4.1}$$

It is easy to see that $a(t) = \dfrac{t^{\frac{7}{3}}}{(1+t)^{\frac{1}{2}}(1-t)^{\frac{1}{6}}}$ is singular at $t=1$, $f(t,y) = y^{-\frac{1}{3}} + y^2$ is singular at $y = 0$, and there exists a $r > 0$, which satisfies

$\dfrac{r^{\frac{4}{3}}}{1+r^{\frac{7}{3}}} > \dfrac{\pi}{8}$ (for example $r = 1$) such that

$$\frac{r}{\left(1+\dfrac{h(r)}{g(r)}\right)\displaystyle\int_0^1 g(rx(1-x))a(x)dx} > \frac{1}{2}.$$

Meanwhile, $\lim\limits_{y\to+\infty} \dfrac{f(t,y)}{y} = +\infty$, $\lim\limits_{y\to 0} f(t,y) = +\infty$. Consequently, the conclution of Theorem 1 and Theorem 2 for BVP (4.1) hold.

Note that $\displaystyle\int_0^1 \frac{x^2}{\sqrt{(1+x)(1-x)}}dx = \frac{\pi}{4}$. Because

$$I_n = \int_0^1 \frac{x^n}{\sqrt{(1+x)(1-x)}}dx = \begin{cases} \dfrac{(2k-1)!!}{(2k)!!}\cdot\dfrac{\pi}{2}, & n = 2k, \\ \dfrac{(2k-2)!!}{(2k-1)!!}, & n = 2k-1. \end{cases}$$

References

[1] R.P.Agarwal & D.O'Regan, A survey of recent results for initial and boundary value problems singular in the dependent variable.

[2] D.R.Anderson, Multiple positive solutions for a three-point boundary value problem, Math.Comput.Modelling.27 (1998) 49-57.

[3] Z.Du,W.Ge,X.Lin, A class of third-order mutiple-point boundary value problems, J.Math.Anal.Appl.294 (2004) 104-112.

[4] D.Guo, V.Lakshmikantham, Nonlinear Problems in Abstract Cones, Academic Press,New York,1988.

[5] Y.Liu, Structure of a class of singular boundary value problem with superlinear effect, J.Math.Anal.Appl.284 (2003) 78-89.

[6] Y.Sun, Positive solutions of singular third-order three-point boundary value problem, J.Math.Anal.Appl. 306 (2005) 589-603.

[7] Q.Yao, Positive solutions of singular third-order three-point boundary value problems, J.Math.Anal.Appl. 354 (2009) 207-212.

MODIFIED NEUMANN INTEGRAL AND ASYMPTOTIC BEHAVIOR IN THE HALF SPACE*

YANHUI ZHANG*[a] GUAN TIE DENG*[b] ZHENZHEN WEI*[a]

[a] *Department of Mathematics, Beijing Technology and Business University, Beijing 100048, China*

[b] *Sch. Math. Sci. & Lab. Math. Com. Sys., Beijing Normal University, Beijing 100875, China*

The main objective is to give the asymptotic behavior at infinity for the Neumann integral and the modified Neumann integral in the half space of $\mathbb{R}^n$. The asymptotic behavior hold outside an explicitly defined exceptional set as [10].

Keywords: Neumann Integral; Modified Neumann Integral; Asymptotic Behavior.

AMS No: 31B05, 35J05.

1. Introduction and main results

Let R^n $(n \geq 2)$ denote the $n-$dimensional Euclidean space. For x in the upper half-space $H = \{x = (x', x_n) \in \mathbb{R}^n, x_n > 0\}$, where $x' \in \mathbb{R}^{n-1}$ is the project of x onto the hyperplane $\partial H = \mathbb{R}^{n-1}$.

We recall that the Laplace equation

$$\Delta u = \frac{\partial^2 u}{\partial x_1^2} + \frac{\partial^2 u}{\partial x_2^2} + \cdots + \frac{\partial^2 u}{\partial x_n^2} = 0$$

admits the fundamental solution[1]

$$E(x) = \begin{cases} \frac{1}{2\pi} \log |x|, \ n = 2, \\ \gamma_n \frac{1}{|x|^{n-2}}, \ n > 2. \end{cases}$$

*Project supported by PHR(IHLB 201008257 and IHLB 201106206) and Scientific Research Common Program of Beijing Municipal Commission of Education(KM 200810011005) and College Students Scientific Research and Undertaking Action Project(PXM 2012-014213-000067)

$\gamma_n = \frac{1}{(2-n)\omega_n}$, and ω_n is the surface area of the unit sphere in $\mathbb{R}^n$. It satisfies

$$\Delta E(x-y) = \delta(x-y), x, y \in \mathbb{R}^n$$

when δ is the Dirac distribution; and the Laplace operator has the property[2]

$$-\widehat{\Delta(f)}(\xi) = 4\pi^2|\xi|^2\hat{f}(\xi), \tag{1.1}$$

for any f is a Schwartz function and $\hat{f}(\xi)$ is the Fourier transform. Motivated by (1.1), $(-\Delta)^{\frac{\alpha}{2}}(0 < \alpha < n)$ can be defined for a Schwartz function f by

$$(-\Delta)^{\frac{\alpha}{2}}(f)(x) = \mathfrak{F}(|\xi|^{-\alpha}\hat{f}(x)) = \gamma(\alpha)\int_{\mathbb{R}^n}\frac{f(y)}{|x-y|^{n-\alpha}}dy, \tag{1.2}$$

where $\gamma_{(\alpha)}|x|^{-n+\alpha} \in L_{loc}(\mathbb{R}^n)$ is known as the Reisz kernel and $\gamma(\alpha)$ is a constant [3], $\mathfrak{F}$ is the Fourier inversion transform.

Define the Green function of the half space H as follows

$$\begin{aligned} G(x,y) &= E(x,y) - E(x,\tilde{y}) \\ &= \gamma_n\left(\frac{1}{|x-y|^{n-2}} - \frac{1}{|x-\tilde{y}|^{n-2}}\right), \ x, y \in \bar{H}, \end{aligned}$$

where $\tilde{}$ denotes the reflection in the boundary plane ∂H, that is, $\tilde{y} = (y_1, y_2, \cdots, y_{n-1}, -y_n)$, if $y = (y_1, y_2, \cdots, y_{n-1}, y_n)$. The Poisson kernel $P(x,y')$ of H is just defined by

$$P(x,y') = \frac{\partial G(x,y)}{\partial y_n}|_{y_n=0} = \gamma_n\frac{x_n}{|x-(y',0)|^n},$$

for $x = (x', x_n), y = (y', y_n) \in \mathbb{R}^n$ and $y' \in \partial H$.

The Dirichlet problem is to find a function u satisfying

$$u(x) \in C^2(H), \Delta u(x) = 0; x \in H, u(x)|_{x_n=0} = f(x'); x' \in \partial H.$$

where f is a measurable function of $\mathbb{R}^{n-1}$. The solution of Dirichlet problem can be written as

$$P[f](x) = \int_{\partial H} P(x,y')f(y')dy', \tag{1.3}$$

(1.3) is called the Poisson integral. The necessary and sufficient condition for $P[f](x) \not\equiv \infty$ is

$$\int_{\partial H}\frac{|f(y')|}{1+|y'|^n}dy' < \infty. \tag{1.4}$$

There are some results about the growth properties for the solution of Dirichlet problem in the half space H, such as [7][10]. In this paper we will consider the solution and the growth properties of the Neumann problem. As to the Dirichlet problem, the Neumann problem of the Laplace's equation is to find a function u satisfying

$$u(x) \in C^2(H), \Delta u(x) = 0; x \in H, \frac{\partial u(x)}{\partial n} = f(x'), x' \in \partial H. \tag{1.5}$$

where f is a measurable function of ∂H.

Just like the Green function of half space H, taking an odd reflection across ∂H

$$\begin{aligned}\tilde{G}(x,y) &= E(x,y) + E(x,\tilde{y}) \\ &= \gamma_n \left(\frac{1}{|x-y|^{n-2}} + \frac{1}{|x-\tilde{y}|^{n-2}} \right), \ x, y \in \bar{H},\end{aligned}$$

the Neumann kernel is

$$N(x,y') = \tilde{G}(x,y)|_{y_n=0} = 2\gamma_n \frac{x_n}{|x-y'|^{n-2}},$$

and the Neumann integral is

$$N[f](x) = 2\gamma_n \int_{\partial H} \frac{x_n f(y')}{|x-y'|^{n-2}} dy'. \tag{1.6}$$

The Neumann integral (1.6) is harmonic in H if ([2], Theorem 6)

$$\int_{\partial H} \frac{f(y')}{1+|y'|^{n-2}} dy' < \infty. \tag{1.7}$$

If f is continuous, then convergence of (1.7) is sufficient for (1.6) to be a classical solution of the Neumann problem on H [3]. There are some results for Neumann problem [3] [7].

Denote by $B(x,\rho)$ and $\partial B(x,\rho)$ respectively the open ball and the sphere of positive radius ρ with center $x \in \mathbb{R}^n$. Our objective is to establish the asymptotic behavior for Neumann integral and modified Neumann integral.

2. Neumann Integral and the Asymptotic Behavior in the Half Space

Let f be a continuous function defined on ∂H. Armitage proved [3]

Theorem A Let f be a continuous function defined on ∂H satisfied (1.7). Then a solution of the Neumann problem on H

$$\mathfrak{M}(|N[f]| : r) = O(1), \quad (r \to \infty), \tag{2.1}$$

where the mean of g is defined as follows:

$$\mathfrak{M}(g:r)=\gamma_n\int_{H\cap B(0,R)}g(x)d\sigma_x.$$

In this section, we will explicitly give the asymptotic behavior of the Neumann integral, in the same way as [10] did in the case of the Dirichlet problem.

Theorem 2.1 The Neumann integral (1.6) has the asymptotic behavior

$$N[f](x)=o(|x|) \tag{2.2}$$

in $H-D$, where $D=\cup_{j=1}^{\infty}B(x_j,\rho_j)$ and $B(x_j,\rho_j)$ is ball radius ρ_j with center $x_j\in\mathbb{R}^n$.

Proof of Theorem 2.1

Let μ be a positive Borel measure in $\mathbf{R}^n$, $\beta\geq 0$. The maximal function $M(d\mu)(x)$ of order β is defined by

$$M(d\mu)(x)=\sup_{0<r<\infty}\frac{\mu(B(x,r))}{r^\beta}, \tag{2.3}$$

then the maximal function $M(d\mu)(x):\mathbf{R}^n\to[0,\infty)$ is lower semicontinuous, hence it is measurable.

Define the measure

$$dm(y')=\frac{f(y')dy'}{1+|y'|^{n-2}}.$$

Then the Neumann integral now can be represented in the form

$$N[f](x)=\int_{\partial H}(1+|y'|^{n-2})N(x,y')dm(y').$$

For every Lebesgue measurable set $E\subset\mathbf{R}^n$, we introduce some signs as [10]. Set

$$E_1(\varepsilon)=\{x\in\mathbb{R}^n:|x|\geq 2,\exists\, t>0,m^{(\varepsilon)}(B(x,t)\cap\mathbb{R}^{n-1})>\varepsilon(\frac{t}{|x|})^{n-1}\}. \tag{2.4}$$

The measure

$$m^{(\varepsilon)}(\mathbb{R}^{n-1})\leq(\frac{3}{4})^n\varepsilon \tag{2.5}$$

defined by

$$m^{(\varepsilon)}(E)=m(E\cap\{y'\in\mathbb{R}^{n-1}:|y'|\geq R_\varepsilon\}), \tag{2.6}$$

and function

$$m_x^{(\varepsilon)}(t) = \int_{|x-(y',0)|\leq t} dm^{(\varepsilon)}(y'). \tag{2.7}$$

Write

$$N^{(\varepsilon)}[f](x) = \int_{\partial H} (1 + |y'|^{n-2}) N(x, y') dm^{(\varepsilon)}(y').$$

Then

$$N[f](x) = N^{(\varepsilon)}[f](x) + N_3[f](x) = N_1^{(\varepsilon)}[f](x) + N_2^{(\varepsilon)}[f](x) + N_3[f](x),$$

where denote

$$N_1^{(\varepsilon)}[f](x) = \int_{\{y'\in\mathbb{R}^{n-1}:\delta\leq|x-y'|\leq 3|x|\}} (1 + |y'|^{n-2}) N(x, y') dm^{(\varepsilon)}(y');$$

$$N_2^{(\varepsilon)}[f](x) = \int_{\{y'\in\mathbb{R}^{n-1}:|x-y'|\geq 3|x|\}} (1 + |y'|^{n-2}) N(x, y') dm^{(\varepsilon)}(y');$$

$$N_3[f](x) = 2\gamma_n \int_{\{y'\in\mathbb{R}^{n-1}:|y'|<R_\varepsilon\}} \frac{x_n f(y')}{|x - y'|^{n-2}} dy'.$$

For $|x| \geq 2R_\varepsilon$ and $x \notin E_1(\lambda)$, we get

$$\begin{aligned}
|N_1^{(\varepsilon)}[f](x)| &= 2\gamma_n (1 + |x|^{n-2}) x_n \int_{\{y'\in\mathbb{R}^{n-1}:\delta\leq|x-y'|\leq 3|x|\}} \frac{dm^{(\varepsilon)}(y')}{|x - y'|^{n-2}} \\
&\leq 2\gamma_n (1 + |x|^{n-2}) x_n \int_\delta^{3|x|} \frac{dm_x^{(\varepsilon)}(t)}{t^{n-2}} \\
&\leq A|x|\varepsilon.
\end{aligned}$$

Moreover

$$\begin{aligned}
|N_2^{(\varepsilon)}[f](x)| &\leq 2\gamma_n x_n \int_{\{y'\in\mathbb{R}^{n-1}:|x-y'|\geq 3|x|\}} \frac{1+|y'|^{n-2}}{|x - y'|^{n-2}} dm^{(\varepsilon)}(y') \\
&\leq 2\gamma_n x_n \int_{\{y'\in\mathbb{R}^{n-1}:|x-y'|\geq 3|x|\}} \left(\frac{1}{|x - y'|^{n-2}} + (1 + \frac{|x|}{|x - y'|})^{n-2}\right) dm^{(\varepsilon)}(y') \\
&\leq 2\gamma_n x_n \int_{\{y'\in\mathbb{R}^{n-1}:|x-y'|\geq 3|x|\}} \left(\frac{1}{(3|x|)^{n-2}} + (1 + \frac{|x|}{3|x|})^{n-2}\right) dm^{(\varepsilon)}(y') \\
&\leq A|x|\varepsilon,
\end{aligned}$$

$$N_3[f](x) = 2\gamma_n x_n \int_{\{y'\in\mathbb{R}^{n-1}:|y'|<R_\varepsilon\}} \frac{f(y')}{|x - y'|^{n-2}} dy' \leq \infty.$$

Thus

$$|N[f](x)| \leq |N_1^{(\varepsilon)}[f](x)| + |N_2^{(\varepsilon)}[f](x)| + |N_3[f](x)| \leq A|x|\varepsilon.$$

By collecting the estimates of $|N_1^{(\varepsilon)}[f](x)|$ and $|N_2^{(\varepsilon)}[f](x)|$, there exists a positive constant A independent of ε, such that $|N[f](x)| \le A|x|\varepsilon$ for $|x| \ge 2R_\varepsilon$ and $x \notin E_1(\varepsilon)$. Let μ_ε be a measure on $\mathbb{R}^n$ defined by $\mu_\varepsilon(E) = m^{(\varepsilon)}(E \cap \mathbb{R}^{n-1})$ for every measurable set E in $\mathbb{R}^n$. By the proof of [10], there exist $x_{j,p}$ and $\rho_{j,p}$ with $R_{p-1} \le |x_{j,p}| < R_p$, such that

$$\sum_{j=1}^{\infty}(\frac{\rho_{j,p}}{|x_{j,p}|})^{n-1} \le \frac{1}{2^p}.$$

Since $R_{p-1} \le |x| < R_p$ and $x \notin G_p = \cup_{j=1}^{\infty} B(x_{j,p}, \rho_{j,p})$, therefore

$$|N[f](x)| \le A|x|\varepsilon_p,$$

we have

$$\sum_{p=1}^{\infty}\sum_{j=1}^{\infty}(\frac{\rho_{j,p}}{|x_{j,p}|})^{n-\beta} \le \sum_{p=1}^{\infty}\frac{1}{2^p}.$$

Setting $G = \cup_{p=1}^{\infty} G_p$, we see the proof of theorem 2.1 is completed.

3. Modified Neumann Integrals and the Asymptotic Behavior in the Half Space

If $m \ge 0$ is an integer, we define a modified Neumann kernel as [2] [7][8][9] for $x \in H - \{y\}$ by

$$N_m(x, y') = \begin{cases} N(x, y'), & |y'| \le 1, \\ N(x, y') - \sum_{k=0}^{m-1} \frac{x_n |x|^k}{|y'|^{n+k-1}} C_k^{\frac{n-2}{2}}\left(\frac{x \cdot y'}{|x||y'|}\right), & |y'| \ge 1. \end{cases} \tag{3.1}$$

The first m terms of the asymptotic expansion of k in inverse power of $|y'|$ are removed. The coefficients are in terms of Gegenbauer polynomials, $C_k^{\frac{n-2}{2}}$, most of whose properties used herein are derived in [4] and [12].

The expansion (3.1) arises from the generating function for Gegenbauer polynomails [4]

$$(1 - 2tz + z^2)^{-\lambda} = \sum_{k=0}^{\infty} z^k C_k^{\lambda}(t), \qquad \lambda > 0 \tag{3.2}$$

where $C_k^{\lambda}(t) = \frac{1}{k!}\frac{\partial^k}{\partial z^k}(1 - 2tz + z^2)^{-\lambda}|_{z=0}$. If $-1 \le t \le 1$, the series converges absolutely for $|z| < 1$ (the left side of (3.2) is singular at $z =$

$t \pm i\sqrt{1-t^2}$). The majorisation and derivative for formulas

$$|C_k^\lambda(t)| \le C_k^\lambda(1) = \frac{\Gamma(2\lambda+k)}{\Gamma(2\lambda)\Gamma(k+1)};$$

$$\frac{d}{dt}C_k^\lambda(t) = 2\lambda C_{k-1}^{\lambda+1}(t)$$

are proved in [4]. $C_k^\lambda(t)$ is a polynomial in t of degree k. And

$$C_0^\lambda(t) = 1, \quad C_1^\lambda(t) = 2\lambda t.$$

The series

$$N(x,y') = 2\gamma_n \frac{x_n}{|x-y'|^{n-2}} = \gamma_n \sum_{k=0}^{\infty} \frac{x_n|x|^k}{|y'|^{n+k-2}} C_k^{\frac{n-2}{2}}\left(\frac{x\cdot y'}{|x||y'|}\right)$$

converges for $|x| < |y'|$, each term is homogeneous in x of degree $k+1$. The Dirichlet version appears in [5],[6][11][12][13]. The Neumann version is discussed by Gardiner [7] and Arimitage [3], Siegel D. and Talvila E.[12]. In this section, we will give the asymptotic behavior for the modified Neumann integral. Note that the condition (3.4) is necessary and sufficient for $N_m[f](x)$ to exist as a Lebesgue integral on H, see [12].

For any $y' \in \partial H$ and $|x| > 1, x_n > 0$, we can give the following inequalities as [9]

$$|N_m(x,y')| \le \begin{cases} \dfrac{A|x|^{m+1}}{(1+|y'|)^{n+m-2}}, & |y'| \le 2|x|, \\ \dfrac{A|x|^{m+2}}{(1+|y'|)^{n+m-1}}, & |y'| > 2|x|, \end{cases} \tag{3.3}$$

for some constant A, here A will denotes a finite, positive constant depending at most on n and m, not necessarily the same on any two occurrences.

When the integral in (1.6) diverges but

$$\int_{\partial H} \frac{f(y')}{1+|y'|^{n+m-2}} dy' < \infty \tag{3.4}$$

for a positive integer m, we can use the modified Neumann kernel $N_m(x,y')$, then

$$N_m[f](x) = \int_{\partial H} N_m(x,y')f(y')dy' \tag{3.5}$$

is the solution of the classical Neumann problem [8].

Let $\omega : \mathbb{R}^{n-1} \to [0,1]$ be continuous so that $\omega(y) \equiv 0$ for $0 \le |y| \le 1$ and $\omega(y) \equiv 1$ for $|y| \ge 2$. The purpose of the function ω is merely to avoid the singularity of the modified kernel at the origin. Siegel D. and Talvila E. [12] have proved that

Theorem B Let f be continuous functions with the origin not in the closure of its support and satisfy (3.4). Then

$$V(x) = N_m[\omega f](x) + N_m[(1-\omega)f](x)$$

is the solution of the classical half space Neumann problem, and

$$V(x) = o(|x|^m \sec^{n-2}\theta), \qquad (x \to \infty, x \in H) \tag{3.6}$$

where $x_n = |x|\cos\theta$.

We give the asymptotic behavior of the modified Neumann integral in this section.

Theorem 3.1 Let f be continuous functions satisfy (3.4). The modified Neumann integral

$$N_m[f](x) = o(|x|^{m+1}) \tag{3.7}$$

in $H - D$, where D is defined as theorem 2.1.

Proof of Theorem 3.1 Let μ be a positive Borel measure in $\mathbf{R}^n$, $\beta \geq 0$. The maximal function $M(d\mu)(x)$ of order β is defined by Theorem 1 is measurable.

Define the measure

$$dn(y') = \frac{f(y')dy'}{1+|y'|^{n+m-2}}.$$

Then the modified Neumann integral now can be represented in the form

$$N_m[f](x) = \int_{\partial H} N_m(x,y')(1+|y'|^{n+m-2})dn(y').$$

For every Lebesgue measurable set $E \subset \mathbf{R}^n$, the measure $n^{(\varepsilon)}$, set $E_1(\lambda)$ and function $m_x^{(\varepsilon)}(t)$ are defined respectively as Theorem 1. Denote

$$N_m^{(\varepsilon)}[f_1](x) = \int_{\{y' \in \mathbb{R}^{n-1}: 1 \leq |y'| \leq 2|x|\}} N_m(x,y')(1+|y'|)^{n+m-2} dn^{(\varepsilon)}(y'),$$

then

$$\begin{aligned}
|N_m^{(\varepsilon)}[f_1](x)| &\leq A|x|^{m+1} \int_{\{y' \in \mathbb{R}^{n-1}: 1 \leq |y'| \leq 2|x|\}} \frac{1+|y'|^{n+m-2}}{(1+|y'|)^{n+m-2}} dn^{(\varepsilon)}(y') \\
&\leq A|x|^{m+1} \int_{\{y' \in \mathbb{R}^{n-1}: 1 \leq |y'| \leq 2|x|\}} dn_x^{(\varepsilon)}(t) \\
&\leq A\varepsilon|x|^{m+1};
\end{aligned}$$

and

$$N_m^{(\varepsilon)}[f_2](x) = \int_{\{y'\in\mathbb{R}^{n-1}:|y'|\geq 2|x|\}} N_m(x,y')(1+|y'|^{n+m-2})dn^{(\varepsilon)}(y').$$

For $|x| \geq 2R_\varepsilon$ and $x \notin E_1(\lambda)$, we have

$$\begin{aligned}
|N_m^{(\varepsilon)}[f_2](x)| &\leq A|x|^{m+2} \int_{\{y'\in\mathbb{R}^{n-1}:|y'|\geq 2|x|\}} \frac{1+|y'|^{n+m-2}}{(1+|y'|)^{n+m-1}} dn^{(\varepsilon)}(y') \\
&\leq A|x|^{m+2} \int_{\{y'\in\mathbb{R}^{n-1}:|y'|\geq 2|x|\}} \frac{dn_x^{(\varepsilon)}(t)}{1+|y'|} \\
&\leq A\varepsilon|x|^{m+1},
\end{aligned}$$

$$\int_{\{y'\in\mathbb{R}^{n-1}:|y'|\leq R_\varepsilon\}} \frac{f(y')}{|x-y'|^{n-2}} dy' \leq \infty.$$

There exists a positive constant A independent of ε, such that if $|x| \geq 2R_\varepsilon$ and $x \notin E_1(\varepsilon)$, we have

$$|N_m[f](x)| \leq A|x|^{m+1}\varepsilon.$$

By collecting the estimates of $|N_m^{(\varepsilon)}[f_1](x)|$ and $|N_m^{(\varepsilon)}[f_2](x)|$, there exists a positive constant A independent of ε, such that $|N_m[f](x)| \leq A|x|^{m+1}\varepsilon$ for $|x| \geq 2R_\varepsilon$ and $x \notin E_1(\varepsilon)$. Let μ_ε be a measure on $\mathbb{R}^n$ defined by $\mu_\varepsilon(E) = n^{(\varepsilon)}(E \cap \mathbb{R}^{n-1})$ for every measurable set E in $\mathbb{R}^n$. By the proof of theorem 2.1, there exist $x_{j,p}$ and $\rho_{j,p}$ with $R_{p-1} \leq |x_{j,p}| < R_p$, such that

$$\sum_{j=1}^{\infty} (\frac{\rho_{j,p}}{|x_{j,p}|})^{n-1} \leq \frac{1}{2^p}.$$

Since $R_{p-1} \leq |x| < R_p$ and $x \notin G_p = \cup_{j=1}^{\infty} B(x_{j,p}, \rho_{j,p})$, therefore

$$|N_m[f](x)| \leq A|x|^{m+1}\varepsilon_p,$$

we have

$$\sum_{p=1}^{\infty}\sum_{j=1}^{\infty} (\frac{\rho_{j,p}}{|x_{j,p}|})^{n-1} \leq \sum_{p=1}^{\infty} \frac{1}{2^p}.$$

Setting $G = \cup_{p=1}^{\infty} G_p$, we see the proof of theorem 3.1 is completed.

References

[1] Axler S., Bourdon P., Ramey W.*Harmonic Function Theory Second Edition.* New York: Springer-Verlag, 1992.

[2] Flett T.M.*On the rate of growth of mean values of holomorphic and harmonic functions.* Proc.London Math.Soc. 20(3)(1970),749-768.

[3] Armitage D.H..*The Neumann problem for a function harmonic in* $\mathbb{R}^n \times (0, \infty)$. Arch.Rational Mech. Anal., 63(1976),89-105.

[4] Szego. G.,*Othogonal Polynomials* American Mathematical Society, Providence, 1975.

[5] Armitage D.H..*Representation of harmonic functions in half space* Proc.London Math.Soc.38(3)(1979),53-71.

[6] Mizuta. Y.,*On the behavior of potentials near a hyperplane.*Hiroshima Math. J. 13(1983), 529-542.

[7] Gardiner.S.J.,*The Dirichlet and Neumann problems for harmonic functions in half-spaces.* J. London Math. Soc., 24(2)(1981), 502-512.

[8] Shu F. and Tanaka M. and Minoru Y.,*Neumann problems on a half-spaces.* Proceedings of the American Mathematical Society, (4)(2011), 1333-1345.

[9] Deng G. T.,*Integral representation of harmonic functions in half space.* Bull Sci. Math.,131 (2007), 53-59.

[10] Zhang Y. H. and Deng G. T., Growth properties for a class of subharmonic functions in half space. *Acta Mathematica Sinica*, 51(2008), 319-326.

[11] Mizuta Y. and Shimomura T., Growth properties for modified Poisson integrals in a half space. *Pacific Journal of Mathematics*, 212(2)(2003), 333-346.

[12] Siegel D. and Talvila E., Sharp growth estimates for modified Poisson integrals in a half space. *Potential Analysis*, 15(2001), 333-360.

[13] Siegel D. and Talvila E., Uniqueness for the n-dimensional half space Dirichlet problem, *Pacific. J. Math.,*, 175(1996), 571-587.

[14] Liwinski S., On the Neumann problem in an n-dimensional half -space.*Prace Math.*
11(1967),179-182.

[15] Levin B.Y. *Lectures on Entire Functions. Transl. Math. Monographs, Vol.150.* Amer. Math. Soc., Providence, RI. 1996.

[16] Hörmander L. *Notions of Convexity.* Boston · Basel · Berlin: Birkhauser, 1994.

PIECEWISE TIKHONOV REGULARIZATION SCHEME TO RECONSTRUCT DISCONTINUOUS DENSITY IN COMPUTERIZED TOMOGRAPHY

JIN CHENG*

School of Mathematical Sciences, Fudan University,
Shanghai, 200433, P. R. China
**E-mail: jcheng@fudan.edu.cn*

YU JIANG

Department of Applied Mathematics, Shanghai University of Finance and Economics,
Shanghai, 200433, P. R. China
E-mail: jiang.yu@mail.shufe.edu.cn

KUI LIN

School of Mathematical Sciences, Fudan University,
Shanghai, 200433, P. R. China
E-mail: 10110180008@fudan.edu.cn

JIAWEI YAN

Guotai Junan Securities Co.,Ltd.,
Shanghai, 200120, P. R. China
E-mail: JWYan@fudan.edu.cn

In this paper, we proposed a piecewise Tikhonov regularization scheme to overcome the disadvantage of the classical Tikhonov regularization for solving the Radon integral equation whose solution is piecewise continues arising in CT imaging . In our new scheme, we firstly detect the location of discontinuous interface by applying the classical Tikhonov regularization. Then a so-called piecewise Tikhonov regularization method is applied to reconstruction the smooth solution in each piece. Some numerical examples has also be presented to demonstrate the effectiveness of this two-steps scheme.

Keywords: Computerized Tomography, Radon Integral Equation, Ill-posed problems, Tikhonov Regularization Method.

1. Introduction

Computed tomography (CT) is an important tool in medical imaging that utilizes computer-processed X-rays to produce tomographic images or "slices" of specific areas of the body. As mentioned in many monographs, for example F. Natterer's book Ref. 4, the mathematical tool for tomographic imaging is the Radon transform.

Let $f(x)$ be the space dependent X-ray attenuation coefficient of the human body. After X-ray went through the human body along a straight line L, the intensity I_1 measured by be the detector is

$$I_1 = I_0 \exp\left\{-\int_L f(x)\,\mathrm{d}x\right\},$$

where I_0 is the initial intensity from the source. This is actually the *Radon transform.* Let's use the parameters s and θ to represent a certain line L in $\mathbb{R}^n$. $s \in \mathbb{R}$ denotes the signed distance from the origin to the line L, while $\theta \in S^{n-1}$ denotes the angles of the line, where S^{n-1} is the unit sphere in $\mathbb{R}^n$. Therefore, we can parameterize L as $x \cdot \theta = s$. Assuming that $f \in \mathscr{S}(\mathbb{R}^n)$, where $\mathscr{S}$ is the Schwartz space, the n-dimensional *Radon transform operator* $\mathbf{R}$ is defined as follow:

$$\mathbf{R}f(\theta, s) := \int_{x\cdot\theta=s} f(x)\,\mathrm{d}x = \int_{\theta^\perp} f(s\theta + y)\,\mathrm{d}y, \quad \forall\,\theta \in S^{n-1}, s \in \mathbb{R},$$

where $\theta^\perp$ is the hyperplanes perpendicular to θ. Given the data measured by the detector from different angels and different slices, i.e. $g(\theta, s)$, the inverse problem of CT consists of solving the *Radon integral equation*:

$$\mathbf{R}f = g. \tag{1}$$

It is well-known that solving the Radon integral equation is ill-posed, since the inverse operator $\mathbf{R}^{-1}$ is not continuous. We usually turn to solve such an minimization problem that

$$f = \arg\min_f \|\mathbf{R}f - g\|,$$

by applying the Tikhonov regularization method (see for example Ref. 1–3, 5). Taking the practice into account, the solution f we need to reconstruct is usually a piecewise discontinuous function in two dimensions. To enhance the resolution near the the discontinuities of f, which is useful in medical diagnosis, we proposed a new regularization scheme, i.e. the *piecewise Tikhonov regularization scheme.* Our reconstruction scheme contains two steps: firstly, we apply an adaptive detecting method to locate the discontinuous interface of the function f; secondly, we solve the Radon equation piece-wisely in each part where the function f is continuous by applying the usual Tikhonov regularization method.

The rest of this paper is organized as follows. After introducing the details of our scheme in section 2, we will present some numerical examples in section 3 to illustrate the effectiveness of our scheme. We can see that this new regularization scheme works better both in theory and in numerical tests.

2. Piecewise Tikhonov Regularization Scheme

Our goal is to reconstruct a discontinuous function $f(x)$ which satisfies (1) by giving $g \in L^2(Z)$, where $Z = S^1 \times \mathbb{R}^1$ is a unit cylinder. In this paper, we always assume that $f(x)$ is bounded and piecewise continues function in Z. It has a compact support $D = \operatorname{supp} f$ in Z, and is discontinues at the C^∞-class smooth boundary $\Gamma := \partial D$. For f and any order of its derivative, any points at Γ is jump discontinuity.

2.1. *Tikhonov Regularization Method*

In order to obtain a stable reconstruct of f, we apply the Tikhonov regularization method. The problem is to find a minimum f^α of the functional

$$J_\alpha(f) = \|\mathbf{R}f - g\|_Z^2 + \alpha\|f\|_{H^1(\Omega)}^2, \tag{2}$$

where $\Omega \subset Z$ is a bounded domain contains $D = \operatorname{supp} f$. The L^2 norm $\|\cdot\|_Z$ is defined as

$$\|g(\theta, s)\|_Z = \left(\int_{S^1}\int_{\mathbb{R}^1} |g(\theta, s)|^2 \,\mathrm{d}\theta \,\mathrm{d}s\right)^{1/2},$$

and the H^1 norm $\|f\|_{H^1(\Omega)}$ is defined as

$$\|f(x)\|_{H^1(\Omega)} = \left(\int_\Omega |f(x)|^2 \,\mathrm{d}x + \int_\Omega \|\nabla f\|^2 \,\mathrm{d}x\right)^{1/2}.$$

By denoting the *adjoint Radon operator* through the usual definition of adjoint operator for any $g(\theta, s) \in L^2(Z)$ as follows:

$$\mathbf{R}^* g(x) := \int_{S^{n-1}} g(\theta, x \cdot \theta) \,\mathrm{d}\theta,$$

we transform our minimization problem

$$f^\alpha = \arg\min_f J_\alpha(f),$$

into a problem solving an Euler equation according to following theorem:

Theorem 2.1. *If f^α is the minimum of $J_\alpha(f)$ defined in (2), then f^α is the unique solution of the Euler equation*

$$\mathbf{R}^*\mathbf{R}f + \alpha f - \alpha\Delta f = \mathbf{R}^* g, \qquad in\ \Omega. \tag{3}$$

Proof. Let f^α minimize the functional $J_\alpha(f)$, then the Frechét derivative of $J_\alpha(f)$ at f^α equal zero, i.e.

$$\lim_{\beta\to 0} \frac{J_\alpha(f^\alpha + \beta v) - J_\alpha(f^\alpha)}{\beta} = 0, \quad \forall v \in H_0^1(\Omega).$$

We can compute the Frechét derivative of $J_\alpha(f)$ as follows

$$\begin{aligned}
&\frac{J_\alpha(f^\alpha + \beta v) - J_\alpha(f^\alpha)}{\beta} \\
&= \frac{1}{\beta}\Big(\|\mathbf{R}f^\alpha - g + \beta\mathbf{R}v\|^2_{L^2(Z)} + \alpha\|f^\alpha + \beta v\|^2_{H^1(\Omega)} \\
&\qquad - \|\mathbf{R}f^\alpha - g\|^2_{L^2(Z)} - \alpha\|f^\alpha\|^2_{H^1(\Omega)}\Big) \\
&= \frac{1}{\beta}\Big(\langle\mathbf{R}f^\alpha - g, \beta\mathbf{R}v\rangle_{L^2(Z)} + \beta^2\|\mathbf{R}v\|^2_{L^2(Z)} \\
&\qquad + \alpha\langle f^\alpha, \beta v\rangle_{H^1(\Omega)} + \alpha\beta^2\|v\|^2_{H^1(\Omega)}\Big) \\
&= \langle\mathbf{R}^*\mathbf{R}f^\alpha - \mathbf{R}^*g, v\rangle_{L^2(\Omega)} + \beta\langle\mathbf{R}^*\mathbf{R}v, v\rangle_{L^2(\Omega)} \\
&\quad + \alpha\langle f^\alpha, v\rangle_{H^1(\Omega)} + \alpha\beta\langle v, v\rangle_{H^1(\Omega)} \\
&\to \langle\mathbf{R}^*\mathbf{R}f^\alpha - \mathbf{R}^*g, v\rangle_{L^2(\Omega)} + \alpha\langle f^\alpha, v\rangle_{H^1(\Omega)} \qquad (\text{as } \beta \to 0).
\end{aligned} \tag{4}$$

By the definition of H^1 norm above, we can computer the inner product in (4) as follows:

$$\begin{aligned}
\langle f^\alpha, v\rangle_{H^1(\Omega)} &= \int_\Omega f^\alpha v\,\mathrm{d}x + \int_\Omega \nabla f^\alpha \cdot \nabla v\,\mathrm{d}x \\
&= \int_\Omega f^\alpha v\,\mathrm{d}x + \int_{\partial\Omega} v\frac{\partial f^\alpha}{\partial n}\,\mathrm{d}s(x) - \int_\Omega v\Delta f^\alpha\,\mathrm{d}x \\
&= \langle f^\alpha - \Delta f^\alpha, v\rangle_{L^2(\Omega)}.
\end{aligned}$$

Therefore, we have

$$\langle \mathbf{R}^*\mathbf{R}f^\alpha + \alpha f^\alpha - \alpha\Delta f^\alpha - \mathbf{R}^*g, v\rangle_{L^2(\Omega)} = 0, \quad \forall v \in H_0^1(\Omega),$$

i.e. f^α is the solution to the Euler equation

$$\mathbf{R}^*\mathbf{R}f + \alpha f - \alpha\Delta f = \mathbf{R}^*g, \qquad \text{in } \Omega. \qquad \square$$

We can make this equation more explicit by using the following formula:

$$\mathbf{R}^*\mathbf{R}f(x) = 2\int_{\mathbb{R}^2} \frac{1}{|x-y|} f(y)\,\mathrm{d}y, \qquad x, y \in \mathbb{R}^2.$$

That is, we have to solve a differential-integral equation as follows:

$$2\int_\Omega \frac{1}{|x-y|} f(y)\,\mathrm{d}y + \alpha f(x) - \alpha\Delta f(x) = \int_{S^1} g(\theta, x\cdot\theta)\,\mathrm{d}\theta. \tag{5}$$

The first term in left hand side of (5) is an improper integral with singular kernel, which actually is an improper integrals. This makes the diagonal will be infinite when we discretize it by a uniform mesh. This is a big difficulty when we solve the (5) numerically. Moreover, the regularization solution f^α obtained from this equation always has a big error near the discontinuous interface when f is piecewise continuous. This can be showed by following numerical example.

Example 2.1. Let $\Omega = [0,1]\times[0,1]$, D is a circle of radius 0.2, with a center $(0.5, 0.5)$.

$$f(x) = \begin{cases} 0 & x \in D^c, \\ 1 & x \in D. \end{cases}$$

The picture of the true solution f is showed in fig.1.

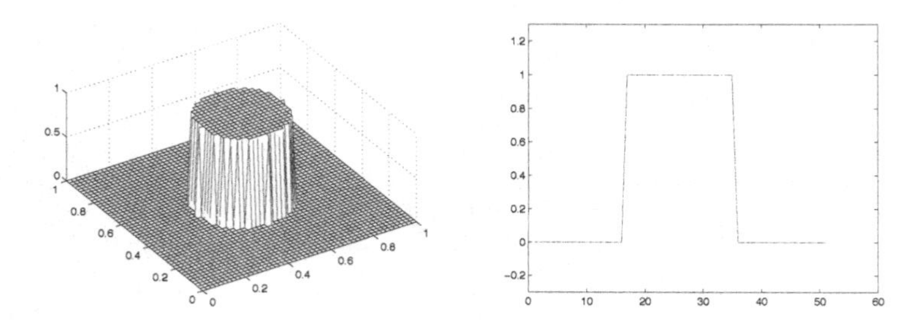

Fig. 1. The true solution $f(x)$.

By solving the above Euler equation 5 with a regularization parameter $\alpha = 0.0001$, the regularization solution f^α is showed in fig.2. From fig.2, we

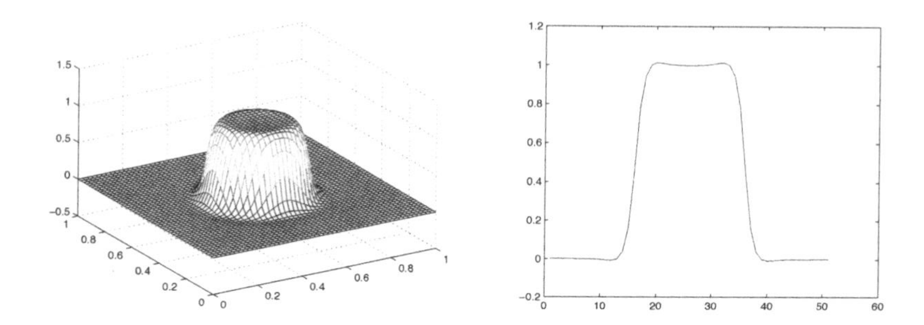

Fig. 2. Tikhonov regularization solution f^α with $\alpha = 0.0001$.

can see that reconstruction f^α of f is quite good in D and D^c, while it is bad near the discontinuous interface, i.e. the circle. Indeed, this Tikhonov regularization method will smooth the solution because of the penalty term $\|f\|_{H^1(\Omega)}$ in (2).

2.2. *Piecewise Tikhonov Regularization Method*

However, the discontinuous interface is very important for medical diagnosis. It requests us to have a explicit reconstruction of the discontinuous interface to separate each continuous piece. We propose a new piecewise Tikhonov regularization method to overcome the disadvantage of the above Tikhonov regularization method.

Let $\{D_i\}_{i=1}^k$ be a set of k open domains, and $\Omega = \cup_{i=1}^k \bar{D}_i$. For every $i = 1, 2, \ldots, k$, we always have $f(x) \in C^1(D_i)$. We define the *piecewise Tikhonov functional* as follows:

$$\tilde{J}_\alpha = \|\mathbf{R}f - g\|^2_{L^2(Z)} + \alpha \|f\|^2_{H^1(\cup_{i=1}^k D_i)}, \tag{6}$$

where α is the regularization parameter, the norm $\|\cdot\|_{H^1(\cup_{i=1}^k D_i)}$ is defined as

$$\|f\|_{H^1(\cup_{i=1}^k D_i)} = \left(\sum_{i=1}^k \|f \cdot \chi_{D_i}\|^2_{H^1(D_i)} \right)^{1/2}$$

and χ_{D_i} is a characteristic function of D_i. We also define a norm $\|\cdot\|_{L^2(\cup_{i=1}^k D_i)}$ as

$$\|f\|_{L^2(\cup_{i=1}^k D_i)} = \left(\sum_{i=1}^{k} \|f \cdot \chi_{D_i}\|^2_{L^2(D_i)}\right)^{1/2}.$$

Instead of minimizing the Tikhonov functional $J_\alpha(f)$, we minimize the piecewise Tikhonov functional $\tilde{J}_\alpha$. We call this the *piecewise Tikhonov regularization method.*

Remark 2.1. Compared to the classical Tikhonov regularization method, piecewise Tikhonov regularization gets rid of the effect coming from the discontinuous interface because of the definition of the penalty term $\alpha\|f\|^2_{H^1(\cup_{i=1}^k D_i)}$ in (6). Suppose we have $\Omega = D \cup \bar{D}^c$ and $f_1 = c\chi_D$ where c is a positive constant. The penalty term $\alpha\|f\|^2_{H^1(\Omega)}$ in functional $J_\alpha(f_1)$ of (2) will obtain a large value near the discontinuous interface ∂D coming from the differentiation of discontinues function f_1. However, we can avoid this by using the modified penalty term $\alpha\|f_1\|^2_{H^1(D\cup\bar{D}^c)}$ because it equals zero no matter what α is. This is quit important and the origin of our piecewise Tikhonov regularization method.

Remark 2.2. The idea of *piecewise Tikhonov regularization method* can be applied for any problems whose true solution is piecewise continuous. To solve the Radon integral equation with piecewise continuous solution in this paper is just an example.

Similarly, we transform our minimization problem

$$f^\alpha = \arg\min_f \tilde{J}_\alpha,$$

into a problem solving an Euler equation according to the following theorem:

Theorem 2.2. *Let f^α is the minimum of the piecewise Tikhonov regularization functional*

$$\tilde{J}_\alpha = \|\mathbf{R}f - g\|^2_{L^2(Z)} + \alpha\|f\|^2_{H^1(\cup_{i=1}^k D_i)},$$

then f^α is the unique solution that satisfies the Euler equation

$$\begin{cases} \mathbf{R}^*\mathbf{R}f + \alpha f - \alpha\Delta f = \mathbf{R}^*g, & \text{in } \cup_{i=1}^k D_i, \\ \mathbf{R}^*\mathbf{R}f = \mathbf{R}^*g, & \text{on } \cup_{i=1}^k \partial D_i. \end{cases} \tag{7}$$

Proof. Let f^α minimize $\tilde{J}_\alpha(f)$. Then the Frechét derivative of $\tilde{J}_\alpha(f)$ is zero at f^α, i.e.

$$\lim_{\beta\to 0}\frac{\tilde{J}_\alpha(f^\alpha+\beta v)-\tilde{J}_\alpha(f^\alpha)}{\beta}=0, \quad \forall v\in H^1(\Omega),\ \operatorname{supp} v\subset \cup_{i=1}^k D_i.$$

We can compute the Frechét derivative of $\tilde{J}_\alpha(f)$ as follows

$$\begin{aligned}
&\frac{\tilde{J}_\alpha(f^\alpha+\beta v)-\tilde{J}_\alpha(f^\alpha)}{\beta}\\
&=\frac{1}{\beta}\Big(\|\mathbf{R}f^\alpha-g+\beta\mathbf{R}v\|^2_{L^2(Z)}+\alpha\|f^\alpha+\beta v\|^2_{H^1(\cup_{i=1}^k D_i)}\\
&\qquad -\|\mathbf{R}f^\alpha-g\|^2_{L^2(Z)}-\alpha\|f^\alpha\|^2_{H^1(\cup_{i=1}^k D_i)}\Big)\\
&=\frac{1}{\beta}\Big(\langle\mathbf{R}f^\alpha-g,\beta\mathbf{R}v\rangle_{L^2(Z)}+\beta^2\|\mathbf{R}v\|^2_{L^2(Z)}\\
&\qquad +\alpha\langle f^\alpha,\beta v\rangle_{H^1(\cup_{i=1}^k D_i)}+\alpha\beta^2\|v\|^2_{H^1(\cup_{i=1}^k D_i)}\Big)\\
&=\langle\mathbf{R}^*\mathbf{R}f^\alpha-\mathbf{R}^*g,v\rangle_{L^2(\Omega)}+\beta\langle\mathbf{R}^*\mathbf{R}v,v\rangle_{L^2(\Omega)}\\
&\quad +\alpha\langle f^\alpha,v\rangle_{H^1(\cup_{i=1}^k D_i)}+\alpha\beta\langle v,v\rangle_{H^1(\cup_{i=1}^k D_i)}\\
&\to\langle\mathbf{R}^*\mathbf{R}f^\alpha-\mathbf{R}^*g,v\rangle_{L^2(\Omega)}+\alpha\langle f^\alpha,v\rangle_{H^1(\cup_{i=1}^k D_i)} \qquad (\text{as } \beta\to 0). \qquad (8)
\end{aligned}$$

By the definition of H^1 norm above, we can computer the inner product in (8) as follows:

$$\begin{aligned}
\langle f^\alpha,g\rangle_{H^1(\cup_{i=1}^k D_i)}&=\sum_{i=1}^k\int_{D_i}(f^\alpha\cdot\chi_{D_i})v\,\mathrm{d}x+\sum_{i=1}^k\int_{D_i}\nabla(f^\alpha\cdot\chi_{D_i})\cdot\nabla v\,\mathrm{d}x.\\
&=\sum_{i=1}^k\int_{D_i}(f^\alpha\cdot\chi_{D_i})v\,\mathrm{d}x+\sum_{i=1}^k\int_{\partial D_i}v\frac{\partial(f^\alpha\cdot\chi_{D_i})}{\partial n}\,\mathrm{d}s(x)\\
&\quad -\sum_{i=1}^k\int_{D_i}v\Delta(f^\alpha\cdot\chi_{D_i})\,\mathrm{d}x\\
&=\sum_{i=1}^k\int_{D_i}(f^\alpha\cdot\chi_{D_i})v\,\mathrm{d}x+\int_{\cup_{i=1}^k\partial D_i}v\frac{\partial(f^\alpha\cdot\chi_{D_i})}{\partial n}\,\mathrm{d}s(x)\\
&\quad -\sum_{i=1}^k\int_{D_i}v\Delta(f^\alpha\cdot\chi_{D_i})\,\mathrm{d}x\\
&=\langle f-\Delta f,v\rangle_{L^2(\cup_{i=1}^k D_i)}.
\end{aligned}$$

Therefore, we have

$$\langle \mathbf{R}^*\mathbf{R}f^\alpha - \mathbf{R}^*g, v\rangle_{L^2(\Omega)} + \alpha\langle f^\alpha - \Delta f^\alpha, v\rangle_{L^2(\cup_{i=1}^k D_i)} = 0, \quad \forall v \in H_0^1(\Omega),$$

i.e. f^α is the solution to the Euler equations:

$$\begin{cases} \mathbf{R}^*\mathbf{R}f + \alpha f - \alpha\Delta f = \mathbf{R}^*g, & \text{in } \Omega := \cup_{i=1}^k D_i, \\ \mathbf{R}^*\mathbf{R}f = \mathbf{R}^*g, & \text{on } \partial\Omega := \cup_{i=1}^k \partial D_i. \end{cases}$$

□

Once more, by using the formula (2.1), we obtain a differential-integral equation as follows:

$$\begin{cases} 2\displaystyle\int_\Omega \frac{1}{|x-y|} f(y)\,\mathrm{d}y + \alpha f(x) - \alpha\Delta f(x) = \int_{S^1} g(\theta, x\cdot\theta)\,\mathrm{d}\theta, & \text{in } \Omega, \\ 2\displaystyle\int_\Omega \frac{1}{|x-y|} f(y)\,\mathrm{d}y = \int_{S^1} g(\theta, x\cdot\theta)\,\mathrm{d}\theta, & \text{on } \partial\Omega. \end{cases}$$

Suppose we know the location of discontinuous interface of f in Example.II.1 i.e. the location of the circle, we apply this piecewise Tikhonov regularization method with a regularization parameter $\alpha = 0.001$ numerically. The regularization solution f^α is showed in fig.3, which illustrates the effectiveness of our new method.

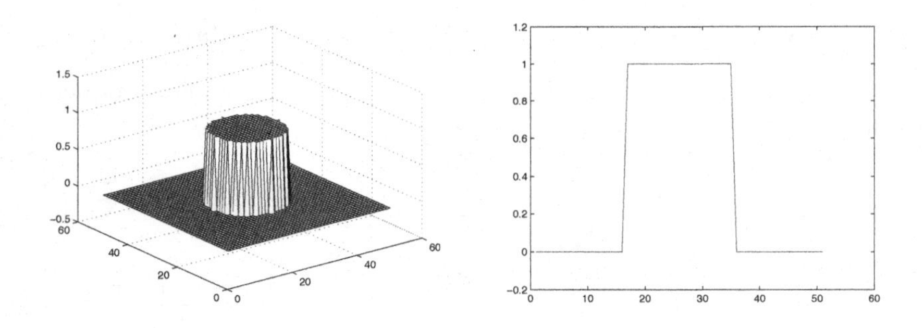

Fig. 3. Piecewise Tikhonov regularization solution f^α with $\alpha = 0.001$.

2.3. *Piecewise Tikhonov Regularization Scheme in Practice*

When we apply the piecewise Tikhonov regularization method above, the location of discontinuous interface $\partial\Omega := \cup_{i=1}^k \partial D_i$ is given as an *a-priori* information. However, we usually do not have such kind of *a-priori* information in practical word, and we need to locate the discontinuous interface. To solve this problem, we introduce a adaptive detecting method.

Firstly, we apply the classical Tikhonov regularization method to solve the Radon integral equation and obtain an intermediate result f^* to (3) even we have illustrated that it is not accurate near the discontinuous interface. Since this regularization solution, though too smooth near the discontinuous interface, will still keep a relative large slope, we can collect the points such that $\|\nabla f\| > \lambda$ for a threshold value λ as the discontinuous interface. Next, we use these as the *a-priori* information in our piecewise Tikhonov regularization method to have the final reconstruction f^α to (7). As a whole, our *reconstruct scheme* contains three steps:

(i) Obtain an intermediate result f^* by applying classical Tikhonov regularization method;
(ii) Detect the location of discontinuous interface by collecting the points satisfying $\|\nabla f_1\| > \lambda$ for a suitable given positive constant λ;
(iii) Applying the piecewise Tikhonov regularization method to reconstruct a piecewise continues solution f^α to (7).

3. Numerical Examples

In this section, we give some numerical examples to demonstrate the effectiveness of our scheme to reconstruct the piecewise continuous solutions.

Example 3.1. We consider $\Omega = [0,1] \times [0,1]$, the true solution $f(x)$ is

$$f(x) = \begin{cases} 0, & x \in D^c, \\ 1, & x \in D, \end{cases}$$

where D is a circle of radius 0.2, with a center $(0.5, 0.5)$. The reconstructed result f^α is showed in fig.4, fig.6, and the points collected as the discontinuous interface is showed in fig.5.

Example 3.2. Nextly, we consider the situation when measured data contains some noise. We add 5% or 10%Gaussian noise to the measured data $g(\theta, s)$ used in Example 3.1, i.e. $g = g + max(max(abs(g))) * randn(size(g)) * 0.05$.

The reconstruction results are showed in fig.7, fig.8.

These figures always showed us the piecewise Tikhonov regularization solution f^α is better than the classical Tikhonov regularization solution f^*.

Example 3.3. Finally, we consider the following more compliable case. We define $\Omega = [0,1] \times [0,1]$,

$$f(x) = \chi_{D_1 \setminus D_2} + 2\chi_{D_2}$$

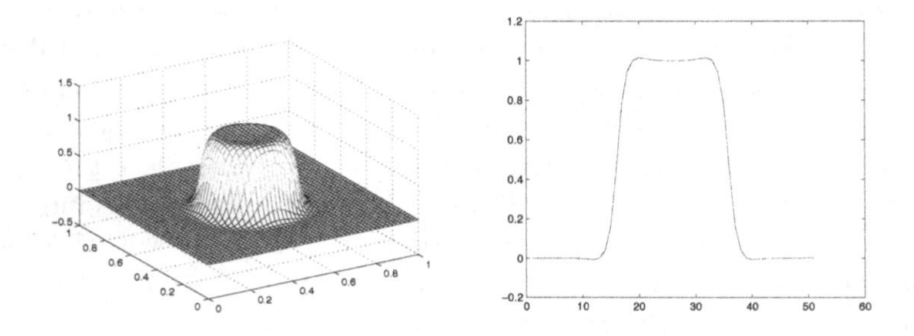

Fig. 4. Tikhonov regularization solution f^* with $\alpha = 0.0001$.

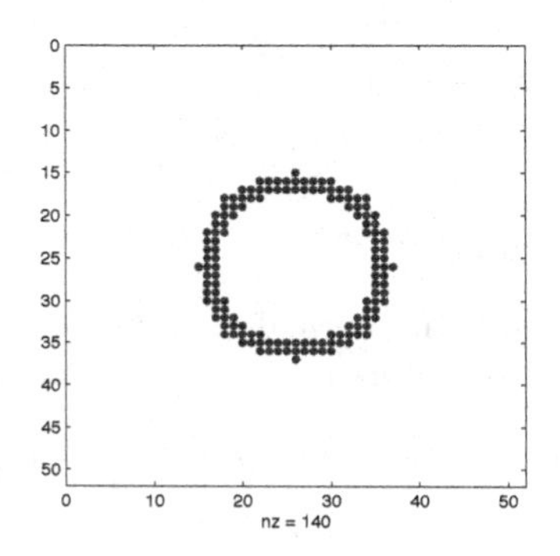

Fig. 5. Points collected as discontinuous interface.

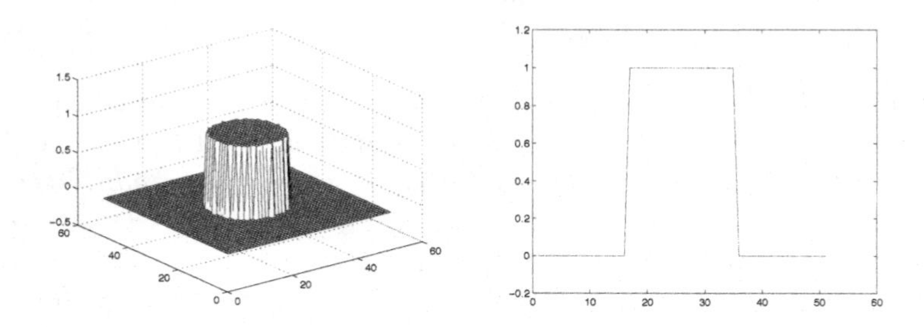

Fig. 6. Piecewise Tikhonov regularization solution f^{α} with $\alpha_2 = 0.001$.

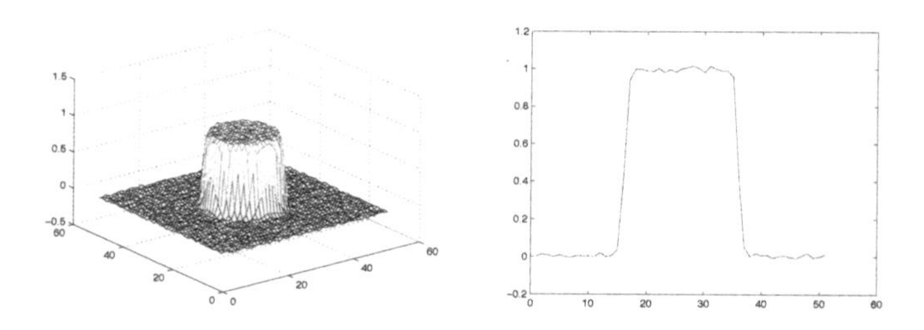

Fig. 7. Tikhonov regularization solution f^* from 5% noisy measured data.

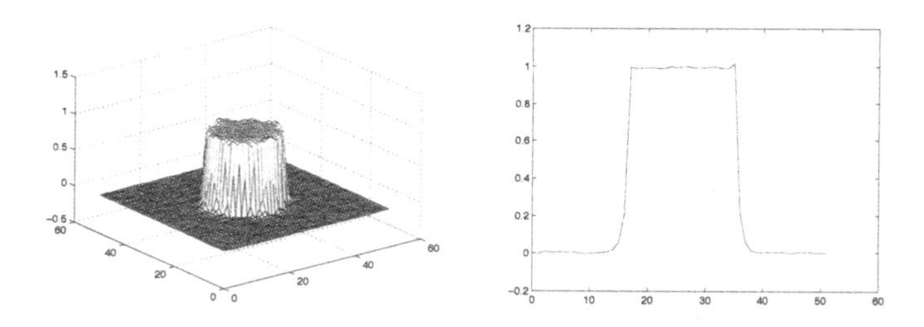

Fig. 8. Piecewise Tikhonov regularization solution f^α from 5% noisy measured data.

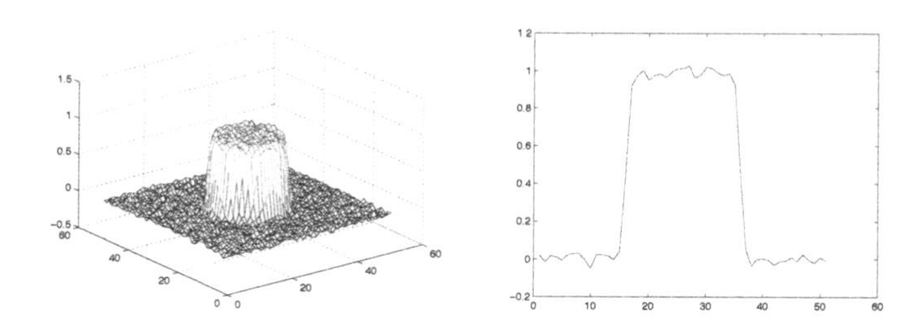

Fig. 9. Tikhonov regularization solution f^* from 10% noisy measured data.

where D_1 is a $[0.25, 0.75] \times [0.25, 0.75]$ square, while D_2 is a circle with radius 0.1, center $(0.45, 0.45)$. The reconstruct result is showed in Fig. 11, which is also illustrate the effectiveness of our scheme.

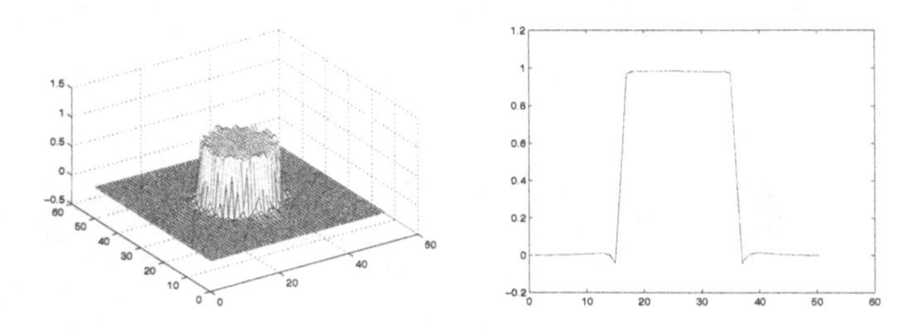

Fig. 10. Piecewise Tikhonov regularization solution f^{α} from 10% noisy measured data.

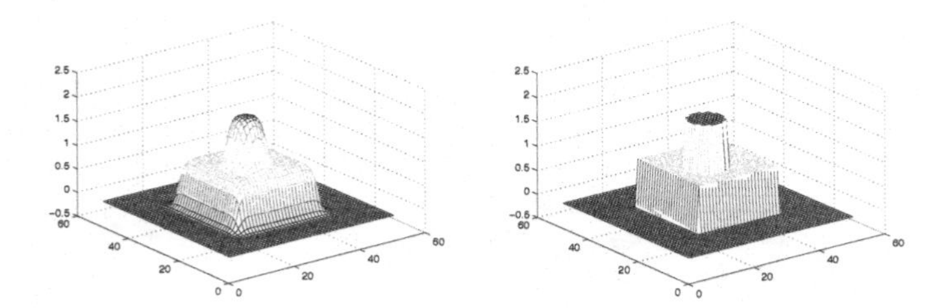

Fig. 11. Left: Tikhonov regularization solution f^{*}, right: Piecewise Tikhonov regularization solution f^{α}.

Acknowledgement

The second author is supported by Training Fund of Shanghai Excellent Young Teachers in Universities and Colleges (Shanghai Municipal Education Commission).

References

1. H. W. Engl, M. Hanke and A. Neubauer, *Regularization of inverse problems* (Springer Netherlands, 1996).
2. C. W. Groetsch, *The theory of Tikhonov regularization for Fredholm equations of the first kind* (Pitman Advanced Pub. Program, 1984).
3. A. Kirsch, *An introduction to the mathematical theory of inverse problems* (Springer Verlag, 1996).
4. F. A. Natterer, *The mathematics of computerized tomography* (SIAM, 2001).
5. C. R. Vogel, *Computational methods for inverse problems* (Society for Industrial Mathematics, 2002).

ABOUT THE QUATERNIONIC JCAOBIAN CONJECTURE

HUA LIU

Department of Mathematics, Tianjin University of Technology and Education
Tianjin, 300222, China
E-mail: daliuhua@163.com

In this paper we consider the Jacobian conjecture in the view of hypercomplex analysis. For a polynomial mapping $P(w) = (p_1(w), p_2(w)) : \mathbb{C}^2 \to \mathbb{C}^2$ we discuss the left holomorphic quaternionic function $f(z_1, z_2, z_3) = p_1(w) + jp_2(w)$ where $w = (x_0 + x_1 i, x_2 + x_3 i)$ and $z_1 = x_1 - x_0 i, z_2 = x_2 - x_0 i, z_3 = x_3 - x_0 i$. Then we give a new approach to the Jacobian conjecture by the use of argument principle for quaternionic holomorphic functions.

Keywords: Quaternionic analysis, left-holomorphic polynomial, Jacobian conjecture.

1. Introduction

Denote by $\mathbb{H}$ the skew field of real quaternion. Recall that it is obtained by endowing $\mathbb{R}^4$ with the multiplication operation defined on the stand basis $1, i, j, k$ —also denoted by e_0, e_1, e_2, e_3 respectively—by

$$\begin{gathered} i^2 = j^2 = k^2 = -1, \\ ij = -ji = k, jk = -kj = i, ki = -ik = j, \\ 1^2 = 1, 1i = i1 = i, 1j = j1 = j, 1k = k1 = k \end{gathered}$$

and extended by distributivity to all quaternions $q = x_0 + x_1 i + x_2 j + x_3 k$. Without confusion we also denote $x = x_0 + x_1 i + x_2 j + x_3 k$ in this paper.

Let Ω be a domain in $\mathbb{H}$ and let $f \to \mathbb{H}$ be a real differentiable function. f is said to be *left-holomorphic* if it satisfies the Cauchy-Riemann equation:

$$Df = \frac{\partial f}{\partial x_0} + \frac{\partial f}{\partial x_1} i + \frac{\partial f}{\partial x_2} j + \frac{\partial f}{\partial x_3} k = 0. \tag{1}$$

In the article we also only say 'holomorphic' since the right case is not involved.

It is unfortunate that the simplest nonconstant function q does not satisfy (1). In order to discuss the differentials we introduce the hypercomplex number—also called Fueter variables—as follows:

$$z_k = -\frac{1}{2}(e_k x + x e_k) = x_k - x_0 e_k, (k = 1, 2, 3). \tag{2}$$

The new variables z_k satisfy the condition $Dz_k = 0$. If we subsstitte them into the formula for df, because of $dx_k = dz_k + dx_0 e_k$, we have

$$df = (Df)dx_0 + \sum_{k=1}^{3} dz_k(\frac{\partial f}{\partial x_k}). \tag{3}$$

So f is holomorphic at x if and only if it can be linearly approximated by hypercomplex variables.

So in general the quaternion polynomial is not holomorphic. Moreover the polynomial of hypercomplex variables need not be holomorphic since $D(z_i z_j) \neq 0$ $(i, j = 1, 2, 3, i \neq j)$.

Delanghe proved that the following Fueter polynomial are left-holomorphic.[3]

Let x be in $\mathbb{H}$. and We call $\mathbf{k} = (k_1, k_2)$ with integer k_i a bi-index, for bi-indices with non-negative components let us take

$$k = |\mathbf{k}| = k_1 + k_2, \ \mathbf{k}! = k_1!k_2!.$$

We call $k = |\mathbf{k}|$ the degree of the multiindex $\mathbf{k}$.

For a multiindex $\mathbf{k}$ with at least one negative component we define

$$\mathcal{P}_{\mathbf{k}}(x) = 0.$$

For the degree $k = 0$ we write shortly $\mathbf{k} = (0, 0) = 0$ and define

$$\mathcal{P}_0(x) = 1.$$

Definition 1.1 (Fuetor polynomials). *For a $\mathbf{k}$ with $k > 0$ we define the Fueter Polynomial $\mathbf{P}_{\mathbf{k}}(x)$ as follows: For each $\mathbf{k}$ let the sequence of indices $j_1, \cdots, j_k$ be given such that the first k_1 indies equal 1 and the second k_2 indices 2. We put*

$$\mathbf{z}^{\mathbf{k}} = z_{j_1} z_{j_2} \cdots z_{j_k} = z_1^{k_1} z_2^{k_2}; \tag{4}$$

this product contains z_1 exactly k_1-times and so on. Then

$$\mathcal{P}(x) = \frac{1}{k!} \sum_{\sigma \in \text{perm}(k)} \sigma(\mathbf{z}^{\mathbf{k}}) = \frac{1}{k!} \sum_{\sigma \in \text{perm}(k)} z_{j_{\sigma(1)}, \cdots, j_{\sigma(k)}}. \tag{5}$$

Here perm(k) is the permutation group with k elements. This symmetrization compensates in some respects the non-commutativity in $\mathbb{H}$.

Fueter polynomials are left and right linearly independent. As the same for complex analysis the quaternion holomorphic functions can also have the Taylor expansion represented by Fueter polynomials, for which we introduce

$$\mathcal{Q}_0(y) = \frac{\overline{y}}{|y|^{n+1}} \tag{6}$$

which is the simplest singular holomorphic function having an isolated singularity at the origin. with the multiindex $\mathbf{k} = (k_0, k_1, k_2, k_3)$ the symbol $\nabla^{\mathbf{k}} = \partial_0^{k_0} \cdots \partial_3^{k_3}$ is defined. With the operator ∇ we again define $\mathcal{Q}_{\mathbf{k}}(x) = \frac{(-1)^{|\mathbf{k}|}}{\mathbf{k}!} \nabla^{\mathbf{k}} \mathcal{Q}_0$. Then we have:([3],p177)

Theorem 1.1. *Let the function f be left-holomorphic for $|x| < R$ in $\mathbb{H}$, where then it can be expanded into the converging Taylor series*

$$f(x) = \sum_{k=0}^{\infty} \sum_{|\mathbf{k}|=k} \mathcal{P}_{\mathbf{k}}(x) a_{\mathbf{k}}, \ \mathbf{k} = (0, k_1, k_2, k_3) \tag{7}$$

with $a_{\mathbf{k}} = \frac{\nabla^{\mathbf{k}} f(0)}{\mathbf{k}!}$.

Especially, by Theorem 1.1 if f is a quaternion polynomial then it can be represented by a sum of Fueter polynomials:

$$f(x) = \sum_{k=0}^{n} \sum_{|\mathbf{k}|=k} \mathcal{P}_{\mathbf{k}}(x) a_{\mathbf{k}} \tag{8}$$

where n is the order of f.

2. Complex polynomials mapping

Let $F = (F_1, \cdots, F_n) : \mathbb{C}^n \to \mathbb{C}^n$ be a polynomial map, i.e. a map of the form

$$(x_1, \cdots, x_n) \mapsto (F_1(x_1, \cdots, x_n), \cdots, F_n(x_1, \cdots, x_n)) \tag{9}$$

where F_i are the n variable polynomial. For $w \in \mathbb{C}^n$ put $F'(w) = \det(JF(w))$ where JF is the complex Jacobian matrix. The famous Jacobian Conjecture is formulated as the following:

Jacobian Conjecture *Let $F : \mathbb{C}^n \to \mathbb{C}^n$ be a polynomial map such that $F'(w) \neq 0$ for all $w \in \mathbb{C}^n$, then F is invertible (i.e. F has an inverse which is also a polynomial map).*

The Jacobian Conjecture was first formulated by O. Keller in the case $n = 2$ for polynomials with integer coefficients. Over the years many people have tried to prove the Jacobian Conjecture. It is still a open problem up to today. But more importantly the study of the Jacobian Conjecture has given rise to several surprising results concerning polynomial maps and many interesting relations with other problems. A well known result is the following:

Theorem 2.1. *The Jacobian Conjecture holds for all $n \geq 2$ and all F with $deg(F) \leq 3$, then the Jacobian Conjecture holds.*

In this little paper we only discuss the conjecture in the case $n = 2$. There appears some interesting correspondence between the Jacobian Conjecture and the quaternion analysis.

Let $p(w) = (p_1(w), p_2(w))$ be a polynomial mapping from $\mathbb{C}^2$ to $\mathbb{C}^2$, where $w = (w_1, w_2) \in \mathbb{C}^2$.

Recall that a quaternion $q = x_0 + x_1 i + x_2 j + x_3 k$ can be represented by two complex numbers:

$$q = (x_0 + x_1 i) + (x_2 + x_3 i)j = (x_0 + x_1 i) + j(x_2 - x_3 i).$$

Let $w_1 = x_0 + x_1 i$ and $w_2 = x_2 + x_3 i$ and consider $\mathbb{C}^2$ to identify with $\mathbb{H}$. Then define a quaternion function f as follows:

$$f(q) = p_1(w) + jp_2(w). \tag{10}$$

Notice that w can be represented by the hypercomplex variables:

$$w_1 = iz_1, \; w_2 = z_2 + iz_3. \tag{11}$$

Theorem 2.2. *$f(q) = f(z_1, z_2, z_3)$ is left-holomorphic for $q \in \mathbb{H}$.*

Proof. Notice that both p_1 and p_2 are complex holomorphic for $(w_1, w_2) \in \mathbb{C}^2$, then

$$\begin{aligned} \frac{\partial p_1(w)}{\partial \overline{w}_1} &= \frac{1}{2}(\frac{\partial p_1(w)}{\partial x_0} + i\frac{\partial p_1(w)}{\partial x_1}) = 0; \\ \frac{\partial p_1(w)}{\partial \overline{w}_2} &= \frac{1}{2}(\frac{\partial p_1(w)}{\partial x_2} + i\frac{\partial p_1(w)}{\partial x_3}) = 0 \end{aligned} \tag{12}$$

and

$$\begin{aligned} \frac{\partial p_2(w)}{\partial \overline{w}_1} &= \frac{1}{2}(\frac{\partial p_2(w)}{\partial x_0} + i\frac{\partial p_2(w)}{\partial x_1}) = 0; \\ \frac{\partial p_2(w)}{\partial \overline{w}_2} &= \frac{1}{2}(\frac{\partial p_2(w)}{\partial x_2} + i\frac{\partial p_2(w)}{\partial x_3}) = 0. \end{aligned} \tag{13}$$

By (2), (12) and (13) we get

$$\begin{aligned} Df(q) &= \frac{\partial f(q)}{\partial x_0} + \frac{\partial f(q)}{\partial x_1} i + \frac{\partial f(q)}{\partial x_2} j + \frac{\partial f(q)}{\partial x_3} k \\ &= \frac{\partial p_1}{\partial x_0} + j\frac{\partial p_2}{\partial x_0} + \frac{\partial p_1}{\partial x_1} i + j\frac{\partial p_1}{\partial x_1} i + j\frac{\partial p_2}{\partial x_1} i \\ &\quad + \frac{\partial p_1}{\partial x_2} j + j\frac{\partial p_2}{\partial x_2} j + \frac{\partial p_1}{\partial x_3} k + j\frac{\partial p_2}{\partial x_3} k \\ &= 2[\frac{\partial p_1}{\partial \overline{w}_1} + j\frac{\partial p_2}{\partial \overline{w}_1} + \frac{\partial p_1}{\partial \overline{w}_2} j + j\frac{\partial p_2}{\partial \overline{w}_2} j] \\ &= 0, \end{aligned} \tag{14}$$

which complete the proof. □

It is obvious that $f = f_0 + f_1 i + f_2 j + f_3 k$ such that

$$f_0 = Re\, p_1,\ f_1 = Im\, p_1,\ f_2 = Re\, p_2,\ f_3 = -Im\, p_2. \tag{15}$$

Let Jf be the Jacobian matrix of map from $\mathbb{R}^4$ to $\mathbb{R}^4$. Then

$$\det(Jf)(x) = -|\det Jp(w_1, w_2)|^2 < 0,\ x \in \mathbb{H}. \tag{16}$$

By [7], Jacobian conjecture is valid if and only if the polynomial mapping is injective. In the view of K-algebra, it is also valid if the map is surjective. So for the proof of conjecture it is enough to prove that $f(q)$ is a surjective mapping from $\mathbb{H}$ to $\mathbb{H}$. And for any $q = \zeta_1 + j\zeta_2 \in \mathbb{H}(\zeta_1, \zeta_2 \in \mathbb{C})$, let $p_1' = p_1 + \zeta_1, p_2' = p_2 + \zeta_2$. Then $P' = (p_1', p_2')$ satisfies the condition of Jacobian Conjecture. So if we can prove there exists at least one zeroes for any f defined as (10) the Jacobian holds.

Proposition 2.1. *The Jacobian conjecture is valid if and only if there exists at least one solution to the quaternion algebraic equation $f(q) = 0$ for any f defined as (10).*

Since $p(w)$ is locally 1-1 mapping from $\mathbb{C}^2$ onto $\mathbb{C}^2$, the zeros of $f(q)$ must be isolated if ever. So it is naturals to generalize the Jacobian conjecture to the case of quaternion holomorphic polynomial that is 1-1 locally mapping.

Open problem *Let $f(q)$ be a holomorphic polynomial function on $\mathbb{H}$ and* $\det Jf = \det(\frac{\partial f_j}{\partial x_i})$ *is non-zero constant function. Then there exists a polynomial function $g(q)$ such that $g(f(q)) = f(g(q)) = q$, i.e., f has the polynomial inverse.*

Remark. The open problem properly generalizes the Jacobian Conjecture since there exist quaternion holomorphic polynomials to not be the composition of two complex polynomials. For example, it is easy to check that $f = iz_1 + jz_2 + 2kz_k$ satisfies the condition of open problem but can not be represented as (10). Moreover the open problem is also different from the real Jacobian Conjecture which there is a counterexample.[6]

3. Some results from quaternion

The single biggest reason for using the complex numbers in preference to any other number field is the so-called— fundamental theorem of algebra— every complex polynomial of degree n has precisely n zeros, counted with multiplicities. The real numbers are not so well-behaved: a real polynomial of degree n can have fewer than n roots.

Quaternion behavior is many degrees freer and predictable. There are many polynomials which display extreme behavior. For example, the cubic $q^2iqi + iq^2iq - iqiq^2 - qiq^2i$ takes the value zero for all $q \in \mathbb{H}$. At the other extreme, since $iq - qi$ is invertible for all $q \in \mathbb{H}$, the equation $iq - qi + 1 = 0$ has no solutions at all.

In order to arrive at any kind of 'fundamental theorem of algebra', we need to restrict our attention considerably. There is still not much information for this problem up today. The following theorem is very old but best in this direction.

We define a monomial of degree n to be an expression of the form

$$a_0qa_1qa_2 \cdots a_{n-1}qa_n, \; a_i \in \mathbb{H} \setminus \{0\}. \tag{17}$$

Then there is the following 'fundamental theorem of algebra for quaternion':[2]

Theorem 3.1. *Let f be a polynomial over $\mathbb{H}$ of degree $n > 0$ of the form $m + g$, where m is a monomial of degree n and g is polynomial of degree $< n$. Then the mapping $f : \mathbb{H} \to \mathbb{H}$ is surjective, and in particular f has zeros in $\mathbb{H}$.*

Since

$$\begin{aligned} x_0 &= \frac{1}{4}(q - iqi - jqj - kqk) \quad x_1 = \frac{1}{4i}(q - iqi + jqj + kqk) \\ x_2 &= \frac{1}{4j}(q + iqi - jqj + kqk) \; x_3 = \frac{1}{4k}(q + iqi + jqj - kqk) \end{aligned} \tag{18}$$

the polynomials of z_1, z_2 and z_3 are also the polynomials of q. So the results from 'fundamental theorem of algebra for quaternion' can be connected with the Jacobian Conjecture.

From (8) we can know that Jacobian Conjecture is connected with the 'fundamental theorem of algebra for holomorphic polynomials'. It is unfortunate that the monomials do not satisfy (1). In other words the Fueter polynomials are not monomials, thus Theorem does not work for any holomorphic polynomial.

In complex analysis the argument principle is a powerful tool from which it is easy to get the fundamental theorem of algebra. Many properties of integrals still work in the quaternion analysis. In the rest of this section we try to give integral approach to Jacobian Conjecture.

For a complex holomorphic function $f(\omega)$ in $\mathbb{H}$ and a simple closed curve $\gamma \subset \mathbb{C}$. Let k be the number of f' zeros surrounded by γ, counted with multiplicities. Then we known that

$$k = \frac{1}{2\pi i} \int_{f(\gamma)} \frac{d\omega}{\omega} \tag{19}$$

where $\omega = f(w)$.

Now we give the definition for the quaternionic holomorphic function by the formula similar to (18).

Definition 3.1. Let f be a left-holomorphic in the domain G and a be an isolated zeroes of f. Moreover let $B_\varepsilon(a)$ be a sufficiently small ball with radius ε around a and the boundary $S_\varepsilon(a)$, so that no additional zero of f is in $B_\varepsilon(a) \cup S_\varepsilon(a)$. Then

$$\operatorname{ord}(f; a) = \frac{1}{\sigma_3} \int_{f(S_\varepsilon(a))} \mathcal{Q}_0(y) dy* \tag{20}$$

denote the *order of the zero of f at the point a.*

In Definition the ball could be replaced by a nullhomologeous 3-dimensional cycle parameterizing a n-dimensional surface of a 4-dimensional simply connected inside G.[4]

It is relatively unpleasant to compute such order integrals explicitly, because one has to perform an integration over the image of a sphere. If f is a local diffeomorphism it can be reformulated as the following theorem:

Theorem 3.2 (Argument principle). *Let $G \subset \mathbb{H}$ be a domain, $f : G \to \mathbb{H}$ holomorphic in G and $a \in G$ an isolate point. Then, under the same* *conditions for $B_\varepsilon(a)$ as in Definition,*

$$\mathrm{ord}(f;a) = \frac{1}{\sigma_4}\int_{\partial B_\varepsilon(a)} \mathcal{Q}_0(x)\mathrm{det}Jf(x)((\mathrm{J}f)^{-1})^\top(x)d\sigma(x). \qquad (21)$$

In contrast to complex analysis, it is possible to have $\mathrm{ord}(f,a) = 0$, even if $f(a) = 0$ for the general holomorphic function.[9]

Now let $f = p_1 + jp_2$ is the holomorphic functions satisfying the condition of Jacobian Conjecture. Then for $B_\varepsilon(a)$ in the above definition $f(\partial B_\varepsilon(a))$ is a closed manifold in $\mathbb{R}^4$. By [8], $\mathrm{ord}(f;a)$ is just the Kronecker index or the winding number of the manifold $f(\partial B_\varepsilon(a))$ with respect to the point a.

Let $B_R(a)$ be a ball with the center of a and radius of R such that there is no zeroes of f on its boundary. Since $\mathrm{det}Jf(x_0,x_1,x_2,x_3) = -|\mathrm{det}P(x_0+x_1i, x_2+x_3i)|^2 < 0)$ for $q = x_0 + x_1i + x_2j + x_3k \in \mathbb{H}$ the Kronecker index of the image of $B_R(a)$

$$I(f(\partial B_R(a));a) = \{|q-a| < R, f(q) = 0\}^{\#}. \qquad (22)$$

Now for $n = 2$ we can give a necessary condition to give Jacobian Conjecture in the version of Cauchy integral of quaternion-valued function.

Theorem 3.3. *Let $p = (p_1,p_2)$ satisfy the condition of Jacobian Conjecture and $f = p_1 + jp_2$ be defined as (10). Then Jacobian conjecture holds if for any $y \in \mathbb{H}$ there exist $a \in \mathbb{H}$ and $R > 0$ such that*

$$\int_{\partial B_R(a)} \frac{\overline{f(x)-y}}{|f(x)-y|^4} det\,Jf(x)((Jf)^{-1})(x)d\sigma(x) > 0. \qquad (23)$$

References

1. H. Bass, E. Connell, and D. Wright, The Jacobian Conjecture: Reduction of Degree and Formal Expansion of the Inverse, Bulletin of the AMS 7(1982), 287-330
2. Ebbinghaus et al: Numbers, springer-Verlag, Graduate Texts in Math. 123(1990)(in particular M.Koecher, R. Remmert: Chapter 7, Hamilton's Qiaternions.)
3. K. Gürlebeck, K. Habetha, S. Wolfgang, Holomorphic functions in the plane and n-dimensional space. Transl. from the German. Basel: Birkhuser. xiii, 394 p. with CD-ROM. (2008).
4. T. Hempfling, Some remarks on zeroes of monogenic functions. Advances in Appl. Clifford Algebras 11(S2), 107-116(2001)
5. H. R. Malonek, Selected topics in hypercomplex function theory. Eriksson, Sirkka-Liisa (ed.), Clifford algebras and potential theory. Proceedings of the

summer school, Mekrijrvi, Finland, June 24–28, 2002. Dedicated to the memorial of Pertti Lounesto. Joensuu: University of Joensuu, Department of Mathematics. Report Series. Department of Mathematics, University of Joensuu 7, 111-150 (2004)

6. S. Pinchuk, A counterexamle to the strong real Jacobian conjecture, Math. Z. 217, 1 4 (1994)
7. W. Rudin, Injective polynomial maps are automorphisms. Am. Math. Mon. 102, No.6, 540-543 (1995).
8. F. Sommen, Monogenic differential forms and homlogy theory. Proc. Royal Irish Acad., Sect A, 84(1984), 87-109.
9. G. Zöll, Ein Residuenkalkül in der Clifford-Analysis und die Möbius transformationen für euklidische Räume. PhD Thesis, Lehrstuhl II für Mathematik RWTH, Aachen 1987.

INTERACTION BETWEEN ANTIPLANE CIRCULAR INCLUSION AND CIRCULAR HOLE OF PIEZOELECTRIC MATERIALS

LIHONG CHANG   XING LI

Department of Mathematics of Baoji University of Arts and Sciences,Shannaxi,721007
School of Mathematics and Computer Science, Ningxia University, Yinchuan,750021

In this paper, by means of Riemann-Schwarz's symmetry priciple integrated with the analysis singularity of complex potentials, The interaction between antiplane cicular inclusion and circular hole of piezoelectric materials under the antiplane line force and inplane line charge is considered, and their complex potential expressions of the electro-elastic field are derived.

Keywords: Piezoelectric Composite Materials, Riemann-Schwarz's symmetry priciple, the round inclousion.

1. Introduction

In the theory of micromechanics of materials, when a finite region with the eigen-strain has the same elastic moduli as the matrix, the region is called an inclusion. If the finite region has the elestic moduli different from those of the matrix, the region is called an inhomogeneity. Esheby [1] developed a general method of solving the elastic field caused by an ellipsoidal inclusion in an infinite homogeneous isotropic medium, when the inclusion has an arbitrary uniform eigenstrain field. Using the method developed by Esheby [1], Chiu [2], Wu and Du [3 − 4], and Li [5 − 8] solved the elastic fields of the infinite honogeneous media with cuboidal and circular cylindrical inclusions,respectively. While Pak [9] analyzed the problem of a piezoelectric circular inclusion in the infinite piezoelectric medium.

Although the inclusion problem has been greatly developed and extensively investigated, relatively little work has been done regarding the elastic field of the infinite homogeneous medium with circular inclusion and circular hole. Lu [10 − 11] investigated the interaction between antiplane circular inclusion and multiple holes and multiple cracks. For the piezoelectric medium, the analytical solution related to the interaction between antiplane circular inclusion and circular hole is not found up to now.

2. Problem formulation

Consider in the infinite homogeneous transversely isotropic piezoelectric medium with a single piezoelectric circular inclusion and a single circular hole in this inclusion, and infinitely along the direction perpendicular to the xy-plane as shown in Fig.1. Here, the xy-plane is a transversely isotropic plane and the circular inclusion and the circular hole radii R and r, respectively. The center of the circular hole at the point z_0, in addition, the interfaces between the circular inclusion and matrix is assumed to be perfectly bonded.

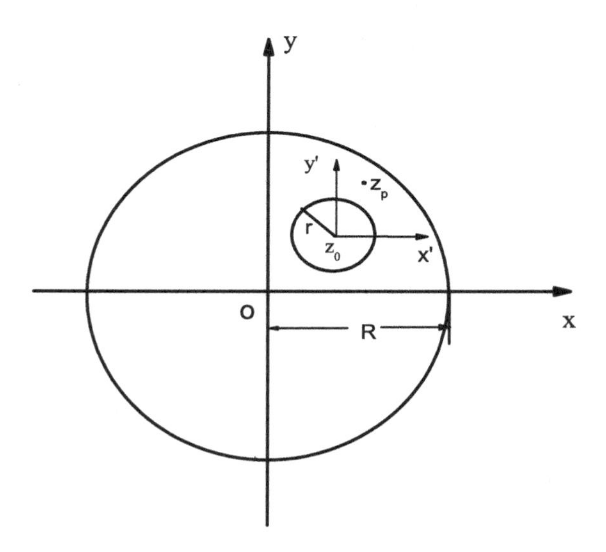

Fig. 1.

As shown in Fig. 1, the circular inclusion is subjected to the antiplane line force and inplane line charge at point z_p near the circular hole. For this problem, the electro-elastic field only is related to the material constants (elastic, piezoelectric and dielectric constants) of the matrix and inclusion when the geometric parameters of system and the antiplane line force and inplane line charge at point z_p are given.Let medium I with C_{441}, e_{151}, ε_{111} occupy the region S^+, while medium II with C_{442}, e_{152}, ε_{112} occupy the region S^-, the subscripts 1 and 2 denote the quantities defined in the region

S^+ and S^-, superscripts $+$ and $-$ used to denote the boundary values of the physical quantities as they approached from S^+ and S^-, respectively.

For the present problem, the antiplane displacement w is coupled with the inplane electric fields E_x and E_y, they are only the functions of coordinates x and y such as $w = w(x, y)$, $E_x = E(x, y)$ and $E_y = E(x, y)$. We can write the field equation in the absence of body forces and free charges, they can be expressed as follows:

Control equation

$$\begin{cases} \sigma_{xz,x} + \sigma_{yz,y} = 0 \\ D_{x,x} + D_{y,y} = 0 \end{cases} \tag{1}$$

Where $\sigma_{iz}, D_i(i = x, y)$ represent stress and electric displacement, and " ' " represent stress and electric displacement taking the derivatives with respect to x and y.

Constitutive equation

$$\begin{bmatrix} \sigma_{xz} \\ D_x \end{bmatrix} = \begin{bmatrix} C_{44} & e_{15} \\ e_{15} & -\varepsilon_{11} \end{bmatrix} \begin{bmatrix} w_{,x} \\ \phi_{,x} \end{bmatrix} \tag{2}$$

$$\begin{bmatrix} \sigma_{zy} \\ D_y \end{bmatrix} = \begin{bmatrix} C_{44} & e_{15} \\ e_{15} & -\varepsilon_{11} \end{bmatrix} \begin{bmatrix} w_{,y} \\ \phi_{,y} \end{bmatrix} \tag{3}$$

Where C_{44}, e_{15}, ε_{11} are the corresponding elasticpiezoelectric and dielectric constants which satisfy the relations $C_{44} > 0$ and $\varepsilon_{11} > 0$, respectively. φ represent the electric potential components. $\phi, i, w, i(i = x, y)$ represent the antiplane displacement w and the electric potential components ϕ taking the derivatives with respect to x and y.

Denote

$$C = \begin{bmatrix} C_{44} & e_{15} \\ e_{15} & -\varepsilon_{11} \end{bmatrix}$$

Substituting Eqs. (2) and (3) into (1) can yield:

$$\begin{bmatrix} \nabla^2 w \\ \nabla^2 \phi \end{bmatrix} = 0 \tag{4}$$

They can be expressed by the analytical function $\Psi_1(z)$ and $\Psi_2(z)$ as follows:

$$w = 2Im\Psi_1(z)$$

$$\phi = 2Im\Psi_2(z)$$

Where $z = x + iy$, is the complex variable and Im refers to the analytical function imaginary part. So, the Eqs. (4) solution can be expressed as follow:

$$U = \begin{bmatrix} w \\ \phi \end{bmatrix} = Imf(z) \tag{5}$$

Where $f(z) = [\Psi_1(z), \Psi_2(z)]^T$, from Eqs. (2) and (3), we can obtain:

$$\begin{cases} \sigma_{yz} + i\sigma_{xz} = C_{44}\Psi_1'(z) + e_{15}\Psi_2'(z) \\ D_y + iD_x = e_{15}\Psi_1'(z) - \varepsilon_{11}\Psi_2'(z) \end{cases} \tag{6}$$

And strain and electric field meet

$$\gamma_{xz} - i\gamma_{yz} = \Psi_1'(z)$$

$$E_x - iE_y = -\Psi_2'(z)$$

where γ_{yz}, γ_{xz}, E_y, E_x are components. Substituting Eqs. (5) into Eqs. (2) and (3) can yield:

$$\begin{bmatrix} \sigma_{yz} + i\sigma_{xz} \\ D_y + iD_x \end{bmatrix} = Cf'(z) \tag{7}$$

$$\begin{bmatrix} \gamma_{yz} + i\gamma_{xz} \\ -E_y - iE_x \end{bmatrix} = f'(z) \tag{8}$$

The resultant force T and resulant hormal component S of the electric displacement along any arc AB can be determined by the formulate [12]

$$T = \int_A^B (D_x dy - D_y dx) = \frac{i}{2}C_{44}\left[\overline{\Psi_1(z)} - \Psi_1(z)\right]_A^B + e_{15}\left[\overline{\Psi_2(z)} - \Psi_2(z)\right]_A^B$$

$$S = \int_A^B (\sigma_{xz} dy - \sigma_{yz} dx) = \frac{i}{2}e_{15}\left[\overline{\Psi_1(z)} - \Psi_1(z)\right]_A^B - \varepsilon_{11}\left[\overline{\Psi_2(z)} - \Psi_2(z)\right]_A^B \tag{9}$$

where $[f(z)]_B^A$ represent the change of function $f(z)$ from point A to point B along the arc. The overbar refers to the complex conjugate.

Introducing generalized potential function $\Phi = [\Phi_1, \Phi_2]^T$, they are met [13]

$$\begin{bmatrix} \sigma_{zx} \\ D_x \end{bmatrix} = -\Phi_{,y} \tag{10}$$

$$\begin{bmatrix} \sigma_{zy} \\ D_y \end{bmatrix} = \Phi_{,x} \tag{11}$$

We can obtain

$$\Phi = CRef(z) \tag{12}$$

Such generalized displacement and generalized stress boundary condition for [14] :

$$C^{-1}\Phi + iU = f(z) \tag{13}$$

3. Solution to the problem

To solve the present problem, in the local coordinate system, for the homogeneous piezoelectric medium, when the antiplane line force P and inplane charge Q are applied in points z_p, respectively. The corresponding potentials can be written as [14] :

$$\phi(z_2) = A_1 \frac{1}{z_2 - z_0}, \psi(z_2) = A_2 \frac{1}{z_2 - z_0} \tag{14}$$

where A_1, A_2 are real constants which need to be determined from the corresponding equilibrium conditions

$$\begin{cases} P + T(z_0 = Re^{i\theta})|_{\theta=0}^{2\pi} = 0 \\ Q + S(z_0 = Re^{i\theta})|_{\theta=0}^{2\pi} = 0 \end{cases} \tag{15}$$

Substituting Eq. (14) into Eqs. (5) and (15) can yield:

$$A_1 = -\frac{P + Q\alpha}{2\pi(1 + \alpha\beta)}, A_2 = -\frac{P\beta - Q}{2\pi(1 + \alpha\beta)} \tag{16}$$

where $\alpha = \frac{e_{15}}{\varepsilon_{11}}, \beta = \frac{e_{15}}{C_{44}}$.

By the application of analysis of singularity, for the problem under consideration in Fig1 in the local coordinate system, the solution of the inclusion can be expressed as follows:

$$\phi_2(z_2) = \phi(z_2) + \phi_{20}(z_2), \psi_2(z_1) = \psi(z_2) + \psi_{20}(z_2) \tag{17}$$

where $\phi_{20}(z_2)$, $\psi_{20}(z_2)$ are the analytical functions out the circular hole. By Eqs.(6)(9)(17), combined with Cauchy integral formula, then after coordinate translation, we can obtain the solutions in the inclusion

$$\phi_2(z) = A_1 \frac{1}{z - z_0} + \frac{1}{2} A_1 \frac{1}{z - z_0 - z_p} + \phi_{20}(z) \tag{18}$$

$$\psi_2(z) = A_2 \frac{1}{z - z_0} + \frac{1}{2} A_2 \frac{1}{z - z_0 - z_p} + \psi_{20}(z) \tag{19}$$

And the solutions in the matrix are $\phi_1(z)$, $\psi_1(z)$, they are the analytical functions.

Denote

$$A = \begin{bmatrix} A_1 \\ A_2 \end{bmatrix}, F_1(z) = \begin{bmatrix} \phi_1(z) \\ \psi_1(z) \end{bmatrix}, F_2(z) = \begin{bmatrix} \phi_2(z) \\ \psi_2(z) \end{bmatrix}, F_{20}(z) = \begin{bmatrix} \phi_{20}(z) \\ \psi_{20}(z) \end{bmatrix}$$

$$F_{22}(z) = A\frac{1}{z - z_0} + \frac{1}{2}A\frac{1}{z - z_0 - z_p} \tag{20}$$

where $F_k(z) = f_k'(z)$, Eq.(18) and Eq.(19) can be expressed :

$$F_2(z) = F_{22}(z) + F_{20}(z) \tag{21}$$

The assumption of perfect bonding between different media implies the continuity of displacement, electric potential, traction and normal component of the electric displacement along the interface. With Eq.(13), we can obtain:

$$U_1^+(t) = U_2^-(t), \Phi_1^+(t) = \Phi_2^-(t), t \in L \tag{22}$$

Express as complex potential function:

$$\left[F_1(t) - \overline{F}_1(t)\right] = \left[F_2(t) - \overline{F}_2(t)\right] \tag{23}$$

$$C_1\left[F_1(t) + \overline{F}_1(t)\right] = C_2\left[F_2(t) + \overline{F}_2(t)\right] \tag{24}$$

Applying *Riemann – Schwarz's* symmetry principle, a new analytic function $\Omega_i(k = 1, 2)$ in $S\pm$ can be introduced

$$\Omega_i = -\frac{R^2}{z^2}\overline{F}_i(\frac{R^2}{z}), (i = 1, 2)\ z \in S^{\pm} \tag{25}$$

With Eqs.(22)–(25) can be obtained as follows:

$$\left[F_1(t) - \Omega_2(t)\right]^+ = \left[F_2(t) - \Omega_1(t)\right]^- \qquad t \in L \tag{26}$$

$$\left[C_1F_1(t) + C_2\Omega_2(t)\right]^+ = \left[C_2F_2(t) + C_1\Omega_1(t)\right]^-, \qquad t \in L \tag{27}$$

So, $F_1(z) - \Omega_2(z)$ and $F_2(z) - \Omega_1(z)$ $C_1F_1(z) + C_2\Omega_2(z)$ and $C_2F_2(z) + C_1\Omega_1(z)$ are analytical continuation function respectively over L .

Form Eqs. (26) and (27), they lead:

$$C_2[F_2(t) + F_{20}(t) - \frac{R^2}{t^2}\overline{F_2(t)} - \frac{R^2}{t^2}\overline{F_{20}(t)}] = C_1[F_1(t) - \frac{R^2}{t^2}\overline{F_1(t)}] \tag{28}$$

$$F_2(t) + F_{20}(t) + \frac{R^2}{t^2}\overline{F_2(t)} + \frac{R^2}{t^2}\overline{F_{20}(t)} = F_1(t) + \frac{R^2}{t^2}\overline{F_1(t)} \tag{29}$$

According to LU [10], the general solution of $F_{20}(z)$ can be expressed as

$$F_{20}(z) = \frac{C_2 - C_1}{C_2 + C_1}\frac{R^2}{z^2}F_{22}(\frac{R^2}{z}) - \frac{C_2 - C_1}{C_2 + C_1}\frac{1}{2\pi z} \tag{30}$$

Substituting Eq.(30) into Eq.(21), we obtain the complex potential solution of the present problem

$$F_2(z) = A\frac{1}{z - z_0} + \frac{1}{2}A\frac{1}{z - z_0 - z_p} + F_{20}(z) \tag{31}$$

$$F_1(z) = \frac{2C_2}{C_2 + C_1}F_{22}(z) + \frac{C_1 - C_2}{C_2 + C_1}\frac{1}{2\pi z} \tag{32}$$

4. Conclusions

Using the *Riemann – Schwarz's* symmetry principle integrated with the analysis singularity of complex potentials, strict derivation and obtained the physical quantity of the closed from solution. And for intelligent materials and manufacturing to provide a theoretical reference.

Acknowledgements

This work is supported by the National Natural Science Foundation of China (51061015, 11261045), the research fund for the doctoral program of higher education of China (20116401110002), and the general program of Baoji University of Arts and Sciences of China(YK1025).

References

1. Eshelby J. D., Progress in solid mechanics, edited by Sneddon L. and Hill R., London Sel, 1957.
2. Chiu Y. P., On the stress field due to initial strains in a cuboid surrounded by an infinite elastic space, J. Appl. Mech. 44(1977), 587.
3. Wu L. Z., Du S. Y., The elastic field caused by a circular cylindrical inclusion. I: Inside the region $x12 + x22 < a2$, $\infty < x3 < \infty$ where the circular cylindrical inclusion is expressed by $x12 + x22 \leqslant a2$, $h \leqslant x3 \leqslant h$, J. Appl. Mech. 62(1995), 579.
4. Wu L. Z., Du S. Y., The elastic field caused by a circular cylindrical inclusion. II: Inside the region $x12 + x22 > a2$, $\infty < x3 < \infty$ where the circular cylindrical inclusion is expressed by $x12 + x22 \leqslant a2$, $h \leqslant x3 \leqslant h$,J. Appl. Mech. 62(1995), 585.
5. Li X., Applications of doubly quasi-periodic boundary value problems in elasticity theory, Shaker Verlag, Aachen, 2001.

6. Li X., Complete plane strain problem of a nonhomogeneous elastic body with a doubly-periodic set cracks, Z. angenw. Math. Mech, 6(2001), 377-391.
7. Li X., The effect of a homogeneous cylindrical inlay on cracks in the doubly-periodic complete plane strain problem, Int. J. Fract. 4(2001), 403-41.
8. Li X., General solution for complete plane strain problem of a nonhomogeneous body with a doubly- periodic set of inlays, Z. angenw. Math. Mech, 9(2006), 682-690.
9. LU J. F., WANG J.H., SHEN W. P., Interaction between antiplane circular inclusion and multiple circular holes and cracks [J].Computational Mecchanics., 2000, 17(2):229-233.
10. LU J. F., SHEN W. P., Antiplane multiple holes and cracks problem of half-plane region[J]. Journal of shanghai jiaotong university., 1998, 32(11):66-70.
11. Pak Y. E. Crack extension force in a piezoelectric material. Jouranl of mechanics, 1990, 57:647-653.
12. K. Gürlebeck, W. Sprössig. Quaternionic analysis and elliptic boundary value problems. Akademie-Verlag, Berlin, 1989.
13. Lu C. K., Complex variables methods in plane elasticity, World Scientific, Singapore, 1995 (Chinese).
14. Wu L. ZH., Interaction of two circular cylindrical inhomogeneities anti-plane shear. Composites Science and Technology, 2000, 60(12-13):2609–2615.

CONVERGENCE OF NUMERICAL ALGORITHM FOR COUPLED HEAT AND MASS TRANSFER IN TEXTILE MATERIALS

MEIBAO GE JIANXIN CHENG DINGHUA XU

School of Sciences, Technology and Arts, Zhejiang Sci-Tech University
Yuhang Zone, Hangzhou 311121, P.R.China
E-mail: gemeibao@126.com
Hangzhou Yuhang High School No. 2228 West Wenyi Road,
Hangzhou 311121, P.R.China
E-mail: chengjianxin1985@163.com
Department of Mathematics, School of Sciences, Zhejiang Sci-Tech University
Xiasha Higher Education Zone, Hangzhou 310018, P.R.China
E-mail: dhxu6708@zstu.edu.cn

In this paper we propose a fundamental numerical method for heat and moisture transfer model, which is a system of coupled ordinary differential equations(ODEs) on temperature, water vapor pressure, water vapor mass flux and condensation rate through parallel pore textiles. By decoupling the coupled ODEs, four explicit formulations are achieved to describe temperature, water vapor pressure, water vapor mass flux and the rate of condensation respectively. We develop a numerical algorithm for the coupled ODEs by finite difference method (FDM) and further prove that the numerical algorithm is convergent with the convergence rate of the first order theoretically. Numerical simulation is implemented for down and polyester material to verify the validity of the algorithm. The numerical results are well matched with the experiment data on the "*Walter*" Manikin.

Keywords: Ordinary Differential Equations; Nonlinear Coupled Equations; Finite Difference Method; Convergence Rate; Heat and Moisture Transfer.

1. Introduction

Motivated by both theoretical interests and practical applications, we focus on numerical algorithms for a steady-state model of heat and moisture transfer through the parallel pore textiles. The model can be formulated into a system of coupled ordinary differential equations on temperature, water vapor pressure, water vapor mass flux and condensation rate through textiles [1-3].

As is well known, textile is certainly of a complex multi-pore structure, which can be described as, for example, the parallel pore structure or pellets accumulation pore structure [1, 4-6]. The textile structure affects the heat and moisture transfer characteristics and further the human body comfort. As for the human body comfort textiles, there are many increasing requirements on human body comfort. Therefore in practical situations we should develop healthier and safer textiles based on heat and mass transfer characteristics. For example, it is hoped that the textiles are of fast decalescence, fast heat radiation, soft or stand-up apparel. Of course, clothing should be light and keeps body warm under lower temperature, meanwhile sweat vaporizes fast and body feels cool under high temperature. This is not only the basic functions of traditional textile products, but also it has a more important significance for functional textiles to adapt to severe environment in aviation, industry, agriculture, national defense and polar region scientific experiments.

The mathematical modeling and numerical simulation can be helpful to study heat and moisture transfer characteristics of textiles [7-16]. But the former researchers usually focused on the physical characteristics of textile materials (such as moisture absorption, condensation characteristics, etc.) and textile structural features (such as porous media, multi-layer structure etc.) in experimental way. In the past half century, different moisture permeability models have been established to describe the heat and moisture transfer process based on the differences of morphological structure of waterproof and moisture-permeable fabric, such as the Hagen-Poiseuill breathable model, Kozeny-Carman breathable models, water vapor permeability model of Kundsen proliferation, and water molecules "Adsorption-proliferation- desorption" model [7]. Some researchers discussed heat and moisture transfer through porous clothing assemblies and porous insulation in experimental way [8-10]. In the last twenty years, some coupled models of simultaneous heat and moisture transfer were established or developed theoretically on the basis of differential equations [11-16]. Based on these models, they have derived different numerical methods to solve these problems, such as finite difference method, finite volume method, finite element method, transfer function method and controllability volume-time-domain recursive method, and the numerical results are well matched with experimental results.

To our best knowledge, the research results both by theoretical and experimental way did not include the mathematical analysis on numerical algorithms, such as stability or convergence results. So in this paper, we shall focus on the mathematical analysis on numerical algorithms for a new well-posed mathematical model of heat and moisture transfer in the parallel pore textile materials, which was first formulated in [1-3]. For concision, we describe the mathematical model as follows

$$\frac{k_1\varepsilon(x)r(x)}{\tau(x)} \cdot \frac{p_v}{T^{3/2}} \cdot \frac{dp_v}{dx} + m_v(x) = 0, \tag{1}$$

$$\frac{dm_v}{dx} + \Gamma(x) = 0, \tag{2}$$

$$\kappa\frac{d^2T}{dx^2} + \lambda\Gamma(x) = 0, \tag{3}$$

$$\Gamma(x) = \frac{-k_2\varepsilon(x)r(x)}{\tau(x)} \cdot (p_{sat} - p_v) \cdot \frac{1}{\sqrt{T}}, \tag{4}$$

where

$T(x)$ is temperature;

$m_v(x)$ is mass flux of water vapor;

$p_v(x)$ is water vapor pressure;

$\Gamma(x)$ is the rate of condensation;

$0 < x < H$, H represents the thickness of textile.

Meanwhile

k_1, k_2 are constants which are related with molecular weight and gas constant;

$\varepsilon(x)$ is porosity of textile surface;

$r(x)$ is radius of cylindrical pore;

$\tau(x)$ is effective tortuosity of the textile;

λ is latent heat of sorption and condensation of water vapor;

κ is thermal conductivity of textiles.

p_{sat} is the saturation vapor pressure within the parallel pore, its empirical formula can be given as follows [4]

$$p_{sat}(T) = 100 \cdot e^{18.956 - \frac{4030}{T+235}}. \tag{5}$$

In order to solve the ODEs (1)-(4) concerning temperature, water vapor pressure, mass flux of water vapor and condensation rate, the boundary conditions are needed. In this case we can give the following conditions

$$\begin{cases} T(0) = T_L, \\ T(H) = T_R, \\ m_v(0) = m_{v,0}, \\ p_v(0) = p_0, \end{cases} \tag{6}$$

where

T_L is the given temperature on the boundary between the thermal underwear and body skin;

T_R is the given temperature on the boundary between the thermal underwear and the environment;

$m_v(0)$ is the given mass flux of water vapor on the boundary between the thermal underwear and body skin;

$p_v(0)$ is the given water vapor pressure on the boundary between the thermal underwear and body skin.

It is difficult to solve the coupled ODEs, since the equations are seriously nonlinear and coupled [2].

In this paper, we derive the distribution of temperature, water vapor pressure, water vapor mass flux and the rate of condensation within the textile materials by means of decoupling and the numerical computation in section 2. We show how to numerically solve the nonlinear integro-differential equation by the finite difference method in section 3. The convergence rate for the numerical algorithm will be given and proved in section 4. In section 5, we shall implement some numerical examples to show the efficiency of the proposed algorithm. Compared with the experimental measurement, the numerical results are well matched with the experimental data on the "*Walter*" Manikin.

2. Decoupling of the Coupled ODEs

We set $A(x) \equiv \frac{\varepsilon(x)\cdot r(x)}{\tau(x)}$, $k_3 = \frac{\kappa}{\lambda}$. According to the eq.(2) and (3), we have

$$m_v(x) = k_3 \cdot T'(x) + m_v(0) - k_3 \cdot T'(0). \tag{9}$$

Let $C_1 = m_v(0) - k_3 \cdot T'(0)$, then $m_v(x) = k_3 \cdot T'(x) + C_1$. According to the eq.(1), we have

$$p_v(x) = \sqrt{p_v^2(0) - 2\int_0^x \frac{T^{3/2}(s)}{k_1 \cdot A(s)} \cdot [k_3 \cdot T'(s) + C_1]ds}. \tag{10}$$

Finally, according to the eq.(3) and (4), we get a two-point boundary value problem of integro-differential equation, the temperature distribution $T(x)$ satisfies

$$\begin{cases} \sqrt{T}T'' = \frac{k_2 \cdot A(x)}{k_3}\left[p_{sat} - \sqrt{p_v^2(0) - 2\int_0^x \frac{T^{3/2}(s)}{k_1 \cdot A(s)}[k_3 T'(s) + C_1]ds}\right], \\ T(0) = T_L, T(H) = T_R, \end{cases} \tag{11}$$

where $0 < x < H$. Consequently we obtain the explicit expression of the functions $m_v(x)$, $p_v(x)$ and $\Gamma(x)$ respectively:

$$m_v(x) = k_3 \cdot T'(x) + m_v(0) - k_3 \cdot T'(0),$$

$$p_v(x) = \sqrt{p_v^2(0) - 2\int_0^x \frac{T^{3/2}(s)}{k_1 \cdot A(s)} \cdot [k_3 \cdot T'(s) + C_1]ds},$$

$$\Gamma(x) = -m_v'(x).$$

3. Presentation of Numerical Algorithm

Step 1. Domain decomposition.

Due to the small thickness of textile and the characteristics of temperature distribution, we employ isometry subdivision:

$$0 = x_0 < x_1 < \cdots < x_N = H,$$

where $h = x_{i+1} - x_i (i = 0, 1, \cdots, N-1)$ is step length.

Step 2. Discretization by FDM and numerical integral methods.

According to the idea that the derivative is approximated by difference quotient, we have

$$T'(x)|_{x=x_i} \approx \frac{T_{i+1} - T_i}{h}, i = 0, 1, \cdots, N-1,$$

$$T''(x)|_{x=x_i} \approx \frac{T_{i+2} - 2T_{i+1} + T_i}{h^2}, i = 0, 1, \cdots, N-2,$$

$$C_1 = m_v(0) - k_3 T'(0) \approx m_{v,0} - k_3 \frac{T_1 - T_0}{h}.$$

Applying the left rectangle formula for the definite integral, we have

$$\int_0^{x_i} \frac{T^{3/2}(s)}{k_1 \cdot A(s)} \cdot [k_3 \cdot T'(s) + C_1]ds$$

$$\approx \sum_{j=0}^{i-1} \frac{T_j^{3/2} h}{k_1 \cdot A(x_j)} \cdot [k_3 \cdot \frac{T_{j+1} - T_j}{h} + C_1^*],$$

where $C_1^* = m_{v,0} - k_3 \frac{T_1 - T_0}{h}$.

Step 3. Iterative algorithm for $T(x)$.

Discretizing the problem (9) by FDM and numerical integral methods, we obtain the iterative scheme:

$$\sqrt{T_i}\cdot\frac{T_{i+2}-2T_{i+1}+T_i}{h^2}=\frac{k_2A(x_i)}{k_3}$$

$$\times\left(p_{sat}(T_i)-\sqrt{p_v^2(0)-2\sum_{j=0}^{i-1}\frac{T_j^{3/2}h}{k_1A(x_j)}[k_3\frac{T_{j+1}-T_j}{h}+C_1^*]}\right),\qquad(12)$$

for $i=1,2,\cdots,N-2$. And for $i=0$, we have

$$\sqrt{T_0}\cdot\frac{T_2-2T_1+T_0}{h^2}=\frac{k_2}{k_3}A(x_0)\left[p_{sat}(T(x_0))-p_v(0)\right].$$

In numerical implementation, we need the approximate value T_1 of $T(x_1)$, for example, by linear interpolation from two point $(0,T_L)$ and (H,T_R):

$$T_1=T_L-\frac{T_L-T_R}{H}\cdot x_1.$$

Using above iterative scheme, we can calculate $T_2,T_3,\cdots,T_{N-1}$ beginning with T_0 and T_1.

Step 4. Numerical computation of $m_v(x),p_v(x)$ and $\Gamma(x)$.

Subsequently we can calculate the distribution of the mass flux of water vapor, water vapor pressure and condensation rate by the eq.(7), (8) and (2):

$$m_{v,i}=k_3\frac{T_i-T_{i-1}}{h}+m_v(0)-k_3\frac{T_1-T_0}{h},$$

$$p_{v,i}=\sqrt{p_{v,0}^2-2\sum_{j=0}^{i-1}\frac{1}{k_1\cdot A(x_j)}\cdot T_j^{3/2}\cdot[k_3\cdot\frac{T_{j+1}-T_j}{h}+C_1^*]\cdot h},$$

$$\Gamma_i=-\frac{m_{v,i+1}-m_{v,i}}{h},i=1,2,\cdots,N-1.$$

4. Convergence Rate of the Proposed Algorithm

First we notice that the constants k_1, k_2, k_3 are all positive, and the functions $\varepsilon(x), r(x), \tau(x)$ are all positive and bounded. We assume that $\varepsilon(x), r(x), \tau(x) \in C[0, H]$, therefore $A(x) = \frac{\varepsilon(x) \cdot r(x)}{\tau(x)} \in C[0, H]$.

Now we conclude the convergence result via the following theorem.

Theorem 4.1. *We assume, a priori, that the temperature $T(x) \in C^3[0, H]$. Then the iterative algorithm (10) is convergent with the convergence rate of the first order.*

Proof. We prove the theorem by two steps.

Step 1. Deduce an equation concerning the term $|T(x_i) - T_i|$.

For simplification, we denote the operator L by

$$LT_i \equiv T_{i+2} - 2T_{i+1} + T_i. \tag{13}$$

Then difference scheme (10) can be written as follows:

$$\frac{\sqrt{T_i} LT_i}{h^2} = \frac{k_2 A(x_i)}{k_3} \times \left(p_{sat}(T_i) - \sqrt{p_v^2(0) - 2\sum_{j=0}^{i-1} \frac{T_j^{3/2} h}{k_1 \cdot A(x_j)} \cdot [k_3 \frac{T_{j+1} - T_j}{h} + C_1^*]} \right) \tag{14}$$

for $i = 1, 2, \cdots, N-2$.

From the problem (9), we obtain

$$\sqrt{T(x_i)} T''(x_i) = \frac{k_2}{k_3} A(x_i) \times \left(p_{sat}(T(x_i)) - \sqrt{p_v^2(0) - 2\int_0^{x_i} \frac{1}{k_1 \cdot A(s)} \cdot T^{3/2}(s) \cdot [k_3 \cdot T'(s) + C_1] ds} \right). \tag{15}$$

Set

$$a_i \equiv \sqrt{T(x_i)} T''(x_i) - \frac{k_2}{k_3} A(x_i) p_{sat}(T(x_i)),$$

$$b_i \equiv \sqrt{T_i} \frac{LT_i}{h^2} - \frac{k_2}{k_3} A(x_i) p_{sat}(T_i),$$

then the equations (11) and (12) imply that

$$b_i^2 = \left(\frac{k_2}{k_3}A(x_i)\right)^2 \cdot$$

$$\left(p_v^2(0) - 2\sum_{j=0}^{i-1}\frac{1}{k_1 \cdot A(x_j)} \cdot T_j^{3/2} \cdot [k_3 \cdot \frac{T_{j+1} - T_j}{h} + C_1^*] \cdot h\right), \tag{16}$$

$$a_i^2 = \left(\frac{k_2}{k_3}A(x_i)\right)^2 \cdot$$

$$\left(p_v^2(0) - 2\int_0^{x_i}\frac{1}{k_1 \cdot A(s)} \cdot T^{3/2}(s) \cdot [k_3 \cdot T'(s) + C_1]ds\right). \tag{17}$$

Subsequently

$$a_i^2 - b_i^2 = 2(\frac{k_2}{k_3}A(x_i))^2 \cdot \left(\sum_{j=0}^{i-1}\frac{1}{k_1 A(x_j)}T_j^{3/2}[k_3\frac{T_{j+1} - T_j}{h} + C_1^*]h - \int_0^{x_i}\frac{1}{k_1 A(s)}T^{3/2}(s)[k_3 T'(s) + C_1]ds\right). \tag{18}$$

We denote $f(x)$ and f_j by

$$f(x) \equiv \frac{1}{k_1 A(x)}T^{3/2}(x)[k_3 T'(x) + C_1],$$

$$f_j \equiv \frac{1}{k_1 A(x_j)}T_j^{3/2}[k_3\frac{T_{j+1} - T_j}{h} + C_1^*],$$

then (15) gives that

$$(a_i + b_i)(a_i - b_i) + 2\left(\frac{k_2}{k_3}A(x_i)\right)^2 \cdot \left(\sum_{j=0}^{i-1} f(x_j) \cdot h - \sum_{j=0}^{i-1} f_j \cdot h\right)$$

$$= 2\left(\frac{k_2}{k_3}A(x_i)\right)^2 \cdot \left(\sum_{j=0}^{i-1} f(x_j) \cdot h - \int_0^{x_i} f(s)ds\right). \tag{19}$$

By mathematical analysis, we can rewrite the left side of identity (19):

The left side of the identity (19)

$$= (a_i + b_i)A_i(T(x_i) - T_i) - (a_i + b_i) \cdot \sqrt{T_i} \cdot O(h) +$$

$$\left(\frac{k_2}{k_3}A(x_i)\right)^2 \cdot \sum_{j=0}^{i-1}\left[2 \cdot B_j(T(x_j) - T_j)h + \frac{k_3 T_j^{3/2} T''(\eta)}{k_1 A(x_j)}h^2 - \right.$$

$$\left.\frac{k_3}{k_1 A(x_j)} T_j^{3/2} \cdot \frac{T''(\nu_j)}{i} x_i \cdot h\right], \tag{20}$$

where

$A_i = \frac{T''(x_i)}{\sqrt{T(x_i)} + \sqrt{T_i}} - \frac{k_2}{k_3}A(x_i) \cdot k \cdot \frac{1}{\varepsilon_i^2} \cdot e^{-\frac{4030}{\varepsilon_i}}$,

$k = 4.03 \times 10^5 e^{18.956}$,

$B_j = \frac{1}{k_1 A(x_j)} \cdot \frac{T(x_j) + T_j + \sqrt{T(x_j)T_j}}{\sqrt{T(x_j)} + \sqrt{T_j}} \cdot (C_1 + k_3 T'(x_j))$,

$\nu_j \in (x_j, x_{j+1}), \eta \in (0, x_1)$,

ε_i lies between $T(x_i) + 235$ and $T_i + 235$.

Now, we calculate the right side of identity (19). Observe that

$$\int_0^{x_i} f(s)ds = \sum_{j=0}^{i-1} \int_{x_j}^{x_{j+1}} f(s)ds = \sum_{j=0}^{i-1} f(x_j) \cdot h + \sum_{j=0}^{i-1} \frac{f'(\zeta_j)}{2} \cdot h^2,$$

where $\zeta_j \in (x_j, x_{j+1})$ and residual item is $R_i[f] = \sum_{j=0}^{i-1} \frac{f'(\zeta_j)}{2} \cdot h^2$. By mathematical analysis, we can prove that there exits $\varsigma \in [\zeta_0, \zeta_{i-1}] \subseteq [0, H]$ such that

$$\int_0^{x_i} f(s)ds - \sum_{j=0}^{i-1} f(x_j) \cdot h = \frac{1}{2} x_i f'(\varsigma)h.$$

Consequently the right side of identity (19) can be rewritten as follows:

The right side of the identity(19)

$$= -\left(\frac{k_2}{k_3}A(x_i)\right)^2 \cdot x_i \cdot f'(\varsigma) \cdot h. \tag{21}$$

Based on the identity (17) and (18), the identity (16) implies that

$$(a_i + b_i)A_i(T(x_i) - T_i) - (a_i + b_i) \cdot \sqrt{T_i} \cdot O(h) +$$

$$\left(\frac{k_2}{k_3}A(x_i)\right)^2 \cdot \sum_{j=0}^{i-1}\left[2 \cdot B_j(T(x_j) - T_j)h + \frac{k_3 T_j^{3/2} T''(\eta)}{k_1 A(x_j)}h^2 - \right.$$

$$\frac{k_3}{k_1 A(x_j)} T_j^{3/2} \cdot \frac{T''(\nu_j)}{i} x_i \cdot h\Bigg] = -\left(\frac{k_2}{k_3} A(x_i)\right)^2 \cdot x_i \cdot f'(\varsigma) \cdot h \tag{22}$$

for $i = 1, 2, \cdots, N-2$.

On the other hand, we know that $f(x) \in C^2[0, H]$ due to $T(x) \in C^3[0, H]$. Therefore there exist a positive constant L, such that $|f'(x)| \leq L$ for any $x \in [0, H]$.

Step 2. Convergence estimates from the identity (19).

First we notice that there exist three positive constants N_1, N_2, N_3 such that

$$|\frac{k_2}{k_3} A(x)|^2 \leq N_1, |\frac{k_3}{k_1 A(x)}| \leq N_2, |\frac{1}{k_1 A(x)}| \leq N_3.$$

Moreover due to $T(x) \in C^3[0, H]$, there exist positive constants m_1, M_1, M_2, M_3, M_4 such that for any $x \in [0, H]$,

$$0 < m_1 \leq |T(x)| \leq M_1, |T'(x)| \leq M_2, |T''(x)| \leq M_3, |T'''(x)| \leq M_4.$$

Hence when $i = 1$,

$$|a_1 + b_1| \cdot |A_1| \cdot |T(x_1) - T_1|$$

$$\leq |a_1 + b_1| \cdot \sqrt{M_1} \cdot O(h) + 2 \cdot N_1 N_2 M_1^{3/2} M_3 H \cdot h + N_1 H L \cdot h.$$

Because $|a_1 + b_1|$ and $|A_1|$ are both bounded, so we have

$$|T(x_1) - T_1| \leq Q_1 \cdot h = O(h)$$

where Q_1 is a constant.

Suppose when $i \leq n-3$, we have

$$|T(x_i) - T_i| \leq O(h).$$

Then when $i = n-2$,we have

$$|a_{n-2} + b_{n-2}| \cdot |A_{n-2}| \cdot |T(x_{n-2}) - T_{n-2}|$$

$$\leq |a_{n-2} + b_{n-2}| \cdot \sqrt{M_1} \cdot O(h) + 2N_1 N_2 M_1^{3/2} M_3 H \cdot h +$$

$$N_1 \cdot \frac{3N_3 M_1 + 6N_2 M_1 M_2}{\sqrt{m_1}} \cdot (n-3) \cdot O(h) \cdot h + N_1 H L \cdot h$$

$$= |a_{n-2} + b_{n-2}| \cdot \sqrt{M_1} \cdot O(h) + 2N_1 N_2 M_1^{3/2} M_3 H \cdot h +$$

$$N_1 \cdot \frac{3N_3 M_1 + 6N_2 M_1 M_2}{\sqrt{m_1}} \cdot (n-3) \cdot C \cdot \frac{H}{n} \cdot h + N_1 H L \cdot h,$$

where C is a constant.

Because $|a_{n-2}+b_{n-2}|$ and $|A_{n-2}|$ are both bounded, so

$$|T(x_{n-2})-T_{n-2}|\le O(h).$$

By mathematical induction, the algorithm is convergence with convergent rate of the first order. This completes the proof of the Theorem 4.1. □

5. Numerical Experiment

In this section, numerical simulation and analysis are carried out for down and polyester textile material respectively.

We consider the case that human body is motionless, and the thermal textile is assumed to be close to the human skin. We obtain the experimental data under the following environmental conditions: the outside air temperature is 20 °C, the relative humidity is 60% and the velocity of wind is 0.16m/s. In our experiment, we lay each four temperature sensors on human body surface, textile inner surface, textile inside points and textile outer surface respectively. The four sensors are located at upside left chest, underside right chest, upside right back and underside left back of human body respectively. So does the experimental data of the rate of condensation.

Using the material parameters of thermal underwear, we get the distribution of the temperature, water vapor mass flux, water vapor pressure and condensation rate through the textile.

5.1. *Computation of the Temperature $T(x)$*

We need solve the following two-point boundary value problem

$$\begin{cases} \sqrt{T}T'' = \frac{k_2A(x)}{k_3}\left[p_{sat}-\sqrt{p_v^2(0)-\int\limits_0^x \frac{2T^{3/2}(s)}{k_1A(s)}[k_3T'(s)+C_1]ds}\right], \\ T(0)=T_L, T(H)=T_R. \end{cases}$$

In the process of computation, we choose $N=50$. $k_1=0.0002$ and $k_2=0.0003$. Meanwhile the values of the material parameters are needed, which are listed as follows:

Case 1. For down material: $r=2\times10^5m$, $\varepsilon=0.8$, $\tau=3$, $\kappa=7.59\times 10^{-3}W/m.K$, $\lambda=2260\times10^3J/kg$, $H=6.71\times10^{-3}m$, $T_L=308.96K$, $T_R=300.96K$, $m_v(0)=3.404\times10^{-5}kg/m^2\cdot s$. The numerical simulation results for down material are shown in the Table 1.

Case 2. For polyester material: $r = 1.4 \times 10^5 m$, $\varepsilon = 0.88$, $\tau = 1.2$, $\kappa = 51.9 \times 10^{-3} W/m.K$, $\lambda = 2522 \times 10^3 J/kg$, $H = 8.71 \times 10^{-3} m$, $T_L = 309.99K$, $T_R = 300.2K$, $m_v(0) = 3.083 \times 10^{-5} kg/m^2 \cdot s$. The numerical simulation results for polyester material are shown in the Table 2.

From the Table 1 and Table 2, it can be seen that the errors between numerical results and experimental data are small, numerical results are well matched with the experimental ones.

On the other hand, the efficiency of the proposed algorithm can be shown by the following Figure 1 for the temperature $T(x)$ in down and polyester materials respectively.

Table 1. Comparison between the numerical results and experimental data for $T(x)$ in the down thermal textile

Thickness (m)	Experimental data (K)	Numerical results (K)	Absolute errors (K)	Relative errors (%)
0.00137	308.46	307.30	1.16	0.38
0.00190	307.96	306.72	1.24	0.40
0.00481	305.56	303.11	2.45	0.81
0.00671	300.96	300.75	0.21	0.07

Table 2. Comparison between the numerical results and experimental data for $T(x)$ in polyester thermal textile

Thickness (m)	Experimental data (K)	Numerical results (K)	Absolute errors (K)	Relative errors (%)
0.00137	309.93	308.45	1.48	0.48
0.00190	310.07	307.83	2.24	0.73
0.00796	303.73	300.92	2.81	0.93
0.00871	300.20	300.06	0.14	0.05

5.2. *Computation of the condensation rate* $\Gamma(x)$

Numerical results of the rate of condensation $\Gamma(x)$ in down and polyester materials respectively, are shown in the Figure 2.

From the numerical simulation results of the temperature distribution within the thermal textile, we can show that the temperatures within the fabric are decreasing, and errors between the numerical results and experimental data are very small. Numerical results are well matched with the

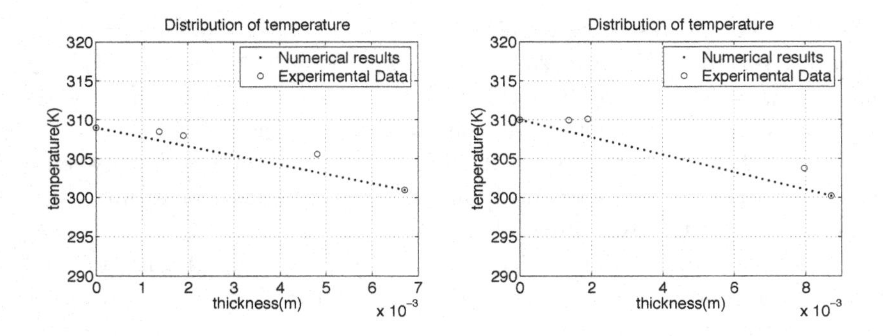

Fig. 1. Temperature distribution $T(x)$ in the down thermal textile (left) and polyester thermal textile (right) respectively.

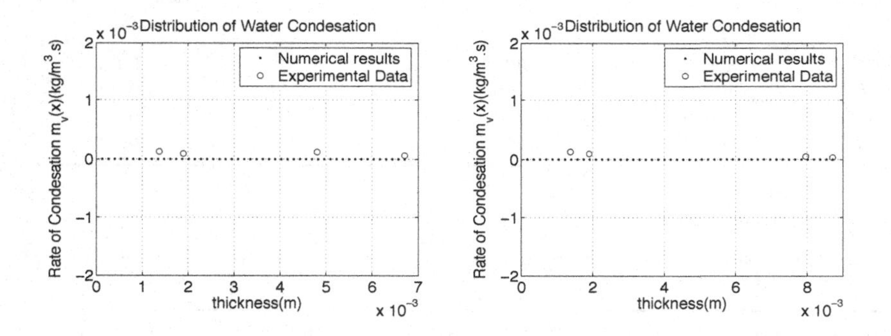

Fig. 2. The rate of condensation $\Gamma(x)$ in the down thermal textile (left) and polyester thermal textile (right) respectively.

experimental ones on the "*Walter*" Manikin. So do the simulation results of the rate of condensation. All above results show that the model and the nonlinear iterative algorithm are effective for heat and mass transfer through textiles.

Acknowledgements

The research is partially supported by National Natural Science Foundation of China (Contract Grant No. 10561001 and 11071221) and Science Foundation of Zhejiang Sci-Tech University(ZSTU) (Contract Grant No. 2010KY009).

References

1. X.H.Zhou, Ph.D. diss., Donghua University, Shanghai(2005).
2. D.H.Xu, J.X.Cheng and X.H.Zhou, Textile Bioengineering and Informatics Symposium Proceeding 2010, TBIS Limited Binary Information Press, Hong Kong, 1151(2010).
3. D.H.Xu, J.X.Cheng and X.H.Zhou, J. Math-for-Industry, **2**, 139(2010).
4. E.J.Frank, *Evaporation of Water*, Lewis Publishers, USA(1992).
5. H.G.Henry, Pro. Roy. Soc. **171A**, 215(1993).
6. J.C.Gretton, Coated Fabric.**25**, 301(1996).
7. H.G.David and P.Nordon, Text. Res. J. **39**, 166(1969).
8. H.Meinander, H.Anttonen, V.Bartels, I.Holmer, R.E.Reinertsen, K.Soltynski and S.Varieras, Eur. J. Appl. Physiol. **92**, 619(2004).
9. G.Phillip, J. Coated Fabrics. **24**, 322(1995).
10. E.L.Ji, Zh.Yu, X.H.Zhou and Sh.Y.Wang, J. Qingdao University(Engineering Edition). **21**, 6(2006).
11. Y.Li and B.V.Holcombe, Text. Res. J. **62**, 211(1992).
12. B.V.Holcombe, Text. Res. J. **68**, 389(1998).
13. A.M.Schneider and B.N.Hoschke, Text. Res. J. **62**, 61(1992).
14. J.T.Fan, Z.Y.Luo and Y.Li, Int. J. Heat Mass Transfer. **43**, 2989(2000).
15. J.T.Fan, X.Cheng and Y.S.Chen, Exp. Therm Fluid Sci. **27**, 723(2003).
16. H.X.Huang, C.H.Ye and W.W.Sun, J. Engineering Math. **61**, 35(2008).

HAVERSIAN CORTICAL BONE WITH A RADIAL MICROCRACK

X. WANG*

Institute of Ethnic Preparatory Educaton,Ningxia University, Yinchuan, China

**E-mail: wangxu685@163.com*

The plane problem for the Haversian cortical bone with a radial microcrack in the interstitial bone under biaxial tension loading is considered. A two-dimensional coated fiber composite model of the Haversian cortical bone is developed correspondingly. By using the solution for the edge dislocation as Green's functions, the problem of a radial microcrack located in the interstitial bone in the vicinity of the osteon is formulated into a system of singular integral equations with Cauchy kernels. And the interaction effect of various material and geometric parameters of the Haversian cortical bone upon the microcrack tip's stress intensity factors is studied. The numerical analysis indicates that the stress intensity factor of the microcrack is dominated by the shear modulus ratio among the Haversian canal, the osteon and the interstitial bone, the distance of the left tip of the microcrack from cement line, the length of the microcrack. The numerical results suggest that when the osteon is softer than the interstitial bone the osteon prompts the microcrack propagation while the osteon is stiffer than the interstitial bone the osteon repels microcrack propagation. However, this interaction effect is limited near the osteon. Some of the numerical results are in accordance with the results obtained. And additional predicted numerical results are given, which need to be confirmed via experiment or computer simulation.

Keywords: Haversian cortical bone; Microcrack; Integral equation; Stress intensity factor.

1. Introduction

Cortical bone is one of the two major types of bone. The hierarchical structure of bone has been described in a number of reviews [1,2]. Haversian cortical bone consists of osteons embedded in interstitial bone, the osteon has a cylindrical shape with a more or less centrally placed neurovascular Haversian canal and the interstitial bone is made of the remnants of the remodeled old osteons or primary bone [3,4]. As the Haversian canal and osteon microstructure have a major influence on these properties, thus the microstructure must be considered in the mathematical model. From a microstructure and mechanical behavior perspective, Haversian cortical bone shares several similarities with coated fiber composite [5]: Haversian canal is analogous to fiber, osteon is analogous to the coating layer, interstitial bone is analogous to matrix, and the cement line acts as a weak interface. Thus, the hierarchical structure of cortical bone may be compared to a coated fiber composite material.

Bone fatigue is well documented [6,7]. This phenomenon is commonly known in the medical community as brittle fracture. Bone matrix accumulates this type of microdamage in the form of microcracks as a result of everyday cyclic loading activities [8,9]. Recently, experimental, computational simulations and theoretical studies focus on the effects of osteons on the fracture mechanics of osteonal cortical bone. Experimental method is one of the most frequently used methods in the study of osteon structure and fracture analysis [10,11,12]. Bone microstructure allows the initiation of microcracks [12], but osteons act as barriers to microcrack propagation in cortical bone and the microstructural barrier effect is dependent on the microcrack length [11,12]. Computational simulations, especially the simulations of microcrack propagation, begin to be used in the study of cortical bone fracture mechanics [13]. Based on linear elastic fracture mechanics, the microcrack propagation near osteons is simulated via finite element method [13]. Theoretical studies, especially the dislocation technique, is used in the fracture mechanics research of cortical bone [14,15]. Using linear elastic fracture mechanics on a composite model of osteonal cortical bone, the interaction effect between a microcrack and an osteon is studied and it shows that a softer osteon promotes microcrack towards the osteon while a stiff one repels the microcrack from the osteon [4]. Based on linear elastic fracture mechanics, the interaction effect between microcracks and an osteon is studied, and it indicates that the interaction effect is limited to the vicinity of the osteon [14,15]. Enhancement and shielding effect of micracracks interactions are also analyzed for various configurations. However, the Haversian cortical bone with a radial microcrack in the interstitial bone has not been theoretically investigated. Thus, it is imperative for us to study the interaction effect between the radial microcrack and the osteon.

The main objective of the present work is to study the interaction effect between a radial microcrack and osteon in Haversian cortical bone and a two-dimensional coated fiber composite model of the Haversian cortical bone is developed correspondingly. A singular integral equation method which uses continuous distributions of edge dislocations as Green's functions to represent the microcrack is used for solving the problem with a radial microcrack in the interstitial bone. Material and geometric parameters of the Haversian cortical bone including shear modulus ratio among the Haversian canal, osteon and the interstitial bone, the distance of the left tip of the microcrack from cement line, the length of the microcrack are involved in the numerical analysis of the microcrack tip's stress intensity factors. The numerical results suggest that when the osteon is softer than the interstitial bone the osteon prompts the microcrack propagation while the osteon is stiffer than the interstitial bone the osteon repels microcrack propagation. And this result is in accordance with the corresponding result given in [13,16]. Additional predicted numerical results are obtained, which need to be confirmed via experiment or computer simulation.

2. Description of the model

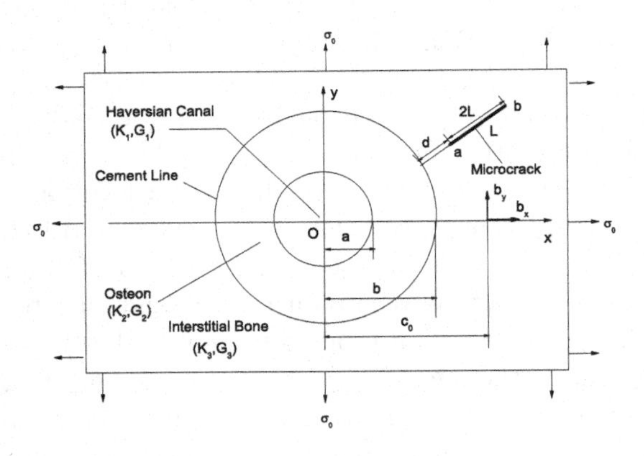

Fig. 1. A coated fiber composite model for the Haversian cortical bone with a radial microcrack.

As illustrated in Figure 1, based on linear elastic fracture mechanics, a two dimensional plane strain model of Haversian cortical bone with a radial microcrack in the interstitial bone near the osteon is developed. The Haversian canal (radius of a) with shear modulus G_1 and constants of $K_1 = 3 - 4v_1$ and the osteon (thickness of $b - a$) with shear modulus G_2 and constants of $K_2 = 3 - 4v_2$ are embedded in the interstitial bone with shear modulus G_3 and constants of $K_3 = 3 - 4v_3$, where v_1, v_2 and v_3 are Poisson's ratios. This model is under a uniform remote biaxial tension loading σ_0.

The solution of the present problem may be obtained through the superposition of two basic problems. In the first problem, an osteon embedded in the interstitial bone without the microcrack and under the uniform remote biaxial tensile loading σ_0; In the second problem, only the stress disturbance due to the existence of the microcrack in the interstitial bone is considered. In this problem the only external loads are the microcrack surface tractions which are equal in magnitude and opposite in sign to the stress obtained in the first problem along the presumed location of the microcrack.

3. Solution of the model

3.1. *The stress fields for solving the first problem*

Solution of the stress fields of the first problem in rectangular coordinate is as follows [17]

$$\sigma_{xx} = \sigma_0 \left[1 - \frac{G_3}{K_3}\frac{b^2}{r^2} a_4 \cos 2\theta\right], \quad (1)$$

$$\sigma_{yy} = \sigma_0 \left[1 + \frac{G_3}{K_3}\frac{b^2}{r^2} a_4 \cos 2\theta\right], \quad (2)$$

$$\sigma_{xy} = -\sigma_0 \left[\frac{G_3}{K_3}\frac{b^2}{r^2} a_4 \sin 2\theta\right], \quad (3)$$

where

$$r^2 = x^2 + y^2, \quad (4)$$

$$\theta = arctg\frac{y}{x}, \quad (5)$$

$$a_4 = -\frac{b^2 (K_2 - K_3)\left(\frac{K_1}{G_3} + m_{23}\right) + a^2 (K_1 - K_2)\left(\frac{K_3}{G_3} + m_{23}\right)}{a^2 (K_1 - K_2)(m_{23} - 1) + b^2\left(\frac{K_1}{G_3} + m_{23}\right)\left(\frac{K_2}{G_3} + 1\right)}, \quad (6)$$

$$m_{13} = \frac{G_1}{G_3}, \quad (7)$$

$$m_{23} = \frac{G_2}{G_3}. \quad (8)$$

In order to provide a solution to the first problem, $p_t(t)$ and $p_n(t)$, which are the tangential and normal stresses in the interstitial bone without the microcrack need to be computed, and t is a point on the location of the imaginary microcrack L.

3.2. *The singular integral equation method for solving the second problem*

Let the interstitial bone contain two edge dislocations at the point($x = c_0, y = 0$) with Burgers vectors b_x and b_y. By using the Muskhelishvili complex variable method [18], the stress fields at a point $P(x, y)$ in the interstitial bone may be expressed as

$$\frac{\pi(K_3+1)}{G_3}\sigma_{xx}(x,y,c_0) = h_{xxx}(x,y,c_0)\,b_x + h_{xxy}(x,y,c_0)\,b_y, \tag{9}$$

$$\frac{\pi(K_3+1)}{G_3}\sigma_{yy}(x,y,c_0) = h_{yyx}(x,y,c_0)\,b_x + h_{yyy}(x,y,c_0)\,b_y, \tag{10}$$

$$\frac{\pi(K_3+1)}{G_3}\sigma_{xy}(x,y,c_0) = h_{xyx}(x,y,c_0)\,b_x + h_{xyy}(x,y,c_0)\,b_y, \tag{11}$$

where $h_{xxx}, h_{xxy}, h_{yyx}, h_{yyy}, h_{xyx}, h_{xyy}$ are given in the Appendix A of the literature [19].

If we assume that b_t and b_n are continuously distributed functions (along and perpendicular) on L, the second problem may be formulated as

$$\sigma_t(t) = \int_{-1}^{1} [h_{t1}(t,t_0)\,b_t(t_0) + h_{t2}(t,t_0)\,b_n(t_0)]dt_0, \tag{12}$$

$$\sigma_n(t) = \int_{-1}^{1} [h_{n1}(t,t_0)\,b_t(t_0) + h_{n2}(t,t_0)\,b_n(t_0)]dt_0, \tag{13}$$

where h_{t1}, h_{t2}, h_{n1} and h_{n2} are formulated as follows

$$\begin{aligned} h_{t1} =& mL\left[-mnh_{xxx} + mnh_{yyx} + \left(m^2-n^2\right)h_{xyx}\right] \\ &+ nL\left[-mnh_{xxy} + mnh_{yyy} + \left(m^2-n^2\right)h_{xyy}\right], \end{aligned} \tag{14}$$

$$\begin{aligned} h_{t2} =& -nL\left[-mnh_{xxx} + mnh_{yyx} + \left(m^2-n^2\right)h_{xyx}\right] \\ &+ mL\left[-mnh_{xxx} + mnh_{yyy} + \left(m^2-n^2\right)h_{xyy}\right], \end{aligned} \tag{15}$$

$$\begin{aligned} h_{n1} =& mL\left[n^2h_{xxx} + m^2h_{yyx} - 2mnh_{xyx}\right] \\ &+ nL\left[n^2h_{xxy} + m^2h_{yyy} - 2mnh_{xyy}\right], \end{aligned} \tag{16}$$

$$h_{n2} = -nL\left[n^2 h_{xxx} + m^2 h_{yyx} - 2mnh_{xyx}\right] + mL\left[n^2 h_{xxy} + m^2 h_{yyy} - 2mnh_{xyy}\right], \tag{17}$$

where

$$x = x_c + mLt, \quad y = y_c + nLt, \quad c_0 = x_c + mLt_0, \quad \mathrm{t}, t_0 \in [-1, 1], \tag{18}$$

and (x_c, y_c) is the midpoint of the microcrack. Here, m and n may be defined as

$$m = \cos\alpha, \quad n = \sin\alpha, \tag{19}$$

where α is the angle between the microcrack and the x axis.

The continuity of displacement requires that

$$\int_{-1}^{1} \mathrm{b_t}(t_0)\, dt_0 = 0, \tag{20}$$

$$\int_{-1}^{1} \mathrm{b_n}(t_0)\, dt_0 = 0. \tag{21}$$

After separating the singularity of the dislocation density functions, $b_t(t_0)$ and $b_n(t_0)$ can be expressed as

$$b_t(t_0) = \frac{F_t(t_0)}{\sqrt{1 - t_0^2}}, \tag{22}$$

$$b_n(t_0) = \frac{F_n(t_0)}{\sqrt{1 - t_0^2}}, \tag{23}$$

where $F_t(t_0)$ and $F_n(t_0)$ are non-singular functions on $[-1, 1]$.

Substituting (22) and (23) into (12),(13),(20) and (21), then

$$\sigma_t(t) = \int_{-1}^{1} \left[h_{t1}(t, t_0) \frac{F_t(t_0)}{\sqrt{1 - t_0^2}} + h_{t2}(t, t_0) \frac{F_n(t_0)}{\sqrt{1 - t_0^2}} \right] dt_0, \tag{24}$$

$$\sigma_n(t) = \int_{-1}^{1} \left[h_{n1}(t, t_0) \frac{F_t(t_0)}{\sqrt{1 - t_0^2}} + h_{n2}(t, t_0) \frac{F_n(t_0)}{\sqrt{1 - t_0^2}} \right] dt_0, \tag{25}$$

$$\int_{-1}^{1} \frac{F_t(t_0)}{\sqrt{1 - t_0^2}} dt_0 = 0, \tag{26}$$

$$\int_{-1}^{1} \frac{F_n(t_0)}{\sqrt{1 - t_0^2}} dt_0 = 0, \tag{27}$$

$$\sigma_t(t) = -p_t(t), \tag{28}$$

$$\sigma_n(t) = -p_n(t). \tag{29}$$

(24)–(29) give the solution to the model that the radial microcrack is located in the interstitial bone near the osteon.

3.3. *The stress intensity factors for the radial microcrack tips*

The stress intensity factors of radial microcracks tips can be calculated as follows [20,21]

$$K_{\mathrm{I}}(\pm 1) = \pm\sqrt{\pi L}\frac{2G_3}{K_3+1}F_n(\pm 1), \tag{30}$$

$$K_{\mathrm{II}}(\pm 1) = \pm\sqrt{\pi L}\frac{2G_3}{K_3+1}F_t(\pm 1). \tag{31}$$

4. Numerical analysis

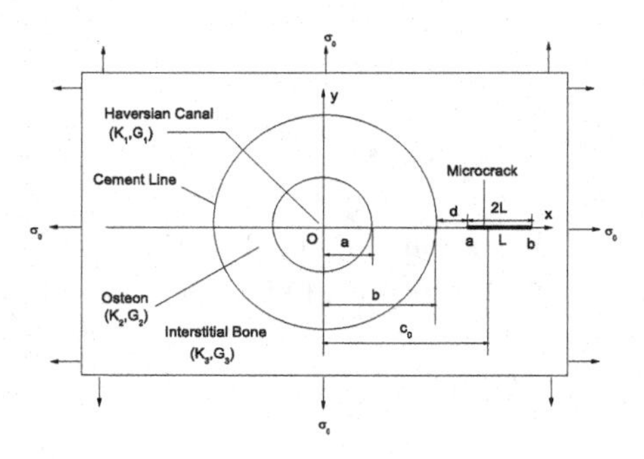

Fig. 2. The Haversian cortical bone model with a radial microcrack under biaxial tension loading.

As the numerical example, the Haversian cortical bone model with a radial microcrack under biaxial tension loading is shown in Figure 2. For the soft osteon and the stiff osteon, the shear modulus $G_3 = 15GPa$ and $10GPa$ respectively. The Poisson's ratio of the Haversian canal, osteon and interstitial bone is $v_1 = v_2 = v_3 = 0.3$. A uniform remote biaxial tension loading $\sigma_0 = 10MPa$. By using the Gauss-Chebyshev Quadrature method [20], the system of the singular integral equation is numerically solved and the stress intensity factors are presented for a variety of material and geometric parameters.

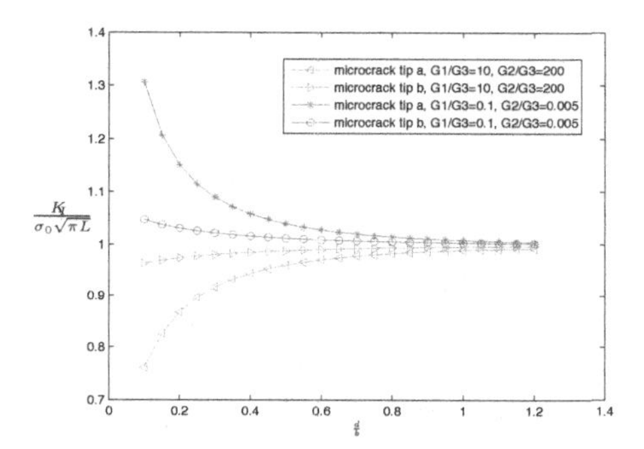

Fig. 3. The normalized stress intensity factors of microcrack tips (a) and (b) versus the normalized distance$\frac{d}{b}$ for various shear modulus ratios.

We assume that the radius of the osteon $b = 100\mu m$, $b/a = 1.5$ and the fixed microcrack length L=100 μm. The normalized stress intensity factors of microcrack tips (a) and (b) are plotted against the normalized distance $\frac{d}{b}$ for various shear modulus ratios is shown in Figure 3. As $\frac{G_1}{G_3} = 0.1$, $\frac{G_2}{G_3} = 0.005$, the stress intensity factor of the microcrack tip (a) increases sharply as the microcrack approaches the osteon and the stress intensity factor of the microcrack tip (b) increases gradually as the microcrack approaches the osteon. As $\frac{G_1}{G_3} = 10$, $\frac{G_2}{G_3} = 200$, the stress intensity factor of the microcrack tip (a) decreases sharply as the microcrack approaches the osteon and the stress intensity factor of the microcrack tip (b) decreases gradually as the microcrack approaches the osteon. However, the influence of the Haversian canal and osteon upon the stress intensity factor is limited in the vicinity of the osteon. The above numerical result is in accordance with the result obtained by finite element method [13].

We assume that the radius of the osteon $b = 100\mu m$, $b/a = 1.5$. The microcrack with fixed tip (b) increases in length, the normalized stress intensity factors of microcrack tips (a) and (b) are plotted against the normalized microcrack length $\frac{L}{b}$ for various shear modulus ratios is shown in Figure 4. Initially, the normalized stress intensity factors of microcrack tip (a) and (b) increase with the increasing microcrack length. While $\frac{G_1}{G_3} = 0.1$,

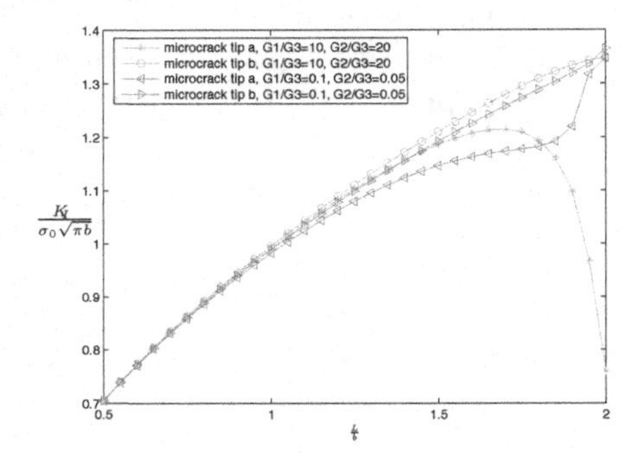

Fig. 4. The normalized stress intensity factors of microcrack tips (a) and (b) versus the normalized microcrack length $\frac{L}{b}$ for various shear modulus ratios.

$\frac{G_2}{G_3} = 0.05$, as the microcrack tip(a) approaches the osteon, the normalized stress intensity factor increases significantly. While $\frac{G_1}{G_3} = 10$, $\frac{G_2}{G_3} = 20$, the stress intensity factor of the microcrack tip (a) decreases sharply. This indicates that when the Haversian canal and osteon is softer than the interstitial bone, the osteon tends to promote microcrack propagation, but it is just the opposite for the stiffer one.

5. Discussion

The result of this study characterizes the fracture mechanics of Haversian cortical bone through a simplified model with a radial microcrack located in the interstitial bone. However, there are several assumptions of this model. First, the linear elastic fracture mechanics is adopted. The plastic zone is involved in the fracture mechanics of ostional cortical bone tissue, which is much smaller than that of an osteon. Thus, it seems justified to adopt the linear elastic fracture mechanics theory. Second, since the longitudinal dimension is quite larger than that of transverse dimension, a plane strain condition is assumed in the mathematical model. Third, the microcrack is located in the interstitial bone and doesn't penetrate the cement line, which is in accordance with the literature [22]. Fourth, a single osteon is assumed. Finally, the interface among the Haversian canal, osteon and the interstitial bone is assumed to be perfectly bonded.

In summary, a coated fiber composite model for the Haversian cortical bone with a radial microcrack located in the interstitial bone near the osteon is developed. The numerical result indicates that the stress intensity factor of microcrack tip is dominated by the shear modulus ratios among the Haversian canal, the osteon and the interstitial bone, the distance of the left tip of the microcrack from the osteon and interstitial bone interface, the length of the microcrack. Some of the results are in accordance with the finite element method given in [13], additional predicted numerical results are obtained, which need to be confirmed via experiment or computer simulation. Further study may provide multiple osteons model with multiple microcracks, which seems a reasonable approximation of Haversian cortical bone.

Acknowledgement

Supported by NNSF(Grant No.10962008, No.51061015, No.11261045 and No.41204041) of P.R. of China, Natural Science Foundation of Ningxia(Grant No.NZ1001) and Natural Science Foundation of Ningxia University(Grant No.ZR1105).

References

1. P. Fratzl, H. S. Gupta, E. P. Paschalis, P. Roschger. Structure and mechanical quality of the collagen-mineral nano-composite in bone. Journal of materials chemistry,2004,14,2115-2123.
2. P. Fratzl, R. Weinkamer. Natures hierarchical materials. Progress in Materials Science,2007,52(8),1263-1334.
3. H.A. Hogan. Micromechanics modeling of Haversian cortical bone properties. Journal of Biomechanics,1992,25(5),549-556.
4. X.E. Guo, L.C. liang, S.A. Goldstein. Micromechanics of Osteonal Cortical Bone Fracture. Journal of Biomechanical Engineerring,1998,120(1),112-117.
5. T. Hoc, L. Henry, M. Verdier, D. Aubry, L. Sedel, A. Meunier. Effect of microstructure on the mechanical properties of Haversian cortical bone. Bone,2006,38,466-474.
6. M.A. Meyers, P.Y. Chen, A.Y.M. Lin, Y. Seki. Biological materials: structure and mechanical properties. Progress in Materials Science. 2008, 53(1), 1-206.
7. J.S. Temenoff, A.G. Mikos. Biomaterials: the intersection of biology and materials science. Pearson/Prentice Hall, Upper Saddle River, NJ, 2008.
8. P. Zioupos. Accumulation of in-vivo fatigue microdamage and its relation to biomechanical properties in ageing human cortical bone. Journal of Microscopy,2001, 201(2), 270-278.
9. V.C. Mow, R. Huiskes. Basic orthopaedic biomechanics and mechano-biology. Lippiincott Williams Wilkins, Philadelphia, PA, 2005.
10. R.K. Nalla, J.J. Kruzic, J.H. Kinney, R.O. Ritchie. Effect of aging on the toughness of human cortical bone: evaluation by R-curves. Bone,2004, 35(6), 1240-1246.

11. F.J. O'Brien, D. Taylor, T.C. Lee. The effect of bone microstructure on the initiation and growth of microcracks. Journal of Orthopaedic Research,2005, 23(2), 475-480.
12. S. Mohsin, F.J. O'Brien, T.C. Lee. Osteonal crack barriers in ovine compact bone. Journal of Anatomy,2006,208(1),81-89.
13. A.R. Najafi, A.R. Arshi, M.R. Eslami, S. Fariborz, M.H. Moeinzadeh. Micromechanics fracture in osteonal cortical bone: A study of the interactions between microcrack propagation,microstructure and the material properties. Journal of Biomechanics,2007,(40),2788-2795.
14. A.R. Najafi, A.R. Arshi, M.R. Eslami, S. Fariborz, M. Moeinzadeh. Haversian cortical bone model with many radial microcracks: An elastic analytic solution. Medical Engineering and Physics,2007,29(6),708-717.
15. A.R. Najafi, A.R. Arshi, K.P. Saffar, M.R. Eslami, S. Fariborz, M.H. Moeinzadeh. A fiber-ceramic matrix composite material model for osteonal cortical bone fracture micromechanics: Soultion of arbitrary microcracks interaction. Journal of the Mechanical Behavior of Biomedical materials,2009,2(3),217-223.
16. F.J. O'Brien, D. Taylor, T.C. Lee. Bone as a composite material: The role of osteons as barriers to crack growth in compact bone. International Journal of Fatigue,2007,29,1051-1056.
17. Z.M. Xiao, B.J. Chen. Stress intensity factor for a Griffith crack interacting with a coated inclusion. International Journal of Fracture,2001,108,193-205.
18. N.I. Muskhelishvili. Some basic problems of the mathematical theory of elasticity. P. NoordhoffLtd, Groningen, Holland,1953.
19. Z.M. Xiao, B.J. Chen. On the interaction between an edge dislocation and a coated inclusion. International Journal of Solids and Structures,2001, 38, 2533-2548.
20. D.A. Hills, P.A. Kelly, D.N. Dai, A.M. Korsunsky. Solution of Crack Problems: the Distributed Dislocation Technique. Kluwer, Dordrecht, 1996.
21. X. Li, Z.X. Li. Effect of a periodic elastic gasket on periodic cracks. Engeering Fracure Mechanics,1993,46(1),127-131.
22. M.B. Schaffler, K.Choi, C. Milgrom. Aging and matrix microdamage accumulation in human compact bone. Bone, 1995,17(6),521-525. ticae Applicatae Sinica, 12(3),284-288.

SPECTRA OF UNITARY INTEGRAL OPERATORS ON $L^2(\mathbb{R})$ WITH KERNELS $k(xy)$

DAOWEI MA

Department of Mathematics, Wichita State University, Wichita, KS 67260-0033, USA
dma@math.wichita.edu

GOONG CHEN

Department of Mathematics, Texas A&M University, College Station, TX 77843, USA, and Science Program, Texas A&M University at Qatar, Education City, Doha, Qatar
gchen@math.tamu.edu

Unitary integral transforms play an important role in mathematical physics. A primary example is the Fourier transform whose kernel is of the form $k(x, y) = k(xy)$, *i.e.*, of the product type. Here we consider the determination of spectrum for such unitary operators as the issue is important in the solvability of the corresponding inhomogeneous Fredholm integral equation of the second kind. A Main Theorem is proven that characterizes the spectral set. Properties of eigenfunctions and eigenspace dimensions are further derived as consequences of the Main Theorem. Concrete examples are also offered as applications.

2010 *Mathematics Subject Classification.* Primary: 42A38
Keywords and phrases: Unitary transformation, convolution operator, Spectrum.

1. Introduction

The authors are honored to dedicate this article to their teacher and friend, Professor Lu Jianke. The first named author is Professor Lu's former student, and both authors have been Professor Lu's friends for decades. Professor Lu has devoted more than sixty years to the research and teaching of mathematics in China. The authors have immense respect for Professor Lu, and great admiration for his work and his enthusiasm and finesse in the study of linear elasticity,[10] integral equations,[11] complex variables, boundary value problems,[8,9] mathematical physics, topology,[6,7] and other areas, just to mention a few. His leadership has produced generations of researchers and teachers at Wuhan University and in many parts of China and the rest of the world. His gracious personality and generosity for helping others have won the authors' hearts as his students and friends.

Integral transforms and the associated integral equations are major lifetime love and interests of Professor Lu and are, aptly, selected as the focal topic of this volume. In this article, we build on our recent results in the study of integral transforms on the real line.[3] Among all integral transforms on the Euclidean space, unitary operators constitute an important class as they often come from the Hamiltonian mechanics of mathematical physics, with a pleasant isometry property. A premiere example of a unitary transform is the powerful *Fourier transform.* Looking at the Fourier transform,[2] we notice that its kernel is of the form $k(x,y) = k(xy) = e^{-ixy}/\sqrt{2\pi}$. (In higher dimensions, it is of the form $k(\mathbf{x}\cdot\mathbf{y})$, where $\mathbf{x}$ and $\mathbf{y}$ are vectors in $\mathbb{R}^n$.) What we have here, for the Fourier transform on $\mathbb{R}$, is

$$\tilde{f}(x) = \mathcal{F}(f)(x) = \int_{\mathbb{R}} k(xy)f(y)\,dy, \quad f(x) = \overline{\mathcal{F}}(\tilde{f})(x) = \int_{\mathbb{R}} \overline{k}(xy)\tilde{f}(y)\,dy,$$

where $k(x) = e^{-ix}/\sqrt{2\pi}$. Thus, $\overline{\mathcal{F}}\mathcal{F} = I$ (= the identity operator), and $\mathcal{F}$ is unitary on $L^2(\mathbb{R})$. The authors asked in [3] the question how to characterize kernels $k(xy)$ of all integral operators K:

$$K:\ L^2(\mathbb{R}) \to L^2(\mathbb{R}),\ \ (Kf)(x) = \int_{\mathbb{R}} k(xy)f(y)\,dy, \tag{1}$$

such that $\overline{K}K = I$. A set of *characterizing conditions* has been obtained in [3, Theorem 2.3] and from there, one can determine an associated abelian group structure for k ([3, Theorem 4.3]). A number of concrete examples, twelve in all, have also been computed [3, Section 3]. Interestingly, the examples also include

(i) "discrete transforms", *i.e.*, those whose kernels k are linear combinations of delta functions; and

(ii) "hybrid" transforms, whose k are a combination of delta functions and point-wise-defined functions.

Integral equations of the type

$$(Kf)(x) = \int_{\mathbb{R}} k(xy)f(y)\,dy = g(x),\ \ g \in L^2(\mathbb{R}), \tag{2}$$

are called an inhomogeneous Fredholm integral equation of the first kind. (Here, (2) is actually a special case because $k(x,y)$ is limited to be $k(xy)$.) Equation (2) is always uniquely solvable here because $f = \overline{K}g$ by the unitarity of K. On the other hand, if we consider an inhomogeneous Fredholm integral equation of the second kind

$$-\alpha f(x) + (Kf)(x) = g(x), \quad g \in L^2(\mathbb{R}), \tag{3}$$

then there is no guarantee that (3) is always solvable, as $(K - \alpha I)^{-1}$ may not exist. If a number $\alpha \in \mathbb{C}$ is such that $(K - \alpha I))^{-1}$ does not exist as a bounded linear operator on $L^2(\mathbb{R})$, we say that $\alpha \in \sigma(K)$, where $\sigma(K)$ is the *spectrum* of the operator K.[16] For any linear operator K, the spectral set $\sigma(K)$ along with the associated eigenfunctions contain some of the most important and useful information concerning the operator K.

In Section 2, we first introduce some prerequisite material from [3]. Then we present the Main Theorem, which provides a necessary and sufficient condition for determining $\sigma(K)$.

In Section 3, we apply the Main Theorem to establish several more theorems about the dimensionality of eigenspaces and the even/odd-ness of eigenfunctions. Then, these theorems are further applied to concrete examples studied in [3].

Section 4 contains concluding remarks.

2. The Main Theorem

Consider a function (or a distribution in a certain class[4]) k and associate to it an operator K_k by

$$K_k f(x) = \int_{\mathbb{R}} k(xy) f(y)\, dy.$$

Its complex-conjugate operator is then

$$\overline{K}_k f(x) = K_{\overline{k}} f(x) = \int_{\mathbb{R}} \overline{k}(xy) f(y)\, dy. \tag{4}$$

Let $\mathscr{K}$ be the family of such k such that $\overline{K}_k K_k = I$ on $L^2(\mathbb{R})$. Let

$$\mathscr{L} = \{K_k : k \in \mathscr{K}\}$$

be the corresponding family of unitary operators on $L^2(\mathbb{R})$.

For the amenability of the problem to subsequent analysis, at this point we perform a natural *splitting* of the kernel k:

$$k(x) = p(x)\chi_{(0,\infty)}(x) + q(-x)\chi_{(-\infty,0)}(x), \tag{5}$$

where p, q are defined on $(0, \infty)$, and χ_S is the characteristic function of a set $S \subset \mathbb{R}$. Thus,

$$p(x) = k(x), \quad q(x) = k(-x), \quad x > 0.$$

The value of k at 0 has no effect on K_k, as an operator on $L^2(\mathbb{R})$.

Let

$$Tf(x) = f(e^x)e^{x/2}. \tag{6}$$

Note that Tp, Tq are defined on $\mathbb{R}$. We call the couple

$$(u, v) = (\sqrt{2\pi}\mathcal{F}Tp, \sqrt{2\pi}\mathcal{F}Tq)$$

the *symbol* of k and that of K_k. By this definition, we have

$$\begin{aligned} u(t) &= \int_{-\infty}^{\infty} (Tp)(x)e^{-itx}\, dx = \int_0^{\infty} k(x)x^{-1/2-it}\, dx, \\ v(t) &= \int_{-\infty}^{\infty} (Tq)(x)e^{-itx}\, dx = \int_0^{\infty} k(-x)x^{-1/2-it}\, dx. \end{aligned} \tag{7}$$

We have the following characterizing conditions for $\mathscr{K}$ (see [3, Corollary 2.4]).

Theorem 2.1. *A function $k \in \mathscr{K}$ if and only if its symbol functions u, v are measurable and satisfy*

$$\overline{u}u + \overline{v}v = 1 \ \ \textit{and} \ \ \overline{u}v + \overline{v}u = 0. \tag{8}$$

□

Let $\mathscr{B} = B(L^2(\mathbb{R}))$ denote the set of bounded linear operators on $L^2(\mathbb{R})$. Then $\mathscr{L} \subset \mathscr{B}$.

Definition 2.2. For $K \in \mathscr{B}$, the spectrum of K is defined to be

$$\sigma(K) = \{\alpha \in \mathbb{C} : (K - \alpha I) \text{ is not invertible}\}.$$ □

Note that if $K - \alpha I$ is invertible, then its inverse is also bounded, by the open mapping theorem.[5] We are concerned with determining the spectra of the operators in $\mathscr{L}$. For general spectral theory, see, *e.g.*, [14,16].

The following proposition is an immediate consequence of the definition.

Proposition 2.3. *If $K \in \mathscr{B}$, then $\sigma(K^n) = \sigma(K)^n$ for each positive integer n. If $K \in \mathscr{B}$ is invertible, then $\sigma(K^n) = \sigma(K)^n$ for each integer n.* □

The following two propositions are well known; we include the proofs here just for the sake of later convenient referencing.

Proposition 2.4. *Let $K \in \mathscr{B}$. Then $\sigma(K) \subset \{\alpha : |\alpha| \leq \|K\|\}$.*

Proof. If $\alpha > \|K\|$, then

$$(K - \alpha I)^{-1} = -\sum_{n=0}^{\infty} \alpha^{-n-1} K^n$$

belongs to $\mathscr{B}$, hence $\alpha \notin \sigma(K)$. $\square$

Proposition 2.5. *Let K be a unitary operator in $\mathscr{B}$, i.e., satisfying $\overline{K}K = I$. Then $\sigma(K) \subset \{\alpha : |\alpha| = 1\}$.*

Proof. By the previous proposition, $\sigma(K) \subset \{\alpha : |\alpha| \leq 1\}$ and $\sigma(K)^{-1} = \sigma(K^{-1}) \subset \{\alpha : |\alpha| \leq 1\}$. Hence $\sigma(K) \subset \{\alpha : |\alpha| = 1\}$. $\square$

For a complex-valued measurable function $\omega(t)$ on $\mathbb{R}$, its *essential range* is defined by

$$R_\omega = \{y \in \mathbb{C} : m(\omega^{-1}(U)) > 0 \text{ for each neighborhood } U \text{ of } y\},$$

where m is the Lebesgue measure. The *positive-measure range* of ω is defined by

$$\Lambda_\omega = \{y \in \mathbb{C} : m(\omega^{-1}(y)) > 0\}.$$

A matrix whose entries are measurable functions on $\mathbb{R}$ is said to be *essentially bounded* if all its entries belong to $L^\infty(\mathbb{R})$.

Let $f \in L^2(\mathbb{R})$. Then

$$f(x) = f_+(x)\chi_{(0,\infty)}(x) + f_-(-x)\chi_{(-\infty,0)}(x),$$

where $f_\pm \in L^2(0,\infty)$. Define $\Gamma : L^2(\mathbb{R}) \to L^2(\mathbb{R})^2$ by

$$\Gamma f = \begin{pmatrix} \mathcal{F}Tf_+ \\ \mathcal{F}Tf_- \end{pmatrix}. \tag{9}$$

Then Γ is an isometry.

Theorem 2.6. ***(Main Theorem)*** *Let $K \in \mathscr{L}$ with symbol (u, v), and let $\alpha \in \mathbb{C}$. Put*

$$W(t) = \begin{pmatrix} u(t) & v(t) \\ v(t) & u(t) \end{pmatrix}, \quad Z(t) = \begin{pmatrix} 0 & W(t) \\ W(-t) & 0 \end{pmatrix}, \quad Q(t) = W(t)W(-t),$$

and

$$w(t) = (u(t) + v(t))(u(-t) + v(-t)), \quad z(t) = (u(t) - v(t))(u(-t) - v(-t)). \tag{10}$$

Then the following are equivalent:

(i) $\alpha \in \sigma(K)$.
(ii) The matrix $(Z(t) - \alpha I)^{-1}$ *is not essentially bounded.*
(iii) The matrix $(Q(t) - \alpha^2 I)^{-1}$ *is not essentially bounded.*
(iv) $\alpha^2 \in \sigma(K^2)$.
(v) $(-\alpha) \in \sigma(K)$.
(vi) The function $\det(Z(t) - \alpha I)^{-1}$ *does not belong to* $L^\infty(\mathbb{R})$.
(vii) The function $\det(Q(t) - \alpha^2 I)^{-1}$ *does not belong to* $L^\infty(\mathbb{R})$.
(viii) $\alpha^2 \in R_w \cup R_z$.

Proof. (i) $\Leftrightarrow$ (ii): Let $K = K_k$, $k \in \mathscr{K}$. Let $f \in L^2(\mathbb{R})$, and $g := Kf$, $g_\alpha := (K - \alpha I)f$ and $h := K^2 f$, $h_\alpha := (K^2 - \alpha^2 I)f$. As usual, let $f_+, f_- : (0, \infty) \to \mathbb{R}$ be defined by

$$f_+(x) = f(x), \quad f_-(x) = f(-x), \quad x > 0,$$

and let

$$F_\pm = \mathcal{F}Tf_\pm. \tag{11}$$

Similarly, $G_\pm = \mathcal{F}Tg_\pm$, $H_\pm = \mathcal{F}Th_\pm$, etc. Then

$$\begin{aligned} g_+(x) &= \int_0^\infty k_+(xy)f_+(y)\,dy + \int_0^\infty k_-(xy)f_-(y)\,dy, \\ Tg_+(x) &= \int_{-\infty}^\infty (Tk_+)(x+y)(Tf_+)(y)\,dy + \int_{-\infty}^\infty (Tk_-)(x+y)(Tf_-)(y)\,dy, \\ G_+(t) &= u(t)F_+(-t) + v(t)F_-(-t). \end{aligned}$$

Similarly,

$$G_-(t) = v(t)F_+(-t) + u(t)F_-(-t).$$

Combining the preceding two equations, we obtain

$$\begin{pmatrix} G_+(t) \\ G_-(t) \end{pmatrix} = W(t) \begin{pmatrix} F_+(-t) \\ F_-(-t) \end{pmatrix}.$$

Replacing t by $-t$ yields

$$\begin{pmatrix} G_+(-t) \\ G_-(-t) \end{pmatrix} = W(-t) \begin{pmatrix} F_+(t) \\ F_-(t) \end{pmatrix}.$$

We then combine the last two equations to obtain

$$\mathbf{G}(t) = Z(t)\mathbf{F}(t),$$

where

$$\mathbf{G}(t) = \begin{pmatrix} G_+(t) \\ G_-(t) \\ G_+(-t) \\ G_-(-t) \end{pmatrix}, \quad \mathbf{F}(t) = \begin{pmatrix} F_+(t) \\ F_-(t) \\ F_+(-t) \\ F_-(-t) \end{pmatrix}. \tag{12}$$

Note that $\mathbf{G}_\alpha(t) = \mathbf{G}(t) - \alpha\mathbf{F}(t)$. Thus,

$$\mathbf{G}_\alpha(t) = (Z(t) - \alpha I)\mathbf{F}(t). \tag{13}$$

Since the map

$$\Gamma : \ f \mapsto \begin{pmatrix} F_+(t) \\ F_-(t) \end{pmatrix}$$

is an isometry, we see that $K - \alpha I$ is invertible if and only if the matrix $(Z(t) - \alpha I)^{-1}$ is essentially bounded. Therefore, (i) and (ii) are equivalent.

(iii) $\Leftrightarrow$ (iv): By the same reasoning as in the last case, we have $\mathbf{G}(t) = Z(t)\mathbf{F}(t)$ and $\mathbf{H}(t) = Z(t)\mathbf{G}(t)$. Hence, $\mathbf{H}(t) = Z^2(t)\mathbf{F}(t)$, or

$$\mathbf{H}(t) = \begin{pmatrix} Q(t) & 0 \\ 0 & Q(-t) \end{pmatrix} \mathbf{F}(t).$$

It follows that

$$\begin{pmatrix} H_+(t) \\ H_-(t) \end{pmatrix} = Q(t) \begin{pmatrix} F_+(t) \\ F_-(t) \end{pmatrix},$$

or

$$\Gamma K^2 f = Q(t)\Gamma f, \tag{14}$$

and hence

$$\Gamma(K^2 - \alpha^2 I)f = (Q(t) - \alpha^2 I)\Gamma f.$$

It is now clear that (iii) and (iv) are equivalent.

(ii) $\Leftrightarrow$ (vi): We have $(Z(t) - \alpha I)^{-1} = \det(Z(t) - \alpha I)^{-1} \cdot M_\alpha(t)$, where $M_\alpha(t)$ is the adjoint matrix of $(Z(t) - \alpha I)$ and hence it is essentially bounded. It follows that (ii) and (vi) are equivalent.

(iii) $\Leftrightarrow$ (vii): The reasoning is the same as in the previous case.

(vi) $\Leftrightarrow$ (vii): It follows from the identity

$$(Z(t) - \alpha I) \begin{pmatrix} I & 0 \\ \alpha^{-1}W(-t) & I \end{pmatrix} = \begin{pmatrix} \alpha^{-1}(Q(t) - \alpha^2 I) & W(t) \\ 0 & -\alpha I \end{pmatrix}, \tag{15}$$

that

$$\det(Z(t) - \alpha I) = \det(Q(t) - \alpha^2 I). \tag{16}$$

Therefore (vi) and (vii) are equivalent.

(iv) $\Leftrightarrow$ (v): By what has been proved, (i) and (iv) are equivalent. It follows that (iv) and (v) are equivalent.

(vii) $\Leftrightarrow$ (viii): It is straightforward to verify that

$$\det(Q(t) - \alpha^2 I) = (w(t) - \alpha^2)(z(t) - \alpha^2), \tag{17}$$

from which the equivalence of (vii) and (viii) follows. □

We call (w, z) defined by (10) the *compound symbol* of k and that of K_k.

Corollary 2.7. *Let $K \in \mathscr{L}$. Then $-\sigma(K) = \sigma(K)$.* □

Corollary 2.8. *Let $K \in \mathscr{L}$ with compound symbol (w, z). Then $\sigma(K)^2 = \sigma(K^2) = R_w \cup R_z$.* □

Corollary 2.9. *Let $K \in \mathscr{L}$. Then K is real if and only if $\sigma(K) = \{\pm 1\}$.*

Proof. Let $K \in \mathscr{L}$ be a real operator. Then $K^2 = I$, hence $\sigma(K)^2 = \{1\}$ by Proposition 2.3. Take an arbitrary nonzero $f \in L^2(\mathbb{R})$. Then $(K+I)(K-I)f = 0$. If $(K-I)f = 0$, then $1 \in \sigma(K)$. If $(K-I)f \neq 0$, then $(-1) \in \sigma(K)$. Either way, $\sigma(K) = \{\pm 1\}$, by the main theorem.

Conversely, suppose that $\sigma(K) = \{\pm 1\}$. Then $\sigma(K^2) = \{1\} = R_w \cup R_z$, by Corollary 2.8. It follows that $w(t) = z(t) = 1$, and hence $Q(t) = I$. By (14), $K^2 = I$. Since $\overline{K}K = I$, we see that $K = \overline{K}$. □

3. Eigenfunctions and Integral Equations

Let $K \in B$ and $\alpha \in \sigma(K)$. Then $K - \alpha I$ is not invertible, hence it is either non-injective or non-surjective. Suppose that $K - \alpha I$ is non-injective. Then there is a nonzero function $f \in L^2(\mathbb{R})$ such that $(K - \alpha I)f = 0$. The function f is called an *eigenfunction* of K associated with the eigenvalue α. The kernel of $(K - \alpha I)$,

$$N(K - \alpha I) = \{f \in L^2(\mathbb{R}) : (K - \alpha I)f = 0\}$$

is called the *eigenspace* of K associated with α.

The following proposition is likely known in the literature; we include a short proof here for the convenience of the reader.

Proposition 3.1. *Let K be a unitary operator in $\mathscr{B}$ and $\alpha \in \mathbb{C}$. Then $(K - \alpha I)$ is invertible if and only if it is surjective.*

Proof. Assume the contrary. Then there is a $K - \alpha I$ that is surjective and non-injective. Hence there is a nonzero $f \in L^2(\mathbb{R})$ with $(K - \alpha I)f = 0$ and an $h \in L^2(\mathbb{R})$ with $(K - \alpha I)h = \overline{f}$. Then

$$0 \neq (\overline{f}, \overline{f}) = ((K - \alpha I)h, \overline{f}) = (h, (\overline{K} - \overline{\alpha} I)\overline{f}) = (h, 0) = 0.$$

This is a contradiction. $\square$

Note that Proposition 3.1 does not hold for a general operator in $\mathscr{B}$. For instance, the "back-shift" operator ι defined by

$$\iota(e_j) = \begin{cases} 0, & \text{if } j = 1, \\ e_{j-1}, & \text{if } j > 1, \end{cases}$$

where $\{e_1, e_2, \dots\}$ is an orthonormal basis of $L^2(\mathbb{R})$, is surjective, but non-injective.

Theorem 3.2. *Let $K \in \mathscr{L}$ with compound symbol (w, z), and $\alpha \in \sigma(K)$. Let*

$$A = A_\alpha = \{t \in \mathbb{R} : w(t) = \alpha^2\}, \quad B = B_\alpha = \{t \in \mathbb{R} : z(t) = \alpha^2\}.$$

Then $N(K - \alpha I)$ is nontrivial if and only if $m(A \cup B) > 0$. Furthermore, the following hold:

(i) If $N(K - \alpha I)$ is nontrivial, then it is infinite dimensional.

(ii) $N(K - \alpha I)$ is nontrivial if and only if $N(K + \alpha I)$ is nontrivial.

(iii) If $m(A) > 0$ and $m(B) = 0$, then every function in $N(K \pm \alpha I)$ is even.

(iv) If $m(A) = 0$ and $m(B) > 0$, then every function in $N(K \pm \alpha I)$ is odd.

Proof. Note that the sets A, B are symmetric about the origin, since w, z are even functions.

Suppose that $m(A \cup B) = 0$. We shall prove that $N(K - \alpha I) = 0$. Let $f \in N(K - \alpha I)$. By (13), we have

$$(Z(t) - \alpha I)\mathbf{F}(t) = 0, \tag{18}$$

where $\mathbf{F}(t)$ is defined by (11) and (12). By the assumption $m(A \cup B) = 0$ and (16), (17), the matrix $(Z(t) - \alpha I)$ is nonsingular almost everywhere on $\mathbb{R}$. It follows that $\mathbf{F} = 0$, and hence $f = 0$.

Now we assume that $m(A) > 0$. Let $A_+ = A \cap \{t : t > 0\}$ and $A_- = A \cap \{t : t < 0\}$. Choose a measurable function $p(t)$ on A_+ such that

$$p(t)^2 = \alpha^{-1}(u(t) + v(t)). \tag{19}$$

For $t \in A_-$, let $p(t) = 1/p(-t)$. The function $p(t)$ is now defined on A and, since $\alpha^2 = w(t)$ on A, equality (19) holds for $t \in A$. It follows that

$$-\alpha p(t) + (u(t) + v(t))p(-t) = 0, \quad t \in A.$$

The last equality is equivalent to

$$(Z(t)) - \alpha I) \begin{pmatrix} p(t) \\ p(t) \\ p(-t) \\ p(-t) \end{pmatrix} = 0, \quad t \in A. \tag{20}$$

Thus, for each function $q(t) \in L^2_e(A)$, the set of *even*, complex-valued, square-integrable, measurable functions on A, we have

$$(Z(t)) - \alpha I) \begin{pmatrix} p(t)q(t) \\ p(t)q(t) \\ p(-t)q(-t) \\ p(-t)q(-t) \end{pmatrix} = 0, \quad t \in A. \tag{21}$$

By (8), $|u(t) + v(t)| = 1$, and hence $|p(t)| = 1$. Recall that Γ is the operator defined by (6) and (9). Define $\tau_A : L^2_e(A) \to L^2(\mathbb{R})$ by

$$\tau_A(q) = \Gamma^{-1}(\chi_A pq \begin{pmatrix} 1 \\ 1 \end{pmatrix}).$$

Then $\|\tau_A(q)\| = \sqrt{2}\|q\|$, hence τ_A is injective. By (21), $\tau_A(L^2_e(A)) \subset N(K - \alpha I)$. Therefore, $N(K - \alpha I)$ is infinite dimensional. Note that each function in $\tau_A(L^2_e(A))$ is even.

Suppose that $m(A) > 0$ and $m(B) = 0$. Then $\text{rank}(Z(t) - \alpha I) = 4$ *a.e.* (almost everywhere) on $\mathbb{R} \setminus A$, and $\text{rank}(Z(t) - \alpha^2 I) < 4$ *a.e.* on A. We now show that $\text{rank}(Z(t) - \alpha^2 I) = 3$ *a.e.* on A. We can verify that

$$2Q(t) = \begin{pmatrix} w(t) + z(t) & w(t) - z(t) \\ w(t) - z(t) & w(t) + z(t) \end{pmatrix}.$$

Since $w(t) = \alpha^2 \neq z(t)$, we see that the $(1, 2)$ entry of $Q(t)$ is non-zero *a.e.* on A. Hence,

$$Q(t) - \alpha^2 I \neq 0 \;\; a.e. \text{ on } A. \tag{22}$$

It follows from (15) and (22) that

$$\text{rank}(Z(t) - \alpha I) = 3 \;\; a.e. \text{ on } A. \tag{23}$$

Now let $f \in N(K - \alpha I)$ and

$$\Gamma f = \begin{pmatrix} F_+ \\ F_- \end{pmatrix}.$$

Then

$$(Z(t)) - \alpha I) \begin{pmatrix} F_+(t) \\ F_-(t) \\ F_+(-t) \\ F_-(-t) \end{pmatrix} = 0, \quad t \in A. \tag{24}$$

Since $Z(t) - \alpha I$ is nonsingular *a.e.* on $\mathbb{R} \setminus A$, equality (24) implies that $\Gamma f(t) = 0$ *a.e.* on $\mathbb{R} \setminus A$. By (23), the two vectors in (20) and (24) are proportional. It follows that there exists a $q \in L^2_e(A)$ such that

$$\begin{pmatrix} F_+(t) \\ F_-(t) \\ F_+(-t) \\ F_-(-t) \end{pmatrix} = \begin{pmatrix} p(t)q(t) \\ p(t)q(t) \\ p(-t)q(-t) \\ p(-t)q(-t) \end{pmatrix}.$$

Thus, $f = \tau_A(q)$. Therefore $\tau_A(L^2_e(A)) = N(K - \alpha I)$, and every function in $N(K \pm \alpha I)$ is even.

The case when $m(B) > 0$ is similar. We define a function $r(t)$ on B such that

$$r(t)r(-t) = 1, \;\; r(t)^2 = \alpha^{-1}(u(t) - v(t)).$$

It follows that

$$-\alpha r(t) + (u(t) - v(t))r(-t) = 0, \quad t \in B,$$

and

$$(Z(t)) - \alpha I) \begin{pmatrix} r(t) \\ -r(t) \\ r(-t) \\ -r(-t) \end{pmatrix} = 0, \quad t \in B.$$

Then the map $\tau_B : L^2_e(B) \to L^2(\mathbb{R})$ defined by

$$\tau_B(s) = \Gamma^{-1}(\chi_B r s \begin{pmatrix} 1 \\ -1 \end{pmatrix})$$

is injective and satisfies $\tau_B(L^2_e(B)) \subset N(K - \alpha I)$. Therefore, $N(K - \alpha I)$ is infinite dimensional. Note that each function in $\tau_B(L^2_e(B))$ is odd.

If $m(A) = 0$ and $m(B) > 0$, then $\tau_B(L^2_e(B)) = N(K - \alpha I)$. Hence, every function in $N(K - \alpha I)$ is odd. □

Corollary 3.3. *Let $K \in \mathscr{L}$. If $\sigma(K)$ is finite, then every $\alpha \in \sigma(K)$ is an eigenvalue with an infinite dimensional eigenspace.* □

Corollary 3.4. *Let $K \in \mathscr{L}$ with compound symbol (w, z). If $\Lambda_w \cap \Lambda_z = \emptyset$, then for each eigenvalue of K, the eigenspaces $N(K \pm \alpha I)$ consist entirely of either even functions or odd functions.* □

Proposition 3.5. *Let $K \subset \mathscr{L}$ be a real operator. Then $\sigma(K) = \{\pm 1\}$, $L^2(\mathbb{R}) = N(K-I) \oplus N(K+I)$, and $N(K \pm I)$ are both infinite dimensional.*

Proof. By Corollary 2.9, $\sigma(K) = \{\pm 1\}$. It is clear that

$$L^2(\mathbb{R}) = (I+K)(L^2(\mathbb{R})) \oplus (I-K)(L^2(\mathbb{R})).$$

It is also clear that $(I+K)(L^2(\mathbb{R})) \subset N(K-I)$. If $f \in N(K-I)$, then

$$f = \frac{f+Kf}{2} + \frac{f-Kf}{2} = \frac{f+Kf}{2} \in (I+K)(L^2(\mathbb{R})).$$

Hence $(I+K)(L^2(\mathbb{R})) = N(K-I)$. Similarly, $(I-K)(L^2(\mathbb{R})) = N(K+I)$. Therefore $L^2(\mathbb{R}) = N(K-I) \oplus N(K+I)$. By Theorem 3.2, $N(K \pm I)$ are infinite dimensional. □

An example of a real $k \in \mathscr{K}$ can be found in [3, Example 3.8]:

$$k(x) = \delta(x-1) - \frac{x \sin(b \log|x|)}{\pi |x|^{3/2} \log|x|}.$$

By applying the preceding theorems, we obtain the following.

Theorem 3.6. *If $\alpha \neq \pm 1$, the Fredholm equation of second kind*

$$-\alpha f(x) + |x|^{-1} f(1/x) - \int_{\mathbb{R}} \frac{xy \sin(b \log|xy|)}{\pi |xy|^{3/2} \log|xy|} f(y)\, dy = g(x)$$

has a unique solution, which is given by

$$(1-\alpha^2) f(x) = \alpha g(x) + |x|^{-1} g(1/x) - \int_{\mathbb{R}} \frac{xy \sin(b \log|xy|)}{\pi |xy|^{3/2} \log|xy|} g(y)\, dy.$$

The solution spaces of the homogeneous Fredholm equations

$$\pm f(x) + |x|^{-1} f(1/x) - \int_{\mathbb{R}} \frac{xy \sin(b \log|xy|)}{\pi |xy|^{3/2} \log|xy|} f(y)\, dy = 0$$

are infinite dimensional. The inhomogeneous Fredholm equations

$$\pm f(x) + |x|^{-1} f(1/x) - \int_{\mathbb{R}} \frac{xy \sin(b \log|xy|)}{\pi |xy|^{3/2} \log|xy|} f(y)\, dy = g(x)$$

are solvable if and only if the right-side function $g(x)$ satisfies the companion homogeneous Fredholm equations

$$\mp g(x) + |x|^{-1} g(1/x) - \int_{\mathbb{R}} \frac{xy \sin(b \log|xy|)}{\pi |xy|^{3/2} \log|xy|} g(y)\, dy = 0,$$

respectively. □

Theorem 3.7. *Let $n \geq 2$ be an integer and let $K \in \mathscr{B}$ be such that $K^n = I$. Then*

$$L^2(\mathbb{R}) = \oplus_{j=1}^n N(K - \beta^j I),$$

where $\beta = e^{2\pi i/n}$.

Proof. Let $P_j = I + \beta^{-j}K + (\beta^{-j}K)^2 + \cdots + (\beta^{-j}K)^{n-1}$ for $j = 1, \ldots, n$. Then $\sum_{j=1}^n P_j = nI$, hence

$$L^2(\mathbb{R}) = P_1(L^2(\mathbb{R})) + \cdots + P_n(L^2(\mathbb{R})).$$

It suffices to show that $N(K - \beta^j I) = P_j(L^2(\mathbb{R}))$. Since $(K - \beta^j I)P_j = 0$, it follows that $N(K - \beta^j I) \supset P_j(L^2(\mathbb{R}))$. Suppose that $f \in N(K - \beta^j I)$. Then $P_\nu f = 0$ for $\nu \neq j$, since $P_\nu = \Pi_{\lambda \neq \nu}(I - \beta^{-\lambda}K)$. It follows that $nf = \sum P_\nu f = P_j f$ and $f \in P_j(L^2(\mathbb{R}))$. Thus $N(K - \beta^j I) = P_j(L^2(\mathbb{R}))$ and the proof is complete. □

An operator $K \in \mathscr{L}$ is said to be *Fourier-like* if $K^2 f(x) = f(-x)$ for each $f \in L^2(\mathbb{R})$. We now determine the spectra and eigenspaces of Fourier-like operators. Some of our work here is inspired by [15], where the spectrum of the Fourier transform is determined, and methods to find eigenfunctions are described.

Let $K \in \mathscr{L}$ be a Fourier-like operator with symbol (u, v). Then (see [3, Theorem 5.2])

$$\overline{u}(-t) = v(t), \quad \overline{v}(-t) = u(t). \tag{25}$$

Equalities (8) and (25) imply that

$$w(t) = 1, \quad z(t) = -1. \tag{26}$$

By Theorem 3.2, Proposition 3.7, and (26), we have the following.

Theorem 3.8. *Let $K \in \mathscr{L}$ be a Fourier-like operator. Set, for $\nu = 1, 2, 3, 4$,*

$$P_\nu = I + i^{-\nu}K + i^{-2\nu}K^2 + i^{-3\nu}K^3.$$

Then the following hold:

(i) $\sigma(K) = \{\pm 1, \pm i\}$.
(ii) $N(K - i^\nu I) = P_\nu(L^2(\mathbb{R}))$, for $\nu = 1, 2, 3, 4$.
(iii) $L^2(\mathbb{R}) = \oplus_{\nu=1}^4 N(K - i^\nu I)$.
(iv) The eigenspaces $N(K - i^\nu I)$ are infinite dimensional.
(v) $N(K \pm I)$ consists of even functions.
(vi) $N(K \pm iI)$ consists of odd functions. □

For Fourier-like kernels, an example is given in [3, Example 3.12]):

$$k(x) = \frac{1-i}{2}\frac{1+\operatorname{sgn}\, x}{2}(-\delta(x-1) + \frac{1-i\operatorname{sgn}(\ln|x|)}{\sqrt{|x|}} \cdot \min(|x|,|x|^{-1}))$$
$$+ \frac{1+i}{2}\frac{1-\operatorname{sgn}\, x}{2}(-\delta(x+1) + \frac{1+i\operatorname{sgn}(\ln|x|)}{\sqrt{|x|}} \cdot \min(|x|,|x|^{-1})).$$

There are plenty of other Fourier-like kernels; but for most of them, finding their explicit forms is intractable. For each positive integer n the following is a Fourier-like kernel:

$$k_n(x) = \frac{1-i}{2\pi\sqrt{2|x|}} \int_{-\infty}^{\infty} e^{it\log|x|} \begin{Bmatrix} \cos(t|t|^{n-1}+\pi/4), & \text{for } x>0 \\ i\sin(t|t|^{n-1}+\pi/4), & \text{for } x<0 \end{Bmatrix} dt.$$

We now turn to an operator $K \in \mathscr{L}$ with full spectrum: $\sigma(K) = \{\alpha : |\alpha| = 1\}$. Let (see [3, Example 3.2])

$$k(x) = \frac{i^{(1-\operatorname{sgn} x)/2}}{2\sqrt{\pi|x|}} \cos\frac{\log^2|x| - \operatorname{sgn}(x)\pi}{4}.$$

For this kernel we have

$$u(t) = \cos(t^2), \quad v(t) = i\sin(t^2),$$

and

$$w(t) = e^{2it^2}, \quad z(t) = e^{-2it^2}.$$

Thus $R_w = R_z = \{\alpha : |\alpha| = 1\}$ and $\Lambda_w = \Lambda_z = \emptyset$. It follows that $\sigma(K_k) = \{\alpha : |\alpha| = 1\}$. We state these results in terms of integral equations.

Theorem 3.9. *If $|\alpha| \neq 1$, then the Fredholm equation of second kind*

$$-\alpha f(x) + \int_{\mathbb{R}} \frac{i^{(1-\operatorname{sgn}(xy))/2}}{2\sqrt{\pi|xy|}} \cos\frac{\log^2|xy| - \operatorname{sgn}(xy)\pi}{4} f(y)\, dy = g(x) \quad (27)$$

has a unique solution. If $|\alpha| = 1$, then the homogeneous Fredholm equation

$$-\alpha f(x) + \int_{\mathbb{R}} \frac{i^{(1-\operatorname{sgn}(xy))/2}}{2\sqrt{\pi|xy|}} \cos\frac{\log^2|xy| - \operatorname{sgn}(xy)\pi}{4} f(y)\, dy = 0$$

does not have nontrivial solutions, and the inhomogeneous Fredholm equation (27) is not always solvable.

Proof. Let $\alpha \in \sigma(K) = \{\alpha : \|\alpha\| = 1\}$. Since $\Lambda_w = \Lambda_z = \emptyset$, we see that $m(A_\alpha) = m(B_\alpha) = 0$. By Theorem 3.2, $N(K - \alpha I) = 0$, hence the corresponding homogeneous Fredholm equation has no nontrivial solutions. By Proposition 3.1, $K - \alpha I$ is non-surjective, hence the non-homogeneous Fredholm equation (27) is not always solvable. □

4. Concluding Remarks

This paper studies the spectra of unitary integral transforms with kernels of the type $k(xy)$. We have obtained interesting properties about the spectra of such operators and their eigenspaces. The analysis and proofs are based mainly on the *compound symbols*. The methodology seems to be quite novel.

Concrete examples given in Section 3 are selected from those given in our prior paper [3], which were obtained by the availability of *exact, explicit* expressions of Fourier transforms of certain functions. Nevertheless, such a catalog of available exact, explicit expressions is quite small from sources such as [1,12,13], for example. We wish to provide a much larger class of concrete examples, but are limited by the above difficulty.

There also exist several unitary integral transforms whose kernels are not exactly of the form $k(xy)$, but "very close to it". The spectral properties of them also appear to satisfy the theorems given in Section 3. So far, they must be studied on a case by case basis. This calls for some further study of generalization and extension in the future.

Acknowledgment

The authors gratefully acknowledge partial financial support by the Texas Norman Hackman Advanced Research Program Grant #010366-0149-2009 from the Texas Higher Education Coordinating Board, and Qatar National Research Fund (QNRF) National Priority Research Program (NPRP) Grant #NPRP4-1162-1-181.

References

1. M. Abramowitz, I.A. Stegun, *Handbook of Mathematical Functions: with Formulas, Graphs, and Mathematical Tables,* Dover Publications, New York, 1972.
2. R.N. Bracewell, *The Fourier Transform and Its Applications,* McGraw-Hill, Boston, 2000.
3. G. Chen, D. Ma, Linear transforms resembling the Fourier transform, preprint.
4. I.M. Gelfand, G.E. Shilov, *Generalized functions. V. 1: Properties and operations,* Translated by E. Saletan, Academic Press, Boston, 1964.
5. E. Hewitt, K. Stromberg, *Real and Abstract Analysis,* Springer-Verlag, New York, 1965.
6. C.K. Lu, Classification of 2-manifolds with singular points. *Bull. Amer. Math. Soc.,* **55** (1949), 1093–1098.
7. C.K. Lu, Classification of 2-manifolds with a singular segment, *Acta Math. Sinica,* **1** (1951), 281–295.

8. J. Lu, On compound boundary problems, *Scientia Sinica,* **14** (1965), 1545–1555.
9. J. Lu, *Boundary value problems for analytic functions,* World Scientific, Singapore, 1993.
10. J. Lu, *Complex variable methods in plane elasticity,* World Scientific, Singapore, 1995.
11. J. Lu, S.-G. Zhong, *Theory of integral equations,* Higher Education Press of China, Beijing, 1995.
12. W. Magnus, F. Oberhettinger, *Formulas and theorems for the special functions of mathematical physics,* Chelsea Pub., New York, 1972.
13. F. Oberhettinger, *Tables of Fourier Transforms and Fourier Transforms of Distributions,* Springer, Berlin, 1990.
14. M. Taylor, Spectral theory, Chapter 8 in *Partial Differential Equations, Vol. 2,* 2nd Ed, Springer-Verlag, New York, 2011.
15. P.P. Vaidyanathan, Eigenfunctions of the Fourier transform, *IETE J. Education,* **49** (2008), 51–58.
16. K. Yosida, *Functional analysis,* Springer-Verlag, New York, 1980.

THE NUMERICAL SIMULATION OF LONG-PERIOD GROUND MOTION ON BASIN EFFECTS *

YIQIONG LI

Institute of Geophysics, China Earthquake Administration
Beijing, 100081, China

XING LI

Ningxia University
Yinchuan, 750021, China

Studies have shown that long-period ground motion can be amplified due to basin structure, the amplification effects generate such as the amplitude increased, the duration lengthened. The basin effects of ground motion are extremely detrimental to engineering structures of long natural vibration period, such as high-rise buildings and bridges. The paper carries out the numerical simulation of long-period ground motion on basin effects with far-field earthquake input, which uses previous developed by the applicant structure-preserving algorithm, symplectic discrete singular convolution differentiator (SDSCD). The preliminary test shows that the algorithm is feasible to simulate long-period ground motion on basin effects. It will lay the foundation on proposing evaluation method to meet the needs of engineering for long-period ground motion on basin effects, to provide more reliable scientific basis for the seismic risk management of major projects.

1. INTRODUCTION

1.1. *Ground motion*

Ground motion is the product of the complex system including earthquake focus, seismic wave propagation in Earth medium and site effects. Ground motion study covers theoretical seismology and engineering seismology, which has not only theoretical significance, but also application value [1]. On the one hand, near-field strong motion records contain the details of source activity; therefore, analyzing near-field strong motion records has been available to research focal process. On the other hand, near-field strong motion generated by the moderate intensity earthquake directly contributes to earthquake disaster. Due to above two reasons, the circles of theoretical seismology and engineering

---

* This work is supported by the National Natural Science Foundation of China (Grant No. 41174047, 41204046), and the Institute of Geophysics, China Earthquake Administration (basic scientific research project, Grant No. DQJB11B10).

seismology pay great attention to ground motion research, which has become a more active research area and has gat great development.

Currently, we can directly obtain ground motion information from strong motion records. However, strong motion records cannot be gained at any time and any places, so it is very useful to evaluate ground motion. The evaluation of ground motion has three kinds of probable approaches [2]: using the correspondence of earthquake intensity and ground motion to convert design earthquake ground motion; using the earthquake attenuation to evaluate ground motion; and using mathematical physics method to simulate ground motion. Here, we focus on the third approach, by which we has achieved great achievements with the development of computing technology. The third approach applies different simulation methods respectively corresponding high-frequency and low-frequency ground motion.

Previous studies based on strong motion records show that low-frequency ground motion is determinate, which can be numerically simulate by certain focus model and velocity structure model; while high-frequency ground motion is random due to the details of focus rupture and the complexity of small scale structure. Therefore, simulating high-frequency ground motion employs the method of random vibration theory, while simulating low-frequency ground motion adopts the numerical method of solving wave equations.

Compared with the random of high-frequency ground motion, the determination of low-frequency ground motion is predominant. We can solve dynamics wave equations to simulate low-frequency wave fields according to seismology. The wave equations simulation consists of seismic wave dynamics, which provide more foundation for the study of seismic wave propagation. The wave equations numerical simulation method has played a significant role in seismic wave fields modeling methods.

1.2. *Numerical modeling methods*

Common numerical modeling methods judging by dealing with the differential of equations have been fairly mature, including finite difference method, finite element method, pseudo-spectral method, boundary element method and spectral element method. These methods are essential to simulate global and regional seismic wave fields.

The finite difference method began in the 1970s. Alterman [3] did pioneering work, who obtained discrete numerical solution of second-order elastic wave equation in layered medium using finite difference scheme. Madariag [4] proposed staggered mesh finite difference to solve velocity-stress elastic wave equations in heterogeneous medium. Zhang and Chen [5] developed the finite difference computation of irregular free surface boundaries model using body-fitted coordinate, which improved the flexibility of finite difference method. Gao [6] simulated strong ground motion in Beijing area with finite difference method, and researched the basin effects. He pointed that the amplification effects of basin and that of surface soil are at the same level. Satol [7] computed strong ground motion in the Sendai basin, Japan, using finite difference method. The finite difference method is based on Taylor series expansion, difference instead of differentiation, differential equations into algebraic equations, which is easy to implement in calculation. However, some difference schemes are attained by virtue of experience according to differentiator, which is lack of mathematical theory guarantee [8].

The finite element method originated in the 1940s, which was first used to solve complex elastic structure analysis in municipal works and aircraft engineering. Seron [9] proposed finite element method to solve elastic wave propagation. Ding [10] simulated long-period ground motion setting virtual earthquake with finite element method in Kunming area. Zhang [11] calculated near-fault basement rock strong ground motion caused by virtual small scale finite source model with modified transmitted artificial boundary using explicit finite element method. Liao [12] summarized the artificial boundary conditions of finite element method. Zhang [13] simulated ground motion in Fuzhou area with finite element method and obtained the distribution characteristics of the surface peak velocity and peak acceleration. Day[14] researched the basin effects of ground motion with finite difference method and finite element method in southern California and pointed that the amplification effects of response spectrum are affected by basin's depth and period. The finite element method has strong mathematical theory to ensure. The finite element method is constructed on Sobolev functional space based on variation principle, which solves the whole wave field by using interpolation function and least potential energy theory. The drawback of the finite element method is complex calculation process and large calculation amount. Moreover, constructing low-order interpolation function leads to low accuracy in finite element method, while high-order interpolation function generates fake information.

The pseudo-spectral method belongs to one of spectral methods. Gazdag [15] and Kosloff [16] first applied pseudo-spectral method to seismic wave problem.

Wang [17] simulated the seismic wave propagation of global model in two-dimensional cylindrical coordinate with pseudo-spectral method. Zhao [18] proposed staggered grid real value Fourier pseudo-spectral method to simulate seismic wave in heterogeneous medium, which improved the accuracy of pseudo-spectral method. Zhao [19] provided the seismic wave propagation and the distribution of ground motion of heterogeneous area, such as alluvial fan and basin. In addition, Zhao's result revealed that seismic wave reflected repeatedly within basin structure after wave arrived at basin from rock structure; therefore, the vibration may repeatedly destruct to the buildings on the ground. The pseudo-spectral method adopts fast Fourier transform to calculate spatial differential and has a swift calculation speed, that its calculation accuracy is higher than classical finite difference method. However, using global variables in computational domain to calculate spatial differential is contrary to local property of differential. It cannot achieve good results in simulation of complex medium.

The boundary element method has natural advantage in simulating irregular interface medium. Several boundary element methods have been developed such as direct integration method, indirect integration method and discrete wave number method, since Wong [20] first used boundary element method to simulate SH wave scattering on irregular terrain. Fujiwara [21] used fast multipole method for solving integral equations of three-dimensional topography and basin problems. Zhou [22] applied local discrete wave number method to simulate ground motion of Zhangbei area. The boundary element method regards analytic basic solution of differential operator as kernel function of boundary integral equation, which combines analytic method and numerical method with high accuracy usually. However, its application demands the existence of basic solution of differential operator, so it is difficultly used in heterogeneous medium.

The spectral element method is a concrete form of weighted residual method, which was introduced to fluid computing by Patera [23], and was first used to simulate seismic wave by Seriani [24]. Komatitsch [25, 26] did a great deal of work on simulating three-dimensional regional and global seismic wave fields, which included simulating ground motion of Los Angeles area. His simulation considered the velocity structure, the density structure and fault and pointed that basin structure affected vitally strong ground motion. The spectral element method has become an important method of seismic simulation with high efficiency, high accuracy and flexibility. The method still has development space, for example, there isn't good scheme of spectral element method when we need adopt triangular grid with complex fault [13].

1.3. *Intention*

The above achievements illustrate that those common numerical methods are effective in the simulation of long-period ground motion. The simulations of Beijing area, Sendai area, Fuzhou area and Los Angeles area show significant basin effects to long-period ground motion, which use several numerical methods based on reliable velocity models and source models. Therefore, the simulation for ground motion of a certain area is time-consuming and is not easy to apply in engineering seismology. The low-frequency components of earthquake contribute to the disaster of engineering structures with long natural vibration period, such as high-rise buildings and bridges. Especially, the basin effects of long-period ground motion are extremely detrimental to engineering structures situated in the basin regions. Nowadays the number of high-rise buildings, urban road overpasses and other lifeline engineering is increasing in many areas, which needs special attention to the seismic safety of such works, particularly in those cities situated in the basin regions. However, the basin effects of ground motion are rarely considered in the building code of China. Under engineering situations, we need first provide the ground motion parameters affected by basin structures, and then adopt detail measures according to the necessity. Therefore, it is necessary to quantitatively study the basin effects of long-period ground motion in characterization models. Through the research, we can directly provide the ground motion parameters for those similar basins, and need not simulate in the every area.

2. SYMPLECTIC METHOD

The above common numerical methods have certain applicable scopes. In the finite difference method and pseudo-spectral method, it is unavoidable to make difference approximation, discretization approximation and finite truncation approximation of series, which destroys the structure of wave equations and solution and makes the numerical solution out of the exact solution. The inherent defect of numerical methods is not prominent in a certain range, so these methods are regarded as reliable and their results are acceptable. When we model seismic wave fields on a large scale, we are concerned about the seismic wave calculation of longer duration. Solving the issue has strong spatial and temporal successions, and obvious accumulative errors in time-domain, thus it need select a high accurate numerical algorithm. The symplectic algorithm, a structure-preserving algorithm, is proposed in such demand. The algorithem is built on Hamilton system. The feature of this algorithm is to keep the structure of equations and conservations unchanged as much as possible during the processes of equation discretization and finite approximation, or make the deviation of numerical solution from exact one within a preset range. It ensures that symplectic algorithm can radically improve the accuracy and credibility of the solution. The calculation experiments show that the symplectic algorithm with the strong stability and long-time tracking capability has important application prospect.

Due to excellent structure-preserving of symplectic algorithm, we adopt non-Tailor expansion explicit symplectic difference scheme in temporal discretization. We use discrete singular convolution differentiator to spatial discretization. Using the differentiator is based on the following considerations: convolution differentiator can be optimized by a localized operator and highlights the local attribute of spatial derivative compared that the pseudo-spectral method needs global variables in computing process; convolution differentiator has higher accuracy than classical finite difference operator and can compensate for the deficiency in modeling complex media using finite difference operator. Thus, convolution differentiator is effective in approximating spatial derivative of wave equations. The SDSCD method combining symplectic scheme and discrete singular convolution differentiator is a good breakthrough of seismic wave simulation method.

2.1. *Symplectic scheme of scalar wave equation*

For a 2D homogeneous isotropic medium, the scalar wave equation can be written as

$$\frac{\partial^2 p}{\partial t^2} = v^2 \left(\frac{\partial^2 p}{\partial x^2} + \frac{\partial^2 p}{\partial z^2} \right) \tag{1}$$

where $p = p(x, z, t)$ is the variable of wave fields, $v = v(x, z, t)$ is the velocity.

Mark the differential operator $v^2\left(\frac{\partial^2}{\partial x^2}+\frac{\partial^2}{\partial z^2}\right)=G$, and $\frac{dp}{dt}=q$, equation (1) is written as

$$\begin{cases} \dfrac{dq}{dt}=G(p) \\ \dfrac{dp}{dt}=q \end{cases} \tag{2}$$

namely

$$\frac{d}{dt}\begin{pmatrix} q \\ p \end{pmatrix}=\begin{pmatrix} 0 & G \\ 1 & 0 \end{pmatrix}\begin{pmatrix} q \\ p \end{pmatrix}=\left(\begin{pmatrix} 0 & G \\ 0 & 0 \end{pmatrix}+\begin{pmatrix} 0 & 0 \\ 1 & 0 \end{pmatrix}\right)\begin{pmatrix} q \\ p \end{pmatrix}=(A+B)\begin{pmatrix} q \\ p \end{pmatrix} \tag{3}$$

From the derivation of symplectic scheme in Appendix A, we get

$$\begin{pmatrix} q^{(n+1)} \\ p^{(n+1)} \end{pmatrix}=\prod_{i=1}^{k}(I+c_i\Delta tA)(I+d_i\Delta tB)\begin{pmatrix} q^{(n)} \\ p^{(n)} \end{pmatrix} \tag{4}$$

where the superscripts of q, p show the index of time axis, Δt is the interval of time, I is the identity matrix with the same order of matrix A and B.

We need determine the coefficients of equation (4), $c_i, d_i, i=1,\cdots,k$; and here we adopt the Iwatsu's coefficients of 3-stage 3-order explicit symplectic scheme[27].

$$c_1=\frac{1}{12}\left(-7+\sqrt{\frac{209}{2}}\right), c_2=\frac{11}{12}, c_3=\frac{1}{12}\left(8-\sqrt{\frac{209}{2}}\right),$$

$$d_1=\frac{2}{9}\left(1+\sqrt{\frac{38}{11}}\right), d_2=\frac{2}{9}\left(1-\sqrt{\frac{38}{11}}\right), d_3=\frac{5}{9}.$$

2.2. *Discrete singular convolution differentiator*

According to distribution theory, the function $f(x)$ in given function space can be defined as the convolution of a kernel function and a test function,

$$f(x)=(T*\eta)(x)=\int_{-\infty}^{+\infty}T(x-t)\eta(t)\,dt \tag{5}$$

where $T(x)$ is a singular kernel function, $\eta(x)$ is a test function in function space.

Singular kernel functions include Hilbert function, Abel function and Delta function. Hilbert function and Abel function often appear in theory of analytic functions. Delta function is used to numerically solve partial differential equation. Here, Delta function is required,

$$T(x)=\delta^{(q)}(x), \quad q=0,1,\cdots, \tag{6}$$

q denotes q th order generalized derivative.

We need replace Delta function with approximate function in numerical calculation. Here, regularized Shannon kernel function is chosen.

$$\delta_{\sigma,\Delta}(x)=\frac{\sin\left(\frac{\pi}{\Delta}x\right)}{\frac{\pi}{\Delta}x}\exp\left(-\frac{x^2}{2\sigma^2}\right) \tag{7}$$

In this formula, Δ is the grid spacing and σ determines the width of the Gaussian envelop. For a given $\sigma \neq 0$, the limit of $\Delta \to 0$ reproduces Delta function.

With regularized Shannon kernel, a function u and its q th order derivative can be approximated by a discrete convolution

$$u^{(q)}(x)\approx\sum_{k=-W}^{W}\delta_{\sigma,\Delta}^{(q)}(x-x_k)u(x_k)=\sum_{k=-W}^{W}\delta_{\sigma,\Delta}^{(q)}(k\Delta)u(x-k\Delta), q=0,1,2,\cdots \tag{8}$$

where $2W+1$ is the computational bandwidth. Generally, a larger W will lead to a higher accuracy. When $q=2$, the second order derivative can be discretized as

$$d_2(k\Delta)=\begin{cases}\delta''_{\sigma,\Delta}(k\Delta), k=\pm1,\cdots,\pm W\\ -\frac{1}{\sigma^2}-\frac{\pi^2}{3\Delta^2}, k=0\end{cases} \tag{9}$$

where $\delta''_{\sigma,\Delta}(k\Delta)=\delta''_{\sigma,\Delta}(-k\Delta), k=1,\cdots,W$.

For practical implementation, the differentiator has to be truncated as a short operator, but doing so could lead to the Gibbs phenomenon. To avoid the Gibbs phenomenon, we use a Hanning window function for truncating the differentiator.

$$w(k) = \left[2\alpha - 1 + 2(1-\alpha)\cos^2\left(\frac{k\pi}{2(W+2)} \right) \right]^{\frac{\beta}{2}} \quad (10)$$

The constants α ($0.5 \leq \alpha \leq 1$) and β allow a family of different windows to be considered. A modified convolutional differentiator can be denoted by

$$\tilde{d}_2(k\Delta) = d_2(k\Delta)w(k), k = 0, \pm 1, \cdots, \pm W . \quad (11)$$

Here, $\alpha = 0.54$ and $\beta = 8$ are chosen.

2.3. *Discrete singular convolution differentiator*

Now that we solve the propagation of seismic wave using symplectic discrete singular convolution differentiator (SDSCD) method, we still consider source input of wave incitation. When we are only concerned with the wave propagation on the surface or soil layer, we can simplify the source model adopting plane-wave incidence or far-field seismic records input. Boore[28] researched scattering from a step change in surface topography with P-SV wave vertical incidence. The result showed that even if incident wavelength is several times of the step height, the amplitude of converted Rayleigh wave at stepped terrain corner can reach higher 0.45 times of the displace of the surface in elastic half-space free field. Harmsen[29] simulated the effect of filled basin with SV wave vertical incidence. The result indicated that Rayleigh wave was observably incited at basin edge low-velocity medium. Vidale[30] used finite difference method to model the process of long-period seismic wave propagation in 1971 California earthquake. He found that the surface waves with wide amplitude and long duration can be regarded as the converted wave from S incident wave.

3. NUMERICAL EXPERIMENTS

Here, we just take a numerical contrast experiment as one example to reveal feasibility of the algorithm for the limit of article length, while we still need do a lot of numerical work to explore the regularity of basin effects.

We consider a layered medium model with basin interface and a contrast model without basin interface, while the model parameters are labeled in Fig.1 and Fig.2. The surface receiver is located at the vertical projection of the centre of basin with ★ marked. The SH waves are incited at the line z=500m with a series of single Ricker wavelet of 0.25Hz peak frequency. The time increment is 2ms. The spatial increment is 40 m and the number of grid points is 300×200. Fig.3

and Fig.4 are the snapshots of wave field at the moment of 8.0s in contrast models.

In the contrast experiment, the peak value displacement of the receiver in basin model is about 30 times that of the receiver in layered model. That shows amplification effects are significant, and we need further research the basin effects.

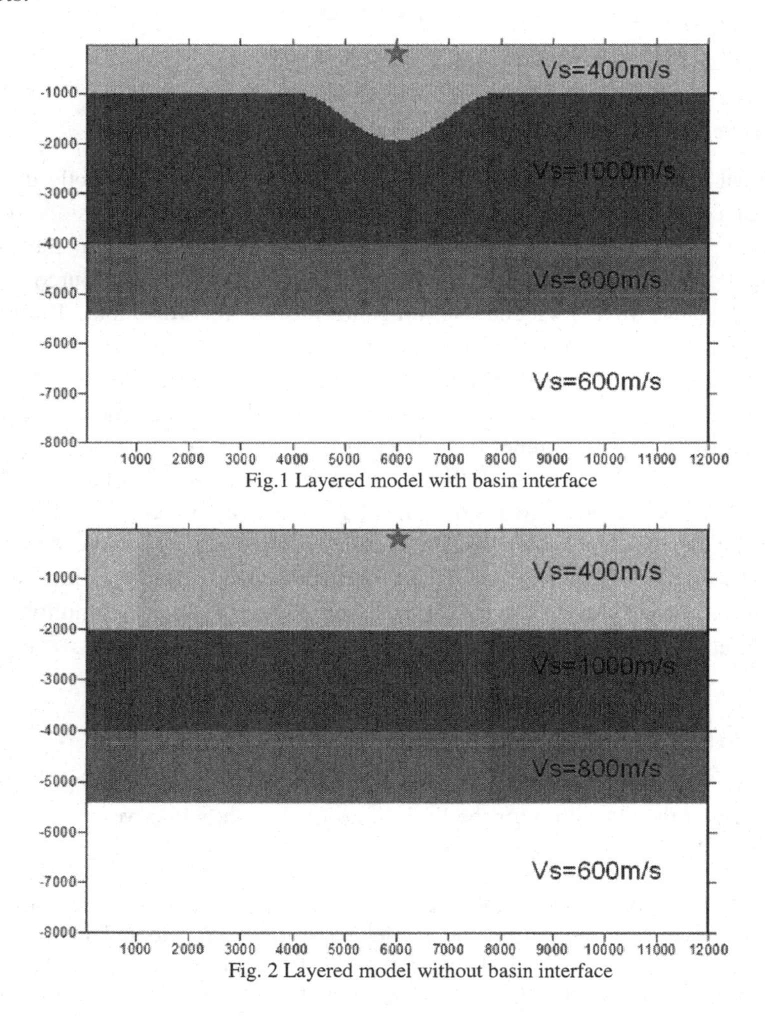

Fig.1 Layered model with basin interface

Fig. 2 Layered model without basin interface

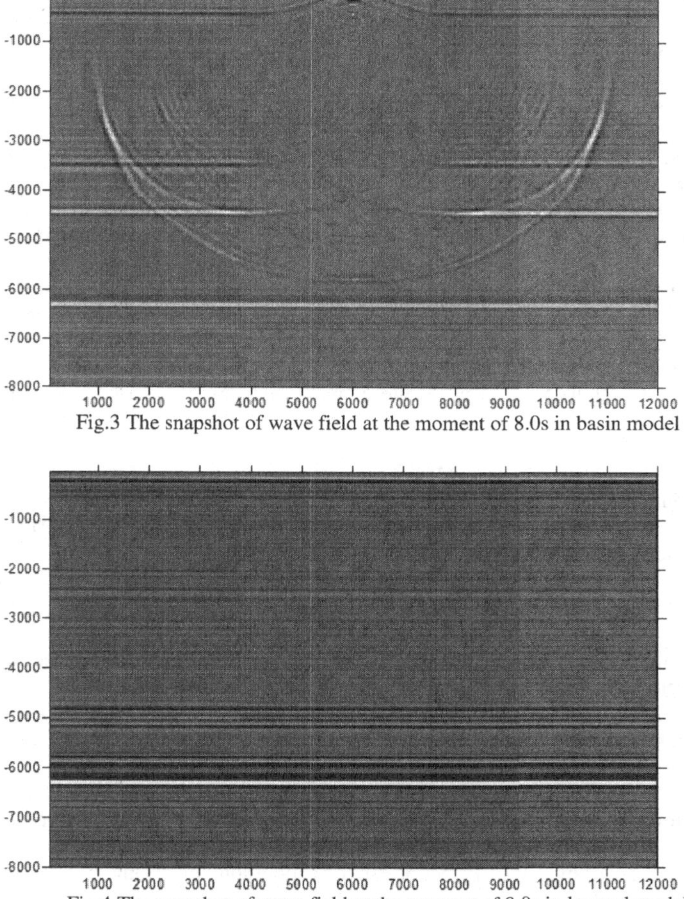

Fig.3 The snapshot of wave field at the moment of 8.0s in basin model

Fig.4 The snapshot of wave field at the moment of 8.0s in layered model

References

[1] Zheng T Y, Yao Z X. Study on strong ground motion in theoretical seismology. Chinese J. Geophys. (in Chinese), 1992, 35(1): 102~110

[2] Hu Y X. Earthquake Engineering (in Chinese). Beijing: Earthquake Press, 2006

[3] Alterman Z, Karal F C. Propagation of elastic waves in layered media by finite difference methods. Bull Seism. Soc. Am., 1968, 58: 367~398

[4] Madaviaga R. Dynamics of an expanding circular fault. Bull Seism. Soc. Am.,1976, 66: 639~666

[5] Zhang W, Chen X F. Traction image method f or irregular free surface boundaries in finite difference seismic wave simulation. Geophy. J Int, 2006, 167(1): 337~353

[6] Gao M T, Yu Y X. Three-dimensional finite-difference simulations of ground motions in the Beijing area. Earthquake Research in China (in Chinese), 2002,

18(4) :356~364
[7] Satol T, et al. Three-dimensional finite-difference waveform modeling of strong motions observed in the Sendai basin, Japan. Bull Seism. Soc. Am., 2001,91(4): 812~825
[8] Zhong W X. Symplectic Solution Methodology in Applied Mechanics (in Chinese). Beijing: Higher Education Press, 2006
[9] Seron F J, Sanz F J, Kindelan M, et al. Finite element method for elastic wave propagation. Comm. Appl. Numerical Methods, 1990, 6(2): 359~368
[10] Ding H P, Liu Q F, et al. A method of numerical simulation for long-period strong ground motion by 3-D FEM. Earthquake Engineering and Engineering Vibration, 2006, 26(5):32~36
[11] Zhang X Z, Hu J J. Simulation of near-fault bedrock strong ground motion field by explicit finite element method. Acta Seismologica Sinica, 2006, 28(6): 80~86
[12] Liao Z P. Introduction to Wave Motion Theories in Engineering (in Chinese). Beijing: Science Press. 2002
[13] Zhang H, Zhou Y Z, et al. Finite element analysis of seismic wave propagation characteristics in Fuzhou basin. Chinese J. Geophys. (in Chines), 2009, 52(5):1270~1279
[14] Day S M, Graves R, et al. Model for basin effects on long-period response spectra in southern California. Earthquake Spectra, 2008,24(1):257~277
[15] Gazdag J. Modeling of acoustic wave equation with transform methods. Geophysics, 1981,46(6):854~859
[16] Kosloff D D, Baysal E. Forward modeling by a Fourier method. Geophysics, 1982,47(10):1402~1412
[17] Wang Yanbing, Takenaka H, Takashi F. Modeling seismic wave propagation in a two-dimensional cylindrical whole-earth model using the pseudospectral method. Geophys. J. Int., 2001,145(3):689~708
[18] Zhao Z X, Xu J R, et al. Staggered grid real value FFT differentiation operator and its application on wave propagation simulation in the heterogeneous medium. Chinese J. Geophys. (in Chines), 2003, 46(2):234~240
[19] Zhao Z X, Xu J R. velocity and acceleration characteristics of seismic ground motions related to basin structure. Chinese J. Geophys. (in Chines), 2005, 48(5):1085~1091
[20] Wong H L, Jenning P C. Effect of canyon topography on strong motion. Bull Seism. Soc. Am. 1975, 65: 1239~1257
[21] Fujiwara H. The fast multipole method for solving integral equations of three-dimensional topography and basin problems. Geophys. J. Int., 2000,140:198~210
[22] Zhou H, Chen X F. Simulation of strong ground motion in Beijing city due to the Zhangbei earthquake. Progress in Geophysics. (in Chinese), 2008, 23(5): 1355~1366
[23] Patera A T. A spectral element method for fluid dynamics: Laminar flow in a channel expansion. Journal of Computational Physics. 1984,54(3): 468~488
[24] Seriani G, Priolo E. Spectral element method for acoustic wave simulation in heterogeneous media. Finite Elements in Analysis and Design. 1994,16: 337~348
[25] Komatitsch D, Tromp J. Introduction to the spectral element method for three-dimensional seismic wave propagation. Geophys. J. Int., 1999,139(3):806~822
[26] Komatitsch D, Liu Q, Tromp J. Simulations of ground motion in the Los Angeles basin based upon the spectral-element method. Bull Seism. Soc. Am., 2004,94(1): 187~206

[27] Iwatsu R. Two new solutions to the third-order symplectic integration method. *Phys. Lett. A,* 2009, **373**(34): 3056~3060
[28] Boore D M, Harmsen S C, Harding S T. Wave scattering from a step change in surface topography. Bull Seism. Soc. Am., 1981, 71(1): 117~125
[29] Harmsen S C. Estimating the diminution of Shear-wave amplitude with distance: Application to the Los Angeles, California, urban area. Bull Seism. Soc. Am., 1997, 87(4): 888~903
[30] Vidale J E, Helmberger D V. Elastic finite-difference modeling of the 1971 San Fernanda, California earthquake. Bull Seism. Soc. Am., 1988, 78(1): 122~141

COMPLETE PLANE STRAIN PROBLEM OF A ONE DIMENSIONAL HEXAGONAL QUASICRYSTALS WITH A DOUBLY-PERIODIC SET OF CRACKS*

XING LI†, PENG-PENG SHI

School of Mathematics and Computer Science, Ningxia University, Yinchuan, 750021 People's Republic of China

The first fundamental complete plane strain (CPS) problem for an infinite one-dimensional hexagonal quasicrystals body containing a doubly-periodic array of cracks in periodical and aperiodical plane is considered. Employing the superposition principle of force, the complete plane strain state, which is a special three-dimensional elastic system, is resolved into two linearly independent two-dimensional (plane) elastic systems, one is the generalized plane strain state, and another is the longitudinal displacement state. Using a technique that based on the complex potential function of Kolosov and Muskhelishvili, solving this problem are transferred into seeking analytic functions which fit certain boundary value problems. Furthermore, the general representation for the solution is constructed, under some general restrictions the boundary value problem is reduced to a normal type singular integral equation with a Weierstrass zeta kernel along the boundary of cracks, and the unique solvability of which is proved.

Keywords: one-dimensional hexagonal quasicrystals, doubly-periodic cracks, complex stress function, singular integral equations

Introduction

The complex potential method in the classical elasticity is effective, in general, mainly for solving some harmonic, quasi-harmonic, harmonic and quasi-harmonic equations which usually appear as governing equations in the study, the solutions can be expressed by analytic functions with respect to the single or multiple complex variables (see, e.g. [2,3,11,12]). In the present work, we wish to use the complex theory to solved the first fundamental complete plane strain (CPS) problem for an infinite one-dimensional hexagonal quasicrystals body containing doubly-periodic array of cracks in aperiodical plane (see, e.g. [5-8]).

---

* This work is supported by the National Natural Science Foundation of China (10962008; 11261045; 51061015) and Research Fund for the Doctoral Program of Higher Education of China (20116401110002).

† Corresponding author: Xing Li, Tel.: +86 951 2061405. E-mail address: li_x@nxu.edu.cn

1. Preliminaries

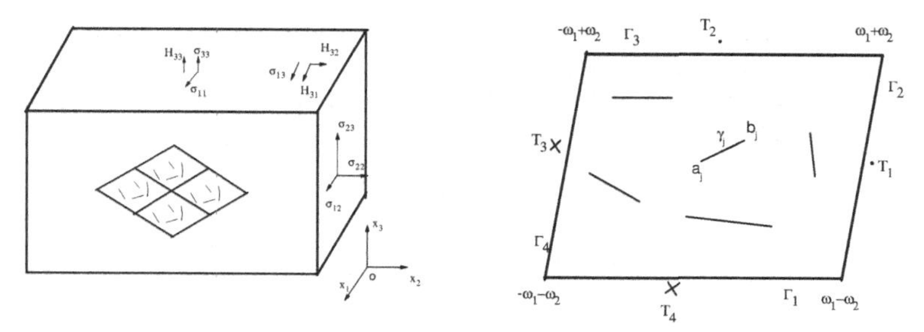

Figure 1. A one-dimensional hexagonal quasicrystals body with doubly-periodic set of cracks in the aperiodical plane

Assume that the x_3 -axis is the quasi-periodic axis, and the (x_1, x_2) -plane the periodic plane of an infinite one-dimensional hexagonal quasicrystals body. According to one-dimensional hexagonal quasicrystals elasticity theory, there are phonon displacement u_1, u_2, u_3 and phanon displacement w_3 (and $w_1 = w_2 = 0$) for the quasicrystals. A schematic diagram of an infinite one-dimensional hexagonal quasicrystals body with a doubly-periodic array of cracks in the aperiodical transverse cross section ($x_2 - x_3$ cross section) is put on the complex plane, $z = x_2 + ix_3$, as shown in figure 1. $2\omega_1$ and $2\omega_2$ are the two primitive periods, with $\operatorname{Im}{}^{\omega_1}\!/_{\omega_2} > 0$. The doubly-periodic fundamental parallelogram in the aperiodical plane of the quasicrystals body will be denoted by P_{00}. Its vertices are $\pm\omega_1 \pm \omega_2$. The boundary of P_{00} will be denoted by Γ with the positive direction taken to be anticlockwise, and $\Gamma = \cup_{i=1}^{4}\Gamma_i$. It will be assumed that there exists a system of m cracks distributed in P_{00}, denoted by $\gamma_j = \overset{\frown}{a_j b_j}(j = 1,\ldots,m)$, which are smooth and non-intersecting curve segments. The positive directions of γ_j are taken to be from a_j to b_j, and we denote $\gamma = \sum_{j=1}^{m}\gamma_j$. The system of cracks inside $P_{mn}(m,n = 0,\pm 1,\ldots)$ is congruent to γ in P_{00}. Without loss of generality, we may assume that the origin is not on the cracks.

If there are only Griffith cracks in the quasicrystals body along the x_1 -axis, the direction of the atom periodic arrangement, assume that the quasicrystals is in complete plane strain (CPS) state, which is a specific case of the three-dimensional elasticity state. Exactly, the quasicrystals subjected to combined antiplane shear and inplane loads in the aperiodical plane, (x_2, x_3 -plane). Obviously, the elastic equilibrium in the aperiodical plane is reduced two-dimensional (plane) elastic problem. Different from classics complete plane strain formula, in our case, considering the contribution of phonon and phanon stress, the generalized complete plane strain (CPS) state is

$$\begin{cases} \sigma_{22} = \sigma_{22}(x_2, x_3), \sigma_{33} = \sigma_{33}(x_2, x_3), \sigma_{23} = \sigma_{23}(x_2, x_3) \\ H_{33} = H_{33}(x_2, x_3), H_{32} = H_{32}(x_2, x_3) \\ \sigma_{12} = \sigma_{12}(x_2, x_3), \sigma_{13} = \sigma_{13}(x_2, x_3) \\ \varepsilon_{11} = e_1, w_{31} = e_2 \end{cases} \tag{1.1}$$

where e_1, e_2 are real constants.

These equations of classics CPS problem similar to the above equations in the absence of phanon physical quantitys were first introduced by Glingauss in reference in 1975. And a creativeness achievement of solving various doubly-periodic CPS problems of elasticity theory attributed the success to Li Xing in reference [5] in 2001. Similarly, we can resolve the special three-dimensional elastic system into two linearly independent plane elastic systems by the superposition principle of forces; one is the generalized plane strain state

$$\begin{cases} \sigma_{22} = \sigma_{22}(x_2, x_3), \sigma_{33} = \sigma_{33}(x_2, x_3), \sigma_{23} = \sigma_{23}(x_2, x_3) \\ H_{33} = H_{33}(x_2, x_3), H_{32} = H_{32}(x_2, x_3) \\ \sigma_{12} = \sigma_{13} = 0, \varepsilon_{11} = e_1, w_{31} = 0 \end{cases} \tag{I}$$

and another is the longitudinal displacement state

$$\begin{cases} \sigma_{22} = \sigma_{33} = \sigma_{23} = H_{33} = H_{32} = 0 \\ \sigma_{12} = \sigma_{12}(x_2, x_3), \sigma_{13} = \sigma_{13}(x_2, x_3), H_{31} = H_{31}(x_2, x_3) \\ \varepsilon_{11} = 0, w_{31} = e_2 \end{cases} \tag{II}$$

For the elastic system (I), the phonon and phanon stress components can be expressed by analytic functions $\varphi_k(z_k), \Phi_k(z_k) = \varphi_k'(z_k), k = 1,2,3$ [9]

$$\begin{aligned} &\sigma_{22} = 2\,\mathrm{Re}\sum_{k=1}^{3}\mu_k^2\Phi_k(z_k), \sigma_{33} = 2\,\mathrm{Re}\sum_{k=1}^{3}\Phi_k(z_k), \sigma_{23} = -2\,\mathrm{Re}\sum_{k=1}^{3}\mu_k\Phi_k(z_k) \\ &H_{32} = 2\,\mathrm{Re}\sum_{k=1}^{3}\vartheta_k\mu_k\Phi_k(z_k), H_{33} = -2\,\mathrm{Re}\sum_{k=1}^{3}\vartheta_k\Phi_k(z_k) \end{aligned} \tag{1.2}$$

and the phonon and phanon displacement components can be expressed as

$$\begin{aligned} u_2 &= 2\,\mathrm{Re}[o_1\varphi_1(z_1) + o_2\varphi_2(z_2) + o_3\varphi_3(z_3)] + c_1 x_2 \\ u_3 &= 2\,\mathrm{Re}[p_1\varphi_1(z_1) + p_2\varphi_2(z_2) + p_3\varphi_3(z_3)] + c_2 x_3 \\ w &= 2\,\mathrm{Re}[q_1\varphi_1(z_1) + q_2\varphi_2(z_2) + q_3\varphi_3(z_3)] + c_4 x_3 \end{aligned} \tag{1.3}$$

For the elastic system (II), taking (1.3) into account, we obtain

$$\begin{aligned} &\frac{\sigma_{12}}{C_{66}} - i\frac{\sigma_{13}}{mC_{44}} = \varphi'(z_m) - \frac{ic}{m}\bar{z}_m - \frac{iR_3}{mC_{44}}e_2 \\ &H_{31} = \frac{R_3}{C_{44}}\sigma_{13} + \frac{C_{44}K_2 - R_3^2}{C_{44}}e_2 \end{aligned} \tag{1.4}$$

The investigation on the doubly-periodic problem will be restricted to doubly-periodic stress distributions in the elastic body (e.g. see [5])

$$\begin{aligned} &\sigma_{ii}(z+2\omega_k) = \sigma_{ii}(z) = \sigma_{ii}(x_2,x_3), i = 1,2,3 \\ &\sigma_{1j}(z+2\omega_k) = \sigma_{1j}(z) = \sigma_{1j}(x_2,x_3), j = 2,3 \\ &\sigma_{23}(z+2\omega_k) = \sigma_{23}(z) = \sigma_{23}(x_2,x_3) \\ &H_{3i}(z+2\omega_k) = H_{3i}(z) = H_{3i}(x_2,x_3), i = 1,2,3 \end{aligned} \tag{1.5}$$

where $z = x_2 + ix_3$.

From (1.2)-(1.4) it now follows immediately that the complex stress function $\varphi_k(z_k),(k=1,2,3)$ and $\varphi(z_m)$ are all doubly quasi-periodic.

2. Kolosov Functions

In order to separate the multi-valued parts of $\varphi_k(z_k),$, we need to construct the Kolosov functions (see, e.g. [5,10,13]). Not that z_1, z_2, z_3 as functions of $z = x_2 + ix_3$ in the aperiodical plane ($x_2 - x_3$ plane), transform Γ, $\gamma_j = \widehat{a_j b_j}$ to $\Gamma', \gamma_j' = \widehat{a_j' b_j'}$, $\Gamma'', \gamma_j'' = \widehat{a_j'' b_j''}$, and $\Gamma''', \gamma_j''' = \widehat{a_j''' b_j'''}$ respectively. The stress functions for the aperiodical plane are denoted by $\varphi_k(z_k),(k=1,2,3)$, which are generally multiple-valued. Furthermore, here we need not only to separate the multi-valued parts but must keep the double periodicity of them. For doing this, we construct the Kolosov functions have the following expressions.

$$\begin{cases} \varphi_1(z_1) = \dfrac{1}{4\pi i}\sum\limits_{j=1}^{m}\Delta_j'[\log\sigma(z_1 - a_j')\sigma(z_1 - b_j') - H_j(z_1)] + \varphi_1^*(z_1) \\ \varphi_2(z_2) = \dfrac{1}{4\pi i}\sum\limits_{j=1}^{m}\Delta_j''[\log\sigma(z_2 - a_j'')\sigma(z_2 - b_j'') - H_j(z_2)] + \varphi_2^*(z_2) \\ \varphi_3(z_3) = \dfrac{1}{4\pi i}\sum\limits_{j=1}^{m}\Delta_j'''[\log\sigma(z_3 - a_j''')\sigma(z_3 - b_j''') - H_j(z_3)] + \varphi_3^*(z_3) \end{cases} \tag{2.1}$$

where $t_j = \mathrm{Re}\,t + \mu_j \,\mathrm{Im}\,t, j = 1,2,3$, $H_j(z_k), k = 1,2,3$ are introduced to eliminate the singularity of $\log\sigma(z_k - a_j')\sigma(z_k - b_j')$, and the expressions as follows [13]

$$\begin{cases} H_j(z_1) = \int_{\gamma'} \frac{2t_1 - a_j' - b_j'}{b_j' - a_j'} \zeta(t_1 - z_1) dt_1 \\ H_j(z_2) = \int_{\gamma''} \frac{2t_2 - a_j'' - b_j''}{b_j'' - a_j''} \zeta(t_2 - z_2) dt_2 \\ H_j(z_3) = \int_{\gamma'''} \frac{2t_3 - a_j''' - b_j'''}{b_j''' - a_j'''} \zeta(t_3 - z_3) dt_3 \end{cases} \tag{2.2}$$

where $\zeta(z)$ is the weierstrass zeta function, namely

$$\zeta(z) = \frac{1}{z} + \sum_{m,n}{}' \{ \frac{1}{z - \Omega_{mn}} + \frac{1}{\Omega_{mn}} + \frac{z}{\Omega_{mn}^2} \} \tag{2.3}$$

where $\Omega_{mn} = 2m\omega_1 + 2n\omega_2$,the prime of $\sum_{m,n}{}'$ above indicates that m and n will not be zero simultaneously. In fact, $\zeta(z)$ is a doubly quai-periodic function with cyclic increments $2\eta_k, k = 1,2$.

$\sigma(z)$ is the weierstrass sigma function, which has a relationship with $\zeta(z)$, namely $\zeta(z) = \frac{\sigma'(z)}{\sigma(z)}$; $\Delta_j', \Delta_j'', \Delta_j''' (j = 1,2,\ldots,m)$ are solutions of a system of linear equations with $X_{2j} + iX_{3j} \neq 0, X_j \neq 0$ which are the phonon and phanon external resultant principal vectors on the two sides of γ_j, respectively.

Then, $\varphi_1^*(z_1), \varphi_2^*(z_2), \varphi_3^*(z_3)$ are single-valued, doubly quasi-periodic functions, and with same singular as $\varphi_1(z_1), \varphi_2(z_2), \varphi_3(z_3)$.

3. Formulation of the First Fundamental CPS Problem

Let us consider the first fundamental CPS problem. The phonon and phanon external stress functions on the positive and the negative sides of γ_j are given, respectively. In addition, the stress resultant principal vectors on Γ_k , $k = 1,2$ in the $x_2 - x_3$ plane, and the shear stress in the x_1 direction are given, too. And the strain real constants e_1, e_2 are given. In fact, paying attention to the kolosov functions (2.1), without loss of generality, we may assume that each $X_2 + iX_3 = 0, X = 0$. Then, we obtain [13]

$$\Delta_j' = \Delta_j'' = \Delta_j''' = 0, j = 1,2,\ldots,m \tag{3.1}$$

$$\varphi_1(z_1) = \varphi_1^*(z_1), \varphi_2(z_2) = \varphi_2^*(z_2), \varphi_3(z_3) = \varphi_3^*(z_3) \tag{3.2}$$

Then, similar to [5,13], we have the boundary conditions on γ_j and heir congruent for the elastic system (I) as

$$\begin{aligned} &\delta_1\varphi_1^\pm(\tau_1) + \delta_2\overline{\varphi}_1^\pm(\tau_1) + \delta_3\varphi_2^\pm(\tau_2) + \delta_4\overline{\varphi}_2^\pm(\tau_2) + \varphi_3^\pm(\tau_3) = \tilde{f}_j^\pm(\tau) + e_{1j} \\ &\lambda_1\varphi_1^\pm(\tau_1) + \overline{\lambda}_1\overline{\varphi}_1^\pm(\tau_1) + \lambda_2\varphi_2^\pm(\tau_2) + \overline{\lambda}_2\overline{\varphi}_2^\pm(\tau_2) = \tilde{g}_j^\pm(\tau) + e_{2j} \end{aligned} \tag{3.3}$$

$$\begin{aligned} &[\delta_1\varphi_1(z_1) + \delta_2\overline{\varphi}_1(z_1) + \delta_3\varphi_2(z_2) + \delta_4\overline{\varphi}_2(z_2) + \varphi_3(z_3)]_{\Gamma_k} = \tilde{F}_k \\ &[\lambda_1\varphi_1(z_1) + \overline{\lambda}_1\overline{\varphi}_1(z_1) + \lambda_2\varphi_2(z_2) + \overline{\lambda}_2\overline{\varphi}_2(z_2)]_{\Gamma_k} = \tilde{G}_k \end{aligned} \tag{3.4}$$

Futhermore, we have the boundary conditions for the elastic system (II)

$$\begin{cases} \varphi^\pm(\tau) - \overline{\varphi}^\pm(\tau) = iC_j^*(m,n), \tau \in \gamma_j(m,n) = \gamma_j \cup \Omega_{mn} \\ [\varphi^\pm(z_m) - \overline{\varphi}^\pm(z_m)]_{\Gamma_k} = |w_k| T_k, k = 1,2,, z \in \Gamma(m,n) = \Gamma \cup \Omega_{mn} \end{cases} \tag{3.5}$$

where $t_m = \mathrm{Re}\, t + im\, \mathrm{Im}\, t$, and $C_j^*(m,n)$ may be arbitrarily fixed constants.

4. Solution, Reduction to Integral Equations

In order to solve the boundary value problems, a new unknown real function $\omega_1(t), t \in \gamma$ will be introduced to describe the contribution of phanon, and another unknown complex function $\omega_2(t), t \in \gamma$ will be introduced to represent that of phonon. Then, we construct the general representation for the solution in the form as (e.g. see [13])

$$\varphi_1(z_1) = \frac{\lambda_2}{2\pi i}\int_{\gamma'}[\omega_1(t) + \omega_2(t)]\zeta(t_1 - z_1)dt_1 + Ar_1z_1 \tag{4.1}$$

$$\begin{aligned} \varphi_2(z_2) = &-\frac{\lambda_1}{2\pi i}\int_{\gamma''}[\omega_1(t) + \omega_2(t)]\zeta(t_2 - z_2)dt_2 \\ &+\frac{1}{\lambda_2 + \overline{\lambda}_2}\frac{1}{2\pi i}\int_{\gamma''}[g_j^+(t) - g_j^-(t)]\zeta(t_2 - z_2)dt_2 + Br_2z_2 \end{aligned} \tag{4.2}$$

$$\begin{aligned} \varphi_3(z_3) = &\frac{1}{2\pi i}\int_{\gamma'''}\left\{\begin{matrix} (-\delta_1\lambda_2 - \delta_2\overline{\lambda}_2 + \delta_3\lambda_1 + \delta_4\overline{\lambda}_1)\omega_1(t) \\ +(\delta_3\lambda_1 - \delta_1\lambda_2)\omega_2(t) + (\delta_4\overline{\lambda}_1 - \delta_2\overline{\lambda}_2)\overline{\omega}_2(t) \end{matrix}\right\}\zeta(t_3 - z_3)dt_3 \\ &-\frac{\delta_3 + \delta_4}{\lambda_2 + \overline{\lambda}_2}\frac{1}{2\pi i}\int_{\gamma'''}[g_j^+(t) - g_j^-(t)]\zeta(t_3 - z_3)dt_3 \\ &+\frac{1}{2\pi i}\int_{\gamma'''}[f_j^+(t) - f_j^-(t)]\zeta(t_3 - z_3)dt_3 + i\,\mathrm{Im}\,Ar_3z_3 + i\,\mathrm{Im}\,Br_3z_3 + Cz_3 \end{aligned} \tag{4.3}$$

where r_1, r_2, r_3 are fixed complex constants, and $\varphi_i^0(z_i) = ikr_iz_i + D_i, i = 1,2,3$ be the solutions of the first fundamental plane strain problem under homogeneous conditions (e.g. see [5,10,13]). And assume temporarily $\omega_1(a_j) = \omega_1(b_j) = \omega_2(a_j) = \omega_2(b_j) = 0, j \in 1,2,\ldots,m$.

Substituting equations (4.1)-(4.3) into the left-hand of the condition (3.4), we get

$$\begin{aligned}
&2A(\delta_1 r_1 \omega_k + \delta_2 \overline{r_1}\overline{\omega}_k) + 2B(\delta_3 r_2 \omega_k + \delta_4 \overline{r_2}\overline{\omega}_k) + 2C\omega_k \\
&= \tilde{F}_k + \delta_1 \lambda_2 \frac{\eta_k}{\pi i}\int_{\gamma'}[\omega_1(t) + \omega_2(t)]dt_1 + \delta_2 \overline{\lambda}_2 \frac{i\overline{\eta}_k}{\pi}\int_{\gamma'}[\omega_1(t) + \overline{\omega}_2(t)]d\overline{t}_1 \\
&+\delta_3 \lambda_1 \frac{i\eta_k}{\pi}\int_{\gamma''}[\omega_1(t) + \omega_2(t)]dt_2 + \delta_4 \overline{\lambda}_1 \frac{\overline{\eta}_k}{\pi \underline{i}}\int_{\gamma''}[\omega_1(t) + \overline{\omega}_2(t)]d\overline{t}_2 \\
&+\frac{\eta_k}{\pi i}\int_{\gamma'''}\left\{\begin{array}{c}(-\delta_1\lambda_2 - \delta_2\overline{\lambda}_2 + \delta_3\lambda_1 + \delta_4\overline{\lambda}_1)\omega_1(t) \\ +(\delta_3\lambda_1 - \delta_1\lambda_2)\omega_2(t) + (\delta_4\overline{\lambda}_1 - \delta_2\overline{\lambda}_2)\overline{\omega}_2(t)\end{array}\right\}dt_3 \\
&+\frac{\delta_3}{\lambda_2 + \overline{\lambda}_2}\frac{\eta_k}{\pi i}\int_{\gamma''}[g_j^+(t) - g_j^-(t)]dt_2 + \frac{\delta_4}{\lambda_2 + \overline{\lambda}_2}\frac{i\eta_k}{\pi}\int_{\gamma''}[\overline{g}_j^+(t) - \overline{g}_j^-(t)]d\overline{t}_2 \\
&+\frac{\delta_3 + \delta_4}{\lambda_2 + \overline{\lambda}_2}\frac{i\eta_k}{\pi}\int_{\gamma'''}[g_j^+(t) - g_j^-(t)]dt_3 + \frac{\eta_k}{\pi i}\int_{\gamma'''}[f_j^+(t) - f_j^-(t)]dt_3
\end{aligned} \tag{4.4}$$

$$\begin{aligned}
&2A[\mathrm{Re}(\lambda_1 r_1 \omega_1) + i\,\mathrm{Re}(\lambda_1 r_1 \omega_2)] + 2B[\mathrm{Re}(\lambda_2 r_2 \omega_1) + i\,\mathrm{Re}(\lambda_2 r_2 \omega_2)] \\
&= \tilde{G}_1 + i\tilde{G}_2 + \mathrm{Re}\{\lambda_1\lambda_2 \frac{\eta_1}{\pi i}\int_{\gamma'}[\omega_1(t) + \omega_2(t)]dt_1 + \lambda_1\lambda_2\frac{i\eta_1}{\pi}\int_{\gamma''}[\omega_1(t) + \omega_2(t)]dt_2 \\
&+\frac{\lambda_2}{\lambda_2 + \overline{\lambda}_2}\frac{\eta_1}{\pi i}\int_{\gamma''}[g_j^+(t) - g_j^-(t)]dt_2\} + i\,\mathrm{Re}\{\lambda_1\lambda_2\frac{\eta_2}{\pi i}\int_{\gamma'}[\omega_1(t) + \omega_2(t)]dt_1 \\
&+\lambda_1\lambda_2\frac{i\eta_2}{\pi}\int_{\gamma''}[\omega_1(t) + \omega_2(t)]dt_2 + \frac{\lambda_2}{\lambda_2 + \overline{\lambda}_2}\frac{\eta_2}{\pi i}\int_{\gamma''}[g_j^+(t) - g_j^-(t)]dt_2\}
\end{aligned} \tag{4.5}$$

Without loss of generality, assume that

$$(\overline{r_1}\delta_2 + \overline{r_2}\delta_4)\,\mathrm{Re}[\lambda_2 r_2(\omega_1 + i\omega_2)] \neq 0 \tag{4.6}$$

Then, the determinant of the system of equations (4.4)-(4.5) is nonzero. Hence, we can obtain A, B, C uniquely from that.

Letting $z \to t_0$, so $z_1 \to t_1^0, z_2 \to t_2^0, z_3 \to t_3^0$, and using the Plemelj formulae, taking equations (4.3), either for the positive or negative boundary, we get the same equations

$$\begin{aligned}
&\mathrm{Re}\{\frac{\lambda_1\lambda_2}{\pi i}\int_{\gamma}[\omega_1(t) + \omega_2(t)][\zeta(t_1 - t_1^0)dt_1 - \zeta(t_2 - t_2^0)dt_2]\} \\
&+A\mathrm{Re}(\lambda_1 A r_1 t^0) + B\mathrm{Re}(\lambda_2 B r_2 t^0) \\
&= \frac{\tilde{g}_j^+(t^0) + \tilde{g}_j^-(t^0)}{2} - \mathrm{Re}\{\frac{\lambda_2}{\lambda_2 + \overline{\lambda}_2}\frac{1}{\pi i}\int_{\gamma''}[g_j^+(t) - g_j^-(t)]\zeta(t_2 - z_2)dt_2\} + e_{2j}
\end{aligned} \tag{4.7}$$

$$
\begin{aligned}
&\frac{\delta_4\overline{\lambda}_1-\delta_2\overline{\lambda}_2}{\pi i}\int_\gamma[\omega_1(t)+\overline{\omega}_2(t)]\zeta(t_3-t_3^0)dt_3\\
&+\frac{\delta_1\lambda_2}{2\pi i}\int_\gamma[\omega_1(t)+\omega_2(t)]\zeta(t_1-t_1^0)dt_1-\zeta(t_3-t_3^0)dt_3]\\
&-\frac{\delta_2\lambda_2}{2\pi i}\int_\gamma[\omega_1(t)+\overline{\omega}_2(t)][\overline{\zeta(t_1-t_1^0)dt_1}-\zeta(t_3-t_3^0)dt_3]\\
&-\frac{\delta_3\lambda_1}{2\pi i}\int_\gamma[\omega_1(t)+\omega_2(t)][\zeta(t_2-t_2^0)dt_2-\zeta(t_3-t_3^0)dt_3]\\
&+\frac{\delta_4\lambda_1}{2\pi i}\int_\gamma[\omega_1(t)+\overline{\omega}_2(t)]\overline{\zeta(t_2-t_2^0)dt_2}-\zeta(t_3-t_3^0)dt_3]\\
&+A(\delta_1 r_1 t^0+\delta_2\overline{r}_1\overline{t}^0)+B(\delta_3 r_2 t^0+\delta_4\overline{B}\overline{r}_2\overline{t}^0)\\
&=\frac{\tilde{f}_j^+(t^0)+\tilde{f}_j^-(t^0)}{2\pi i}-\frac{1}{2\pi i}\int_{\gamma'''}[\tilde{f}_j^+(t)-\tilde{f}_j^-(t)]\zeta(t_3-t_3^0)dt_3+e_{1j}\\
&-\frac{\delta_3}{\lambda_2+\overline{\lambda}_2}\frac{1}{2\pi i}\int_{\gamma''}[\tilde{g}_j^+(t)-\tilde{g}_j^-(t)][\zeta(t_2-t_2^0)dt_2-\zeta(t_3-t_3^0)dt_3]\\
&+\frac{\delta_4}{\lambda_2+\overline{\lambda}_2}\frac{1}{2\pi i}\int_{\gamma''}[\tilde{g}_j^+(t)-\tilde{g}_j^-(t)][\overline{\zeta(t_2-t_2^0)dt_2}+\zeta(t_3-t_3^0)dt_3]
\end{aligned}
\tag{4.8}
$$

In order to solve the boundary value problem (3.5), we construct the modified Sherman transform

$$\phi(z_m)=\frac{1}{2\pi i}\int_\gamma w_3(t)\zeta(t-z_m)dt+Ez \tag{4.9}$$

where $w_3(t)$ is an unknown real function, E is an undermined complex comstant.

Substituting equation (4.9) into the second formulae of (3.5), we get a system of algebraic equations for the unknown real constant $\operatorname{Re}E, \operatorname{Im}E$

$$(w_k-\overline{w}_k)\operatorname{Re}E-i(w_k+\overline{w}_k)\operatorname{Im}E=\frac{|w_k|}{2}T_k+\frac{1}{\pi i}\operatorname{Re}\{\eta_k\int_\gamma w_3(t)dt\} \tag{4.10}$$

Because of the determinant of the system of equations (4.10) is nonzero. we can obtain E uniquely from that.Substituting (4.9) into the first equation of (4.10), either for the positive or negative boundary, we get the same equations as

$$\frac{1}{2\pi i}\int_\gamma w_3(t)d\log[\frac{\overline{\sigma(t-t^0)}}{\sigma(t-t^0)}]+i\operatorname{Im}(Et^0)-iC_j^*=0 \tag{4.11}$$

5. Unique Solvability

It is now be proved that if $C_j, C_{3j}, j=1,2,\ldots,m$ will be suitably chosen (also uniquely), then equations (4.7) and (4.8) are solvable in h_{2m}. Similarly to the work of [5], for this purpose, the homogeneous equation, obtain from equation (4.7) and (4.8) for $T_{2j}^{\pm}(t)=0, T_{3j}^{\pm}(t)=0, T_{hj}^{\pm}(t)=0, e_1=0$ will be considered and it will be shown that it has no non-trivial solutions. This mean if $\omega_1^0(t), \omega_2^0(t)$ be any solution of this equation, after $C_j = C_j^0, C_{3j} = C_{3j}^0$ be taken, the $\omega_1^0(t) = \omega_2^0(t) = 0$ everywhere on γ. It obvious that (2.32) is satisfied in this case. Let $\varphi_1^0(z_1), \varphi_2^0(z_2), \varphi_3^0(z_3), A^0, B^0, C^0$ be the corresponding values of $\varphi_1(z_1)$, $\varphi_2(z_2), \varphi_3(z_3), A, B, C$ determined by equations (3.1), (4.1-4.3), and equations (4.5), (4.6) for $\omega_1(t) = \omega_1^0(t), \omega_2(t) = \omega_2^0(t)$. It is easy to verify that they satisfy the corresponding boundary conditions (4.8) and (4.9), which from the first fundamental problem under homogeneous conditions. By the uniqueness theorem [11], the elastic region may only have a rigid motion as

$$\varphi_1^0(z_1) = ikr_1 z_1 + D_1, \varphi_2^0(z_2) = ikr_2 z_2 + D_2, \varphi_3^0(z_3) = ikr_3 z_3 + D_3 \tag{5.1}$$

Referring to formulae (4.1)-(4.3), we have

$$\begin{aligned}
\varphi_1^0(z_1) &= ikr_1 z_1 + D_1 = \frac{\lambda_2}{2\pi i}\int_{\gamma'}[\omega_1^0(t) + \omega_2^0(t)]\zeta(t_1 - z_1)dt_1 + A^0 r_1 z_1 \\
\varphi_2^0(z_2) &= ikr_2 z_2 + D_2 = -\frac{\lambda_1}{2\pi i}\int_{\gamma''}[\omega_1^0(t) + \omega_2^0(t)]\zeta(t_2 - z_2)dt_2 + B^0 r_2 z_2 \\
\varphi_3^0(z_3) &= ikr_3 z_3 + D_3 \\
&= \frac{1}{2\pi i}\int_{\gamma'''}\left\{\begin{array}{c}(-\delta_1\lambda_2 - \delta_2\overline{\lambda}_2 + \delta_3\lambda_1 + \delta_4\overline{\lambda}_1)\omega_1^0(t) \\ +(\delta_3\lambda_1 - \delta_1\lambda_2)\omega_2^0(t) + (\delta_4\overline{\lambda}_1 - \delta_2\overline{\lambda}_2)\overline{\omega}_2^0(t)\end{array}\right\}\zeta(t_3 - z_3)dt_3 + Cz_3
\end{aligned} \tag{5.2}$$

Considering quasi-periodicity of the two sides of equations, we obtain

$$A^0 = B^0 = C^0 = ik \tag{5.3}$$

$$\begin{aligned}
&\frac{1}{2\pi i}\int_{\gamma'}[\omega_1^0(t) + \omega_2^0(t)]dt_1 = 0 \\
&\frac{1}{2\pi i}\int_{\gamma'''}\left\{\begin{array}{c}(-\delta_1\lambda_2 - \delta_2\overline{\lambda}_2 + \delta_3\lambda_1 + \delta_4\overline{\lambda}_1)\omega_1^0(t) \\ +(\delta_3\lambda_1 - \delta_1\lambda_2)\omega_2^0(t) + (\delta_4\overline{\lambda}_1 - \delta_2\overline{\lambda}_2)\overline{\omega}_2^0(t)\end{array}\right\}dt_3 = 0
\end{aligned} \tag{5.4}$$

Taking equations (5.4) into account, from equations (4.5) and (4.6) we have

$$A^0 = B^0 = C^0 = 0 \tag{5.5}$$

Hence, from equations (5.3) and (5.5) we have $k = 0$, Furthermore, equations (5.2) can be reduced by

$$\begin{aligned}
D_1 &= \frac{\lambda_2}{2\pi i}\int_{\gamma'}[\omega_1^0(t) + \omega_2^0(t)]\zeta(t_1 - z_1)dt_1 \\
D_2 &= -\frac{\lambda_1}{2\pi i}\int_{\gamma''}[\omega_1^0(t) + \omega_2^0(t)]\zeta(t_2 - z_2)dt_2 \\
D_3 &= \frac{1}{2\pi i}\int_{\gamma'''}\left\{\begin{array}{c}(-\delta_1\lambda_2 - \delta_2\overline{\lambda}_2 + \delta_3\lambda_1 + \delta_4\overline{\lambda}_1)\omega_1^0(t) \\ +(\delta_3\lambda_1 - \delta_1\lambda_2)\omega_2^0(t) + (\delta_4\overline{\lambda}_1 - \delta_2\overline{\lambda}_2)\overline{\omega}_2^0(t)\end{array}\right\}\zeta(t_3 - z_3)dt_3
\end{aligned} \tag{5.6}$$

Considering $\omega_1(t)$ be a real function which describe the contribution of phanon, and $\omega_2(t)$ a the complex functions which describe that of phanon on the cracks, and using the Plemelj formulae from (5.6), and we get immediately

$$\omega_1^0(t) = \omega_2^0(t) = 0, t \in \gamma \tag{5.7}$$

We now turn to prove the equation (4.11) to be uniquely solvable. To do this, we must prove $w_3^0(t) = 0$ under homogeneous conditions of the first fundamental problem. In this case, we have (e.g. see [5])

$$\varphi^0(z_m) = \frac{1}{2\pi i}\int_\gamma w_3^0(t)\zeta(t - z_m)dt + E^0 z = E_z \tag{5.8}$$

Considering quasi-periodicity of (5.9), we obtain

$$\frac{1}{2\pi i}\int_\gamma w_3^0(t)dt = 0, E^0 = 0 \tag{5.9}$$

By the Plemelj formulae from (5.8), we get immediately

$$w_3^0(t) = 0, t \in \gamma \tag{5.10}$$

Now, we have proved the system of singular integral equations, which is constituted by equations (4.7) and (4.8), and the Fredholm integral equation of the second kind, which is constituted by equation (4.11), are uniquely solvable, respectively. Thus, after obtaining the unique solution $w_1(t), w_2(t)$ from the above singular integral equation, we may obtain the functions $\varphi_1^*(z_1), \varphi_2^*(z_2), \varphi_3^*(z_3)$ from equations (4.1)-(4.3), furthermore, the stress functions $\varphi_1(z_1), \varphi_2(z_2), \varphi_3(z_3)$ will be obtained from equations (2.1). And after obtaining the unique solution $w_3(t)$ from the above Fredholm equation of the second kind, the functions $\varphi(z_m)$ may be obtained from equations (4.9).

References

1. Fan T. Y., et al., The strict theory of complex variable function method of sextuple harmonic equation and applications, J. Math. Phys., 51(2010), 053519.
2. Fan T. Y., Mathematical theory of elasticity of quasicrystals and its applications, Science Press, Beijing, 2010.
3. Lekhnitskii S. G., Theory of elasticity of an anisotropic body. Holden Day. San Francisco,1963.
4. Li X., On the mathematical problem of composite materials with a doubly periodic set of cracks, Appl. Math. Mech. (PRC) 14(1993), 1143-1150.
5. Li X., Applications of doubly quasi-periodic boundary value problems in elasticity theory, Shaker Verlag, Aachen, 2001.

6. Li X., Complete plane strain problem of a nonhomogeneous elastic body with a doubly-periodic set cracks, Z. angenw. Math. Mech, 6(2001), 377-391.
7. Li X., The effect of a homogeneous cylindrical inlay on cracks in the doubly-periodic complete plane strain problem, Int. J. Fract. 4(2001), 403-41.
8. Li X., General solution for complete plane strain problem of a nonhomogeneous body with a doubly- periodic set of inlays, Z. angenw. Math. Mech, 9(2006), 682-690.
9. Liu G. T., He Q. L. and Guo R. P., The plane strain theory for one-dimensional hexagonal quasicrystals in aperiodical plane, Acta Physica Sinica, 6(2009), S118-S123(Chinese).
10. Lu C. K., On the complex Airy functions in doubly-periodic plane elasticity, J. Math. (PRC), 6(1986), 319-330 (Chinese).
11. Lu C. K., Complex variables methods in plane elasticity, World Scientific, Singapore, 1995 (Chinese).
12. Muskhelishvili N. I., Some basic problems of the mathematical theory of elasticity, Noordhoff, Groningen, 1956.
13. Zheng K., On the fundamental problems in an anisotropic plane with a doubly-periodic set cracks, Comm. On Appl. Math. and Comput. 1(1993), 14-20(Chinese).
14. Zheng K., On the doubly-periodic fundamental problems in plane elasticity, Acta Mathematical Scientia 8(1988), 95-104(Chinese).

THE PROBLEM ABOUT AN ELLIPTIC HOLE WITH III ASYMMETRY CRACKS IN ONE-DIMENSIONAL HEXAGONAL PIEZOELECTRIC QUASICRYSTALS*

HUASONG HUO, XING LI **

(School of Mathematics and Computer Science, Ningxia University, Yinchuan 750021)

In this paper, by introducing the appropriate conformal transformation and using the complex variable function method, the problem about an elliptic hole with III asymmetry cracks in one-dimensional hexagonal piezoelectric quasicrystals is sovlved and the SIFs at the crack tip are obtained in the case of impermeable and permeable boundary conditions. In some special onditions, the cracks can be turned into a Griffith crack. By imitation computation, we can get the SIFs at the tip of the Griffith crack. especially, when the quasicrystals become vestigial the crystal, we can also get the SIFs at the tip of the Griffith crack, which is in accordance with the know results.

Key words: one-dimensional hexagonal piezoelectric quasicrystals, elliptic hole with III asymmetry cracks, complex variable method, SIFs

0. Introduction

Quasicrystal is found in recent years a new structure. It is its special structure led to its characteristics changed, For example, in force, thermal, electrical, magnetic and related physical and chemical properties[1-4]etc. For quasicrystal defects research has made certain achievements[5-8], however, the piezoelectric quasicrystal research are rare, Piezoelectric for quasi crystal material functional widely used has certain practical significance. This paper, by using generalized complex variable method, With the help of the appropriate conformal mapping, the problem about along the quasi-periodic direction through two asymmetric elliptical hole edge cracks in one-dimensional hexagonal piezoelectric quasicrystals is researched, we get the SIFs at the tip of cracks in electric impermeability and electric permeability two circumstances. In extreme cases, it can degenerate into existing relevant conclusion.

---

* This work is supported by the National Natural Science Foundation of China (10962008; 51061015; 11261045) and Research Fund for the Doctoral Program of Higher Education of China (20116401110002).

** Corresponding author, Tel.: +86 951 2061405. E-mail address: li_x@nxu.edu.cn (X. Li).

1. Basic equation

x_1-x_2 is the coordinate cycle plane in one-dimensional hexagonal piezoelectric quasicrystals, the coordinate axis x_3 is the quasi-periodicity direction, establish rectangular coordinate system, Electric polarization direction for x_3 direction, u_1, u_2, u_3 are phonon displacement component, ω is phason displacement, ϕ is the electric potential; $\xi_{11}, \xi_{22}, \xi_{33}, 2\xi_{23}, 2\xi_{31}, 2\xi_{12}, \omega_1, \omega_2, \omega_3$, are the phonon,phason strains and electric fields; $\sigma_{11}, \sigma_{22}, \sigma_{33}, \sigma_{23}, \sigma_{31}, \sigma_{12}, H_1, \quad H_2, H_3, D_1, D_2, D_3$ are the phonon、phason stress components and electric displacements, according to references[7], The generalized Hooke's law for one-dimensional hexagonal piezoelectric quasicrystal with point group 6 mm is given.

$$\begin{cases} \sigma_{11} = C_{11}\varepsilon_{11} + C_{12}\varepsilon_{22} + C_{13}\varepsilon_{33} + R_1\omega_3 - e_{31}^1 E_3, \\ \sigma_{22} = C_{12}\varepsilon_{11} + C_{22}\varepsilon_{22} + C_{13}\varepsilon_{33} + R_1\omega_3 - e_{31}^1 E_3, \\ \sigma_{33} = C_{13}\varepsilon_{11} + C_{13}\varepsilon_{22} + C_{33}\varepsilon_{33} + R_2\omega_3 - e_{33}^1 E_3, \\ \sigma_{23} = \sigma_{32} = 2C_{44}\varepsilon_{31} + R_3\omega_2 - e_{15}^1 E_2, \\ \sigma_{13} = \sigma_{31} = 2C_{44}\varepsilon_{31} + R_3\omega_1 - e_{15}^1 E_1, \\ \sigma_{12} = \sigma_{21} = 2C_{66}\varepsilon_{12}, \\ H_3 = R_1(\varepsilon_{11} + \varepsilon_{22}) + R_2\varepsilon_{33} + k_1\omega_3 - e_{31}^2 E_3, \\ H_1 = 2R_3\varepsilon_{31} + k_2\omega_1 - e_{15}^2 E_1, \\ H_2 = 2R_3\varepsilon_{32} + k_2\omega_2 - e_{15}^2 E_2, \\ D_3 = e_{31}^1(\varepsilon_{11} + \varepsilon_{22}) + e_{33}^1\varepsilon_{33} + e_{33}^2\omega_3 + \in_{33} E_3, \\ D_1 = 2e_{15}^1\varepsilon_{31} + e_{15}^2\omega_1 + \in_{11} E_1, \\ D_2 = 2e_{15}^1\varepsilon_{32} + e_{15}^2\omega_2 + \in_{11} E_2. \end{cases} \quad (1)$$

where C_{ij}, k_i are elastic constants in the phonon and phason fields; R_i are phonon-phason coupling elastic constants; e_{ij}^1, e_{ij}^2 are piezoelectric coefficients; $\in_{11}$, $\in_{33}$ are two dielectric coefficients.

Geometric equation:

$$\begin{cases} \varepsilon_{ij} = 0.5(\partial_j u_i + \partial_i u_j), \\ \omega_j = \partial_j \omega, \\ E_j = -\partial_j \phi, (i, j = 1, 2, 3). \end{cases} \quad (2)$$

Regardless of the energy balance equation:

$$\begin{cases} \partial_1\sigma_{11} + \partial_2\sigma_{12} + \partial_3\sigma_{13} = 0, \\ \partial_1\sigma_{21} + \partial_2\sigma_{22} + \partial_3\sigma_{23} = 0, \\ \partial_1\sigma_{31} + \partial_2\sigma_{32} + \partial_3\sigma_{33} = 0, \\ \partial_1 H_1 + \partial_2 H_2 + \partial_3 H_3 = 0, \\ \partial_1 D_1 + \partial_2 D_2 + \partial_3 D_3 = 0. \end{cases} \tag{3}$$

In one-dimensional hexagonal piezoelectric quasicrystals, When the defect through the x_3 direction, material geometric properties will not with quasi-periodicity direction change, in that way

$$\partial_3 u_i = 0, \partial_3\omega = 0, \partial_3\sigma_{ij} = 0, \partial_3 H_i = 0, \partial_3 D_i = 0, (i, j = 1, 2, 3) \tag{4}$$

Substituting (4) into (1)-(3), we obtain

$$\begin{cases} C_{44}\nabla^2 u_3 + R_3\nabla^2\omega + e_{15}^1\nabla^2\phi = 0, \\ R_3\nabla^2 u_3 + K_2\nabla^2\omega + e_{15}^2\nabla^2\phi = 0, \\ e_{15}^1\nabla^2 u_3 + e_{15}^2\nabla^2\omega - \in_{44} \nabla^2\phi = 0. \end{cases} \tag{5}$$

When $\begin{vmatrix} C_{44} & R_3 & e_{15}^1 \\ R_3 & K_2 & e_{15}^2 \\ e_{15}^1 & e_{15}^2 & -\in_{44} \end{vmatrix} \neq 0$, type (5) can be written as a further

$$\nabla^2 u_3 = 0, \nabla^2\omega = 0, \nabla^2\phi = 0. \tag{6}$$

From (6) we know, this problem can be thought of as three harmonic equation of solving problem, using the complex function theory knowledge, using u_3, ω, ϕ express as three analytic function of the real component.

$$u_3 = \mathrm{Re}\,\varphi_1(z), \omega = \mathrm{Re}\,\varphi_2(z), \phi = \mathrm{Re}\,\varphi_3(z) \tag{7}$$

where Re express as analytic function of the real component, $\varphi_i(z)(i = 1, 2, 3)$ express as any three of the analytic function, $z = x_1 + ix_2, i = \sqrt{-1}$.

Now, using the complex variable function method to solve the above elastic balance issue.

Using the nature of the analytic function and Cauchy—Riemann theorem, from (1)-(4) obtain

$$\begin{cases} \sigma_{23} = \sigma_{32} = -\dfrac{1}{2i}[C_{44}(\varphi_1' - \overline{\varphi_1'}) + R_3(\varphi_2' - \overline{\varphi_2'}) + e_{15}^1(\varphi_3' - \overline{\varphi_3'})], \\ H_2 = -\dfrac{1}{2i}[R_3(\varphi_1' - \overline{\varphi_1'}) + k_2(\varphi_2' - \overline{\varphi_2'}) + e_{15}^2(\varphi_3' - \overline{\varphi_3'})], \\ D_2 = -\dfrac{1}{2i}[e_{15}^1(\varphi_1' - \overline{\varphi_1'}) + e_{15}^2(\varphi_2' - \overline{\varphi_2'}) - \in_{11} (\varphi_3' - \overline{\varphi_3'})]. \end{cases} \tag{8}$$

where $\varphi_i' = d\varphi_i \ / \ dz$, for simplicity, using φ_i express $\varphi_i(z), (i = 1,2,3)$.

2. Two asymmetric elliptical hole edge crack problem (Impermeable problem)

2.1 *The permeability of stress field*

If an one-dimensional hexagonal piezoelectric quasicrystals has two asymmetric elliptical hole edge crack along the quasi-periodic direction penetration (x_3 direction), elliptic semi-major axis for a long, elliptical short half shaft for b long, craze for c_1-a, c_2-a long , elliptical hole and the boundary of the crack for L long, τ is a quasi-cycle the direction of shear force in infinity. The following chart

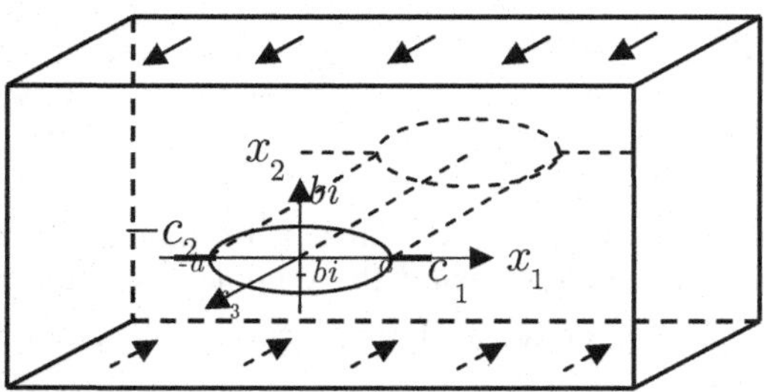

That by linear elastic theory, this problem can be considered only in the crack surface stress $\sigma_{32} = -\tau_1, H_{32} = -\tau_2, D_2 = -\text{T}$, impermeable boundary condition for

$$\begin{cases} \sqrt{x_1^2 + x_2^2} \to \infty : \sigma_{32} = 0, H_2 = 0, D_2 = 0; \\ (x_1, x_2) \in L : \sigma_{32} = -p, H_2 = -q, D_2 = -\text{T}. \end{cases} \tag{9}$$

Substituting (9) into (8), we obtain

$$\begin{cases} C_{44}(\varphi_1' - \overline{\varphi_1'}) + R_3(\varphi_2' - \overline{\varphi_2'}) + e_{15}^1(\varphi_3' - \overline{\varphi_3'}) = 2pi, \\ R_3(\varphi_1' - \overline{\varphi_1'}) + k_2(\varphi_2' - \overline{\varphi_2'}) + e_{15}^2(\varphi_3' - \overline{\varphi_3'}) = 2qi, \\ e_{15}^1(\varphi_1' - \overline{\varphi_1'}) + e_{15}^2(\varphi_2' - \overline{\varphi_2'}) - \in_{11}(\varphi_3' - \overline{\varphi_3'}) = 2Ti. \end{cases} \tag{10}$$

Introducing conformal mapping [9]

$$z_1 = z_2 = z = \omega(\zeta) = \frac{a+b}{2} \cdot \frac{\kappa}{4\zeta} + \frac{a-b}{2} \cdot \frac{4\zeta}{\kappa} \tag{11}$$

where

$$\alpha_1 = \frac{d_1^2 + 1}{2d_1}, \alpha_2 = \frac{d_2^2 + 1}{2d_2},$$

$$d_1 = \frac{c_1 + \sqrt{c_1^2 - a^2 + b^2}}{a+b}, d_2 = \frac{c_2 + \sqrt{c_2^2 - a^2 + b^2}}{a+b},$$

$$\kappa = \alpha_1(1+\zeta)^2 + \alpha_2(1-\zeta)^2 +$$

$$\sqrt{(\alpha_1^2 - 1)(1+\zeta)^4 + 2(\alpha_1\alpha_2 + 1)(1-\zeta^2)^2 + (\alpha_2^2 - 1)(1-\zeta)^4}$$

Using this mapping, the z plane area mapping into unit circle internal of ζ plane, the z plane boundary L map into the circumference of the unit circle η of ζ plane, $\omega^{-1}(c_1) \to 1$, $\omega^{-1}(-c_2) \to -1$, $\omega^{-1}(bi) \to A_1$, $\omega^{-1}(-bi) \to A_2$. The a mirror to B_1 shore, the a next to reflect the B_2 shore.

Define

$$\psi_i(\zeta) = \varphi_i(z) = \varphi_i(w(\zeta)), (i = 1,2,3)$$

where

$$\varphi_i'(\mathrm{z}) = \frac{\psi_i'(\zeta)}{w'(\zeta)}. \tag{12}$$

Because $\psi_i'(\zeta)$ in unit circle analytic, therefore

$$\frac{1}{2\pi i}\int_\eta \frac{\psi_i'(\sigma)}{\sigma - \zeta} d\sigma = \psi_i'(\zeta), \tag{13}$$

Because $\frac{\overline{\omega'(\sigma)}}{\omega'(\sigma)} = -\frac{1}{\sigma^2}$, so $\frac{\overline{\omega'(\sigma)}}{\omega'(\sigma)}\overline{G_i'(\sigma)}$ is the boundary values of the analytical function $-\frac{1}{\zeta^2}\overline{G_i'}(\frac{1}{\zeta})$ in unit circle external. Using the Cauchy formula, so

$$\frac{1}{2\pi i}\int_\eta \frac{\overline{w'(\sigma)}}{w'(\sigma)} \cdot \frac{\overline{\psi_i'(\sigma)}}{\sigma - \zeta} d\sigma = 0, \tag{14}$$

Define

$$\frac{1}{2\pi i}\int_\eta \frac{\omega'(\sigma)}{\sigma - \zeta} d\sigma = \frac{(\alpha_1 + \alpha_2)(a+b)}{4} = F(\zeta). \tag{15}$$

Substituting (12) into (10), both sides are multiplied by $1/(2\pi i(\sigma - \zeta))$, along the boundary η integral, using(13)-(15), we obtai

$$\begin{cases}
\psi_1'(\zeta) = 2\mathrm{i} \cdot F(\zeta) \cdot \\
\dfrac{e_{15}^1 k_2 T + (e_{15}^2)^2 p + \in_{11} k_2 p - R_3 \in_{11} q - R_3 e_{15}^2 T - e_{15}^1 e_{15}^2 q}{C_{44}(e_{15}^2)^2 - 2e_{15}^1 e_{15}^2 R_3 + k_2 (e_{15}^1)^2 - R_3{}^2 \in_{11} + C_{44} \in_{11} k_2}, \\
\psi_2'(\zeta) = 2\mathrm{i} \cdot F(\zeta) \cdot \\
\dfrac{(e_{15}^1)^2 q + C_{44} \in_{11} q - R_3 \in_{11} p + C_{44} e_{15}^2 T - e_{15}^1 e_{15}^2 p - e_{15}^1 R_3 T}{C_{44}(e_{15}^2)^2 - 2e_{15}^1 e_{15}^2 R_3 + k_2 (e_{15}^1)^2 - R_3{}^2 \in_{11} + C_{44} \in_{11} k_2}, \\
\psi_3'(\zeta) = 2\mathrm{i} \cdot F(\zeta) \cdot \\
\dfrac{R_3{}^2 T + e_{15}^1 k_2 p - R_3 e_{15}^1 q - R_3 e_{15}^2 p + C_{44} e_{15}^2 q - C_{44} k_2 T}{C_{44}(e_{15}^2)^2 - 2e_{15}^1 e_{15}^2 R_3 + k_2 (e_{15}^1)^2 - R_3{}^2 \in_{11} + C_{44} \in_{11} k_2}.
\end{cases} \tag{16}$$

Using the above formula, we can get the original plane crack surface displacement field and stress field of expression, due to the complicated situation, it is not listed here.

2.2 *Impermeable stress intensity factor of the crack tip*

By reference [8], using (16), in the ζ plane, we obtain the crack tip of phonon field, phason field and electric displacement intensity factor when $z = c_1$, corresponding $\zeta = 1$

$$\begin{cases} K_{III}^{II} = 2\sqrt{\pi}p\lim\limits_{\zeta\to 1}\dfrac{F(\zeta)}{\sqrt{\omega''(\zeta)}} \\ = \dfrac{p(a+b)\sqrt{\pi(\alpha_1+\alpha_2)}({\alpha_1}^2-1)^{1/4}\sqrt{\alpha_1+\sqrt{{\alpha_1}^2-1}}}{\sqrt{(a+b)(\alpha_1+\sqrt{{\alpha_1}^2-1})^2-(a-b)}}, \\ K_{III}^{\perp} = \lim\limits_{\zeta\to 1}2\sqrt{\pi}q\dfrac{F(\zeta)}{\sqrt{w''(\zeta)}} \\ = \dfrac{q(a+b)\sqrt{\pi(\alpha_1+\alpha_2)}({\alpha_1}^2-1)^{1/4}\sqrt{\alpha_1+\sqrt{{\alpha_1}^2-1}}}{\sqrt{(a+b)(\alpha_1+\sqrt{{\alpha_1}^2-1})^2-(a-b)}}, \\ K_{III}^{D} = 2\sqrt{\pi}\mathrm{T}\lim\limits_{\zeta\to 1}\dfrac{F(\zeta)}{\sqrt{\omega''(\zeta)}} \\ = \dfrac{\mathrm{T}(a+b)\sqrt{\pi(\alpha_1+\alpha_2)}({\alpha_1}^2-1)^{1/4}\sqrt{\alpha_1+\sqrt{{\alpha_1}^2-1}}}{\sqrt{(a+b)(\alpha_1+\sqrt{{\alpha_1}^2-1})^2-(a-b)}}. \end{cases} \tag{17}$$

(1) When $a \to b$, model is becoming an one-dimensional hexagonal piezoelectric quasicrystals has two asymmetric round hole edge cracks problem when $z = c_1$, corresponding $\zeta = 1$

$$\begin{cases} K_{III}^{II} = 2\sqrt{\pi}p\lim\limits_{\zeta\to 1}\dfrac{F(\zeta)}{\sqrt{\omega''(\zeta)}} \\ = p\sqrt{2\pi a(\alpha_1+\alpha_2)\sqrt{{\alpha_1}^2-1}(\alpha_1-\sqrt{{\alpha_1}^2-1})}, \\ K_{III}^{\perp} = \lim\limits_{\zeta\to 1}2\sqrt{\pi}q\dfrac{F(\zeta)}{\sqrt{w''(\zeta)}} \\ = q\sqrt{2\pi a(\alpha_1+\alpha_2)\sqrt{{\alpha_1}^2-1}(\alpha_1-\sqrt{{\alpha_1}^2-1})}, \\ K_{III}^{D} = 2\sqrt{\pi}\mathrm{T}\lim\limits_{\zeta\to 1}\dfrac{F(\zeta)}{\sqrt{\omega''(\zeta)}} \\ = \mathrm{T}\sqrt{2\pi a(\alpha_1+\alpha_2)\sqrt{{\alpha_1}^2-1}(\alpha_1-\sqrt{{\alpha_1}^2-1})}. \end{cases} \tag{18}$$

(2) When one-dimensional hexagonal piezoelectric quasicrystals becoming a crystals, and $b \to 0, c_1 \to c_2$

When $z = c_1$, corresponding $\zeta = 1$

$$\begin{cases} K_{III}^{II} = p\sqrt{\pi c}, \\ K_{III}^{D} = \mathrm{T}\sqrt{\pi c}. \end{cases} \tag{19}$$

This conclusion and references [10] the conclusion consistent.

3. Two asymmetric elliptical hole edge crack problem (Permeability problem)

3.1 *The permeability of stress field*

Permeable boundary condition

$$\begin{cases} \sqrt{x_1^2 + x_2^2} \to \infty : \sigma_{32} = 0, H_2 = 0, D_2 = 0; \\ (x_1, x_2) \in L : \sigma_{32} = -p, H_2 = -q, \phi\text{=constant}. \end{cases} \tag{20}$$

Substituting (20) into (8), we obtain

$$\begin{cases} C_{44}(\varphi_1' - \overline{\varphi_1'}) + R_3(\varphi_2' - \overline{\varphi_2'}) = 2pi, \\ R_3(\varphi_1' - \overline{\varphi_1'}) + k_2(\varphi_2' - \overline{\varphi_2'}) = 2qi, \\ e_{15}^1(\varphi_1' - \overline{\varphi_1'}) + e_{15}^2(\varphi_2' - \overline{\varphi_2'}) \\ = -\dfrac{(e_{15}^1 k_2 - e_{15}^2 R_3)p + (e_{15}^2 C_{44} - e_{15}^1 R_3)q}{C_{44}k_2 - R_3^2}. \end{cases} \tag{21}$$

Substituting (12) into (21), both sides are multiplied by $1 / (2\pi i(\sigma - \zeta))$, along the boundary η integral, using(13)-(15),we obtain

$$\begin{cases} \psi_1'(\zeta) = 2\mathrm{i}\dfrac{k_2 p - R_3 q}{C_{44}k_2 - R_3^2} \cdot F(\zeta), \\ \psi_2'(\zeta) = 2\mathrm{i}\dfrac{C_{44}q - R_3 p}{C_{44}k_2 - R_3^2} \cdot F(\zeta). \end{cases} \tag{22}$$

Similarly, using the above formula, We can get the original plane crack surface displacement field and stress field of expression, also, we can available permeability crack boundary electric displacement, however, due to the complicated situation, it is not listed here.

3.2 *Permeable stress intensity factor of the crack tip*

In the same way, by reference [8], using (22), in the ζ plane, we obtain the crack tip of phonon field, phason field and electric displacement intensity factor

When $z = c_1$, corresponding $\zeta = 1$

$$\begin{cases} K_{III}^{II} = 2\sqrt{\pi} p \lim\limits_{\zeta \to 1} \dfrac{F(\zeta)}{\sqrt{\omega''(\zeta)}} \\ = \dfrac{p(a+b)\sqrt{\pi(\alpha_1+\alpha_2)}(\alpha_1{}^2-1)^{1/4}\sqrt{\alpha_1+\sqrt{\alpha_1{}^2-1}}}{\sqrt{(a+b)(\alpha_1+\sqrt{\alpha_1{}^2-1})^2-(a-b)}}, \\ K_{III}^{\perp} = \lim\limits_{\zeta \to 1} 2\sqrt{\pi} q \dfrac{F(\zeta)}{\sqrt{w''(\zeta)}} \\ = \dfrac{q(a+b)\sqrt{\pi(\alpha_1+\alpha_2)}(\alpha_1{}^2-1)^{1/4}\sqrt{\alpha_1+\sqrt{\alpha_1{}^2-1}}}{\sqrt{(a+b)(\alpha_1+\sqrt{\alpha_1{}^2-1})^2-(a-b)}}, \\ K_{III}^{D} = \lim\limits_{\zeta \to 1} 2\sqrt{\pi} \dfrac{(e_{15}^1 k_2 - e_{15}^2 R_3)p + (e_{15}^2 C_{44} - e_{15}^1 R_3)q}{C_{44}k_2 - R_3{}^2} \dfrac{F(\zeta)}{\sqrt{w''(\zeta)}} \\ = \dfrac{(e_{15}^1 k_2 - e_{15}^2 R_3)p + (e_{15}^2 C_{44} - e_{15}^1 R_3)q}{C_{44}k_2 - R_3{}^2} \cdot \\ \dfrac{(a+b)\sqrt{\pi(\alpha_1+\alpha_2)}(\alpha_1{}^2-1)^{1/4}\sqrt{\alpha_1+\sqrt{\alpha_1{}^2-1}}}{\sqrt{(a+b)(\alpha_1+\sqrt{\alpha_1{}^2-1})^2-(a-b)}}. \end{cases} \tag{23}$$

(1) When $a \to b$, model is becoming an one-dimensional hexagonal piezoelectric quasicrystals has two asymmetric round hole edge cracks problem

When $z = c_1$, corresponding $\zeta = 1$

$$\begin{cases} K_{III}^{II} = 2\sqrt{\pi} p \lim\limits_{\zeta \to 1} \dfrac{F(\zeta)}{\sqrt{\omega''(\zeta)}} \\ = p\sqrt{2\pi a(\alpha_1+\alpha_2)\sqrt{\alpha_1{}^2-1}(\alpha_1-\sqrt{\alpha_1{}^2-1})}, \\ K_{III}^{\perp} = \lim\limits_{\zeta \to 1} 2\sqrt{\pi} q \dfrac{F(\zeta)}{\sqrt{w''(\zeta)}} \\ = q\sqrt{2\pi a(\alpha_1+\alpha_2)\sqrt{\alpha_1{}^2-1}(\alpha_1-\sqrt{\alpha_1{}^2-1})}, \\ K_{III}^{D} = \lim\limits_{\zeta \to 1} 2\sqrt{\pi} \dfrac{(e_{15}^1 k_2 - e_{15}^2 R_3)p + (e_{15}^2 C_{44} - e_{15}^1 R_3)q}{C_{44}k_2 - R_3{}^2} \dfrac{F(\zeta)}{\sqrt{w''(\zeta)}} \\ = \dfrac{(e_{15}^1 k_2 - e_{15}^2 R_3)p + (e_{15}^2 C_{44} - e_{15}^1 R_3)q}{C_{44}k_2 - R_3{}^2} \cdot \\ \sqrt{2\pi a(\alpha_1+\alpha_2)\sqrt{\alpha_1{}^2-1}(\alpha_1-\sqrt{\alpha_1{}^2-1})}. \end{cases} \tag{24}$$

(2) When one-dimensional hexagonal piezoelectric quasicrystals becoming a crystals, and $b \to 0, c_1 \to c_2$

When $z = c_1$, corresponding $\zeta = 1$

$$\begin{cases} K_{III}^{II} = p\sqrt{\pi c}, \\ K_{III}^{D} = \dfrac{e_{15}^{1}}{C_{44}} \cdot \sqrt{\pi c} p. \end{cases} \tag{25}$$

This conclusion is consistent with reference [11].

4. Conclusion

This paper, by using generalized complex variable method, considering of the electric field on the one-dimensional hexagonal piezoelectric quasicrystals influence, it has two asymmetric and along the quasi-periodic direction penetration of elliptical hole edge cracks, and through the use of appropriate conformal mapping, obtained the field under the action of electric impermeable and permeable situation III type crack stress intensity factor of the crack tip. We can be seen in the electric impenetrating cases, the phonon field, phason field stress and electric displacement at the crack tip are singular and are dependent on the magnitude of stress and crack size; but in point penetration cases, the electrical load can be seen on the crack tip singularity no effects. In the model tends to the limit, we get the stress intensity factor results and classical results are consistent. This paper solves problems help rich quasicrystal material functional, have certain practical value.

References

1. Shechtman D, Blech I, Gratias D, Cahn J W. Metallic phase with long-rangeorientational order and no translational symmetry [J]. Phys. Rev. Lett., 1984, 53: 1951-1953.
2. Chen J Z. Structure and symmetry new theory of quasicrystals [M]. Wuhan: Huazhong University of Science and Technology, 1996. (in Chinese)
3. Liu Y Y, Fu X J, Dong X X. Physical properties of one-dimensional quasicrystals [J]. Physics progress, 1997, 17: 1-63.
4. Ding D H, Wang R H, Yang W J, etc. Elasticity, plasticity and dislocations of quasicrystals [J]. Progress in Physics, 1998, 18: 223-260.
5. Li X F, Fan T Y. A straight dislocation in one dimensional hexagonal quasicrystals [J]. Phy Stat B, 1999, 212: 19-26.
6. Fan T Y. Mathematical theory of elasticity and defects of quasicrystals[J]. Advances in Mechanics, 2000(30): 161-174.

7. Wang X, Pan E. Analytical solutions for some defect problems in 1D hexagonal and 2D octagonal quasicrystals [J]. Pramana, 2008, 70: 911-933.
8. Fan T Y. Mathematical Theory of Elasticity of Quasicrystals and Its Applications [M]. Beijing: Science Press, 2010.
9. Guo J H, Liu G T. Analytic solutions of problem about an elliptic hole with two straight cracks in one-dimensional hexagonal quasicrystals [J]. Applied Mathematics and Mechanics, 2008, 439-446.
10. Wang B L. Piezoelectric material and structure of fracture mechanics [M]. Beijing: National defence industry press, 2002.
11. Gao C F, Wang M Z. Periodical cracks in piezoelectric media [J]. Mechanics Research Communications, 1999, 26(4): 427-638.
12. Wang S Z, Yi X L. The elastic problems complex-variable function method of application and development [J]. Mechanics in Engineering, 2009: 110-113.
13. Lu J K. Elastic plane complex variable method [M]. Wuhan: Wuhan University press, 2005. (in Chinese)
14. Wang W S, Li X. Anti-plane problems of periodic permeable cracks in piezoelectric medium [J]. Chinese Journal of Engineering Mathematics, 2007, 24(2): 209-214.
15. Muskhelishvili, write. Zhao H Y, translate. Mathematical theory of elastic mechanics, several basic problems [M]. Science press, 1958.

THE SECOND FUNDAMENTAL PROBLEM OF PERIODIC PLANE ELASTICITY OF A ONE-DIMENSIONAL HEXAGONAL QUASICRYSTALS*

JIANG-YAN CUI, PENG-PENG SHI, XING LI†

School of Mathematics and Computer Science, Ningxia University, Yinchuan 750021,

People's Republic of China

In this paper, we disscuss one-dimensional hexagonal quasicrystals and infinite elasticity plane with single periodic of second fundamental problem by the condition basical hypothesis. We use the plane elasticity of complex variables method and Hilbert kernel integral formula proving the one-dimensional hexagonal quasicrystals and infinite elasticity plane with single periodic of second fundamental problem, and obtaining the solution is not only existence but also unique. And in the proving process, we provide the specific arithmetic. At last, we offer an example to verify.

Keywords: one-dimensional hexagonal quasicrystals; elasticity plane; periodic; fundamental problem

1. *The one-dimensional hexagonal quasicrystals and infinite elasticity plane with single periodic of stress function*

1.1. *Basic assumptions*

In this paper, we disscuss the one-dimensional hexagonal quasicrystals and infinite elasticity plane with single periodic of second fundamental problem in the following basic assumptions.

* This work is supported by the National Natural Science Foundation of China (10962008; 11261045; 51061015) and Research Fund for the Doctoral Program of Higher Education of China (20116401110002).

† Corresponding author: Xing Li, Tel.: +86 951 2061405. E-mail address: li_x@nxu.edu.cn

In the infinite elasticity plane, we assume the stress and displacement to be the periodic of $a\pi$, and the stress is bounded, but the displacement is not necessary bounded, and we discuss the boundary conditions are the periodic of $a\pi$.

On the boundary of a periodic segment, the principal stress vector is $X+iY$, and by the equilibrium conditions, we have

$$\sigma_{33}(-\infty i)=\frac{Y}{a\pi},\sigma_{23}(-\infty i)=\frac{X}{a\pi},H_3(-\infty i)=\frac{Z}{a\pi} \tag{1}$$

So, we just need to discuss elastomer within one periodic.

For the periodic problem in this paper, mostly with elastomer has $z=x_2+ix_3$ plane of the lower half plane S^-, this time, in the $z=-\infty i$, the stress is existence, and the stress is understood as the limit when $z\to-\infty i$, according to the basic assumptions, this limit is finite.

1.2. *The periodic of stress function*

For the quasicrystals plane in infinite elastic, the stress components $\sigma_{22},\sigma_{33},\sigma_{23},H_3$ can be described as $f_1'(z_1), f_2'(z_2), f_3'(z_3)$,and the displacement components U,V,W can be described as $f_1(z_1), f_2(z_2), f_3(z_3)$,that is

$$\sigma_{22} = u_1^2 f_1'(z_1) + \overline{u_1^2 f_1'(z_1)} + u_2^2 f_2'(z_2) + \overline{u_2^2 f_2'(z_2)} + u_3^2 f_3'(z_3) + \overline{u_3^2 f_3'(z_3)} \tag{2}$$

$$\sigma_{33} = f_1'(z_1) + \overline{f_1'(z_1)} + f_2'(z_2) + \overline{f_2'(z_2)} + f_3'(z_3) + \overline{f_3'(z_3)} \tag{3}$$

$$\sigma_{23} = -[u_1 f_1'(z_1) + \overline{u_1 f_1'(z_1)} + u_2 f_2'(z_2) + \overline{u_2 f_2'(z_2)} + u_3 f_3'(z_3) + \overline{u_3 f_3'(z_3)}] \tag{4}$$

$$H_3 = -[\eta_1 f_1'(z_1) + \overline{\eta_1 f_1'(z_1)} + \eta_2 f_2'(z_2) + \overline{\eta_2 f_2'(z_2)} + \eta_3 f_3'(z_3) + \overline{\eta_3 f_3'(z_3)}] \tag{5}$$

$$U = o_1 f_1(z_1) + \overline{o_1 f_1(z_1)} + o_2 f_2(z_2) + \overline{o_2 f_2(z_2)} + o_3 f_3(z_3) + \overline{o_3 f_3(z_3)} \tag{6}$$

$$V = p_1 f_1(z_1) + \overline{p_1 f_1(z_1)} + p_2 f_2(z_2) + \overline{p_2 f_2(z_2)} + p_3 f_3(z_3) + \overline{p_3 f_3(z_3)} \tag{7}$$

$$W = q_1 f_1(z_1) + \overline{q_1 f_1(z_1)} + q_2 f_2(z_2) + \overline{q_2 f_2(z_2)} + q_3 f_3(z_3) + \overline{q_3 f_3(z_3)} \tag{8}$$

With $f_1(z_1), f_2(z_2), f_3(z_3)$ are the holomorphic function of z_1, z_2, z_3, and $z = x_2 + ix_3, z_k = x_2 + u_k x_3, k = 1,2,3$. Also

$$\begin{gathered}\eta_k = \frac{-(b_1 + b_3)\mu_k^2 - b_2}{c_1\mu_k^2 + c_2}, o_k = a_1 u_k^2 + a_2 - b_1\eta_k \\ p_k = a_2 u_k + \frac{a_3}{u_k} - \frac{b_2\eta_k}{u_k}, q_k = b_1 u_k + \frac{b_2}{u_k} - \frac{c_2\eta_k}{u_k}\end{gathered} \tag{9}$$

Here, the $u_1, \overline{u}_1, u_2, \overline{u}_2, u_3, \overline{u}_3$ are one-dimensional hexagonal quasicrystals and infinite elasticity plane theory of control equation $(L_1 L_3 + L_2^2)U = 0$, $L_2 U + L_2 V = 0$ corresponding to the characteristic of the three pairs of conjugate roots. Where

$$\begin{cases} L_1 = a_3 \dfrac{\partial^4}{\partial x_2^4} + a_1 \dfrac{\partial^4}{\partial x_3^4} + (2a_2 + a_4)\dfrac{\partial^4}{\partial x_2^2 \partial x_3^2} \\ L_2 = b_2 \dfrac{\partial^3}{\partial x_2^3} + (b_1 + b_3)\dfrac{\partial^3}{\partial x_2 \partial x_3^2} \\ L_3 = c_2 \dfrac{\partial^2}{\partial x_2^2} + c_1 \dfrac{\partial^2}{\partial x_3^2} \end{cases} \tag{10}$$

Furthermore, the $a_1, a_2, a_3, a_4, b_1, b_2, b_3, c_1, c_2$ can be determined by constitutive equation of elastic coefficient with

$$\begin{cases} c_1 = -(a_1 C_{12} + a_2 C_{13} + b_1 R_1)e_1 \\ c_2 = -(a_2 C_{12} + a_3 C_{13} + b_2 R_1)e_1 \\ a_1 = \dfrac{C_{33}K_1 - R_2^2}{\Delta_1}, a_2 = \dfrac{R_1 R_2 - C_{13}K_1}{\Delta_1} \\ a_3 = \dfrac{C_{11}K_1 - R_1^2}{\Delta_1}, a_4 = \dfrac{K_2}{\Delta_2} \\ b_1 = \dfrac{C_{13}R_2 - C_{33}R_1}{\Delta_1}, b_2 = \dfrac{R_1 C_{13} - R_2 C_{11}}{\Delta_1} \\ b_3 = -\dfrac{R_3}{\Delta_2}, c_1 = \dfrac{C_{44}}{\Delta_2}, b_3 = \dfrac{C_{11}C_{33} - C_{13}^2}{\Delta_1} \\ \Delta_1 = C_{11}C_{33}K_1 + 2C_{13}R_1 R_2 - C_{33}R_1^2 - C_{11}R_2^2 - C_{13}^2 K_1 \\ \Delta_2 = C_{44}K_2 - R_3^2 \end{cases} \tag{11}$$

Here the constitutive equation of one-dimensional hexagonal quasicrystals and infinite elasticity plane is

$$\begin{cases} \sigma_{22} = C_{12}\varepsilon_{11} + C_{11}\varepsilon_{22} + C_{13}\varepsilon_{33} + R_1 w_{33} \\ \sigma_{33} = C_{13}\varepsilon_{11} + C_{13}\varepsilon_{22} + C_{33}\varepsilon_{33} + R_2 w_{33} \\ H_{33} = R_1(\varepsilon_{11} + \varepsilon_{22}) + R_2\varepsilon_{33} + K_1 w_{33} \\ \sigma_{23} = C_{44} 2\varepsilon_{32} + R_3 w_{32} \\ H_{32} = R_3 2\varepsilon_{32} + K_2 w_{32} \\ \sigma_{11} = C_{11}\varepsilon_{11} + C_{12}\varepsilon_{22} + C_{13}\varepsilon_{33} + R_1 w_{33} \end{cases} \tag{12}$$

Lemma: On the basic assumption, the stress functions $f_1'(z_1), f_2'(z_2), f_3'(z_3)$ are the periodic functions with $a\pi$.

Proof: In fact, we will do (6)-(8) for the derivation, we can get

$$\frac{\partial U}{\partial x_2} = o_1 f_1'(z_1) + \overline{o_1 f_1'(z_1)} + o_2 f_2'(z_2) + \overline{o_2 f_2'(z_2)} + o_3 f_3'(z_3) + \overline{o_3 f_3'(z_3)} \tag{13}$$

$$\frac{\partial V}{\partial x_2} = p_1 f_1'(z_1) + \overline{p_1 f_1'(z_1)} + p_2 f_2'(z_2) + \overline{p_2 f_2'(z_2)} + p_3 f_3'(z_3) + \overline{p_3 f_3'(z_3)} \tag{14}$$

$$\frac{\partial W}{\partial x_2} = q_1 f_1'(z_1) + \overline{q_1 f_1'(z_1)} + q_2 f_2'(z_2) + \overline{q_2 f_2'(z_2)} + q_3 f_3'(z_3) + \overline{q_3 f_3'(z_3)} \tag{15}$$

Then according to the displacement of periodic assumption (even if the displacement is quasi periodic), the $\frac{\partial U}{\partial x_2}, \frac{\partial V}{\partial x_2}, \frac{\partial W}{\partial x_2}$ are the periodic with $a\pi$.

We combine (2)(3)(5) and (13)(14)(15), because the determinant

$$\begin{vmatrix} u_1^2 & \overline{u}_1^2 & u_2^2 & \overline{u}_2^2 & u_3^2 & \overline{u}_3^2 \\ 1 & 1 & 1 & 1 & 1 & 1 \\ -\eta_1 & -\overline{\eta}_1 & -\eta_2 & -\overline{\eta}_2 & -\eta_3 & -\overline{\eta}_3 \\ o_1 & \overline{o}_1 & o_2 & \overline{o}_2 & o_3 & \overline{o}_3 \\ p_1 & \overline{p}_1 & p_2 & \overline{p}_2 & p_3 & \overline{p}_3 \\ q_1 & \overline{q}_1 & q_2 & \overline{q}_2 & q_3 & \overline{q}_3 \end{vmatrix} \neq 0 \tag{16}$$

Therefore, the $f_1'(z_1), f_2'(z_2), f_3'(z_3)$ can only be described as $\sigma_{22}, \sigma_{33}, H_3$, $\frac{\partial U}{\partial x_2}$, $\frac{\partial V}{\partial x_2}, \frac{\partial W}{\partial x_2}$. So, with the former periodically can get the same periodicity, and the stress functions are the periodic.

Need to point out, generally $f_1(z_1), f_2(z_2), f_3(z_3)$ are more value, but if the elastomer is periodic simply connected domain (half plane or infinite level strip), although it has periodic hole or crack, and in one of its periodic boundary, the stress of the main vector is zero, then they are single value and holomorphic.

2. The one-dimensional hexagonal quasicrystals and infinite elasticity plane with single periodic of second fundamental problem

2.1. *The solution of existence and uniqueness about second fundamental problem*

Assuming the one-dimensional hexagonal quasicrystals are the plane of the lower half plane with $z = x_2 + ix_3$,on the x_2 shaft with $z = t$,which the t is a real number. On x_2 shaft, known displacement of distributions are

$$U = g_1(t), V = g_2(t), W = g_3(t),$$

Then

$$\frac{\partial U}{\partial x_2} = g_1'(t), \frac{\partial V}{\partial x_2} = g_2'(t), \frac{\partial W}{\partial x_2} = g_3'(t) \tag{17}$$

In which they partially satisfy with Holder condition and for period of $a\pi$.

With the basic assumption, known as displacement,how to solve the elastic balance is named second fundamental problem.

Theorem: In the above conditions, the one-dimensional hexagonal quasicrystals and infinite elasticity plane with single periodic of second fundamental problem have the solution which are existence and uniqueness.

Proof: We will provide a detailed solution and calculation method in the process of proving.

First of all, as a linear combination of the component stress

We will eliminate $f_2'(z_2), f_3'(z_3)$ in equation (13)-(15)

$$\frac{\partial U}{\partial x_2} + m_1 \frac{\partial V}{\partial x_2} + n_1 \frac{\partial W}{\partial x_2} = f_1'(z_1)(o_1 + m_1 p_1 + n_1 q_1) + \overline{f_1'(z_1)}(\overline{o_1} + m_1 \overline{p_1} + n_1 \overline{q_1}) \tag{18}$$
$$+ \overline{f_2'(z_2)}(\overline{o_2} + m_1 \overline{p_2} + n_1 \overline{q_2}) + \overline{f_3'(z_3)}(\overline{o_3} + m_1 \overline{p_3} + n_1 \overline{q_3})$$

In which $m_1 = \dfrac{q_3 - q_2}{p_2 q_3 - p_3 q_2}, n_1 = \dfrac{p_2 - p_3}{p_2 q_3 - p_3 q_2}$

We will eliminate $f_1'(z_1), f_3'(z_3)$ in equation (13)-(15)

$$\frac{\partial U}{\partial x_2} + m_2 \frac{\partial V}{\partial x_2} + n_2 \frac{\partial W}{\partial x_2} = f_2'(z_2)(o_2 + m_2 p_2 + n_2 q_2) + \overline{f_1'(z_1)}(\overline{o_1} + m_2 \overline{p_1} + n_2 \overline{q_1}) \quad (19)$$
$$+\overline{f_2'(z_2)}(\overline{o_2} + m_2 \overline{p_2} + n_2 \overline{q_2}) + \overline{f_3'(z_3)}(\overline{o_3} + m_2 \overline{p_3} + n_2 \overline{q_3})$$

In which $m_2 = \dfrac{q_3 - q_1}{p_1 q_3 - p_3 q_1}, n_2 = \dfrac{p_1 - p_3}{p_1 q_3 - p_3 q_1}$

We will eliminate $f_1'(z_1), f_2'(z_2)$ in equation (13)-(15)

$$\frac{\partial U}{\partial x_2} + m_3 \frac{\partial V}{\partial x_2} + n_3 \frac{\partial W}{\partial x_2} = f_3'(z_3)(o_3 + m_3 p_3 + n_3 q_3) + \overline{f_1'(z_1)}(\overline{o_1} + m_3 \overline{p_1} - n_3 \overline{q_1}) \quad (20)$$
$$+\overline{f_2'(z_2)}(\overline{o_2} - m_3 \overline{p_2} - n_3 \overline{q_2}) + \overline{f_3'(z_3)}(\overline{o_3} - m_3 \overline{p_3} - n_3 \overline{q_3})$$

In which $m_3 = \dfrac{q_2 - q_1}{p_1 q_2 - p_2 q_1}, n_3 = \dfrac{p_1 - p_2}{p_1 q_2 - p_2 q_1}$

Then make to z_1, z_2, z_3 tend x_2 with t, by (18)-(20) and using condition (17), we can get

$$g_1'(t) + m_1 g_2'(t) + n_1 g_3'(t) = f_1'(z_1)(o_1 + m_1 p_1 + n_1 q_1) + \overline{f_1'(z_1)}(\overline{o_1} + m_1 \overline{p_1} + n_1 \overline{q_1})$$
$$+\overline{f_2'(z_2)}(\overline{o_2} + m_1 \overline{p_2} + n_1 \overline{q_2}) + \overline{f_3'(z_3)}(\overline{o_3} + m_1 \overline{p_3} + n_1 \overline{q_3})$$
$$g_1'(t) + m_2 g_2'(t) + n_2 g_3'(t) = f_2'(z_2)(o_2 + m_2 p_2 + n_2 q_2) + \overline{f_1'(z_1)}(\overline{o_1} + m_2 \overline{p_1} + n_2 \overline{q_1})$$
$$+\overline{f_2'(z_2)}(\overline{o_2} + m_2 \overline{p_2} + n_2 \overline{q_2}) + \overline{f_3'(z_3)}(\overline{o_3} + m_2 \overline{p_3} + n_2 \overline{q_3})$$
$$g_1'(t) + m_3 g_2'(t) + n_3 g_3'(t) = f_3'(z_3)(o_3 + m_3 p_3 + n_3 q_3) + \overline{f_1'(z_1)}(\overline{o_1} + m_3 \overline{p_1} - n_3 \overline{q_1})$$
$$+\overline{f_2'(z_2)}(\overline{o_2} - m_3 \overline{p_2} - n_3 \overline{q_2}) + \overline{f_3'(z_3)}(\overline{o_3} - m_3 \overline{p_3} - n_3 \overline{q_3})$$

Then, we will multiplied $\dfrac{1}{2a\pi i}\cot\dfrac{t-z_1}{a}, \dfrac{1}{2a\pi i}\cot\dfrac{t-z_2}{a}, \dfrac{1}{2a\pi i}\cot\dfrac{t-z_3}{a}$ by these three equation sides and along $l_0 : -\dfrac{a\pi}{2} \le t \le \dfrac{a\pi}{2}$ integral,and use the Hilbert kernel integral formula

$$\frac{1}{2a\pi i}\int_{l_0} g(t)\cot\frac{t-z}{a}dt = -g(z) + \frac{1}{2}g(-\infty i), \frac{1}{2a\pi i}\int_{l_0} \overline{g(t)}\cot\frac{t-z}{a}dt = \frac{1}{2}\overline{g(-\infty i)}$$

can get

$$\frac{1}{2a\pi i}\int_{l_0}[g_1'(t)+m_1g_2'(t)+n_1g_3'(t)]\cot\frac{t-z_1}{a}dt$$
$$=-f_1'(z_1)(o_1+m_1p_1+n_1q_1)+\frac{1}{2}f_1'(-\infty i)(o_1+m_1p_1+n_1q_1)$$
$$+\frac{1}{2}\overline{f_1'(-\infty i)}(\overline{o_1}+m_1\overline{p_1}+n_1\overline{q_1})+\frac{1}{2}\overline{f_2'(-\infty i)}(\overline{o_2}+m_1\overline{p_2}+n_1\overline{q_2})$$
$$+\frac{1}{2}\overline{f_3'(-\infty i)}(\overline{o_3}+m_1\overline{p_3}+n_1\overline{q_3})$$

$$\frac{1}{2a\pi i}\int_{l_0}[g_1'(t)+m_2g_2'(t)+n_2g_3'(t)]\cot\frac{t-z_2}{a}dt=$$
$$-f_2'(z_2)(o_2+m_2p_2+n_2q_2)+\frac{1}{2}f_2'(-\infty i)(o_2+m_2p_2+n_2q_2)$$
$$+\frac{1}{2}\overline{f_1'(-\infty i)}(\overline{o_1}+m_2\overline{p_1}+n_2\overline{q_1})+\frac{1}{2}\overline{f_2'(-\infty i)}(\overline{o_2}+m_2\overline{p_2}+n_2\overline{q_2})$$
$$+\frac{1}{2}\overline{f_3'(-\infty i)}(\overline{o_3}+m_2\overline{p_3}+n_2\overline{q_3})$$

$$\frac{1}{2a\pi i}\int_{l_0}[g_1'(t)+m_3g_2'(t)+n_3g_3'(t)]\cot\frac{t-z_3}{a}dt$$
$$=-f_3'(z_3)(o_3+m_3p_3+n_3q_3)+\frac{1}{2}f_3'(-\infty i)(o_3+m_3p_3+n_3q_3)$$
$$+\frac{1}{2}\overline{f_1'(-\infty i)}(\overline{o_1}+m_3\overline{p_1}+n_3\overline{q_1})+\frac{1}{2}\overline{f_2'(-\infty i)}(\overline{o_2}+m_3\overline{p_2}+n_3\overline{q_2})$$
$$+\frac{1}{2}\overline{f_3'(-\infty i)}(\overline{o_3}+m_3\overline{p_3}+n_3\overline{q_3})$$

So

$$f_1'(z_1)=\frac{\int_{l_0}[g_1'(t)+m_1g_2'(t)+n_1g_3'(t)]\cot\frac{t-z_1}{a}dt}{2a\pi i(o_1+m_1p_1+n_1q_1)}+\gamma_1 \tag{21}$$

$$f_2'(z_2)=\frac{\int_{l_0}[g_1'(t)+m_2g_2'(t)+n_2g_3'(t)]\cot\frac{t-z_2}{a}dt}{2a\pi i(o_2+m_2p_2+n_2q_2)}+\gamma_2 \tag{22}$$

$$f_3'(z_3)=\frac{\int_{l_0}[g_1'(t)+m_3g_2'(t)+n_3g_3'(t)]\cot\frac{t-z_3}{a}dt}{2a\pi i(o_3+m_3p_3+n_3q_3)}+\gamma_3 \tag{23}$$

in which

$$\gamma_1=\frac{1}{2(o_1+m_1p_1+n_1q_1)}[f_1'(-\infty i)(o_1+m_1p_1+n_1q_1)$$
$$+\overline{f_1'(-\infty i)}(\overline{o_1}+m_1\overline{p_1}+n_1\overline{q_1})+\overline{f_2'(-\infty i)}(\overline{o_2}+m_1\overline{p_2}+n_1\overline{q_2}) \quad (24)$$
$$+\overline{f_3'(-\infty i)}(\overline{o_3}+m_1\overline{p_3}+n_1\overline{q_3})]$$

$$\gamma_2=\frac{1}{2(o_2+m_2p_2+n_2q_2)}[f_2'(-\infty i)(o_2+m_2p_2+n_2q_2)$$
$$+\overline{f_1'(-\infty i)}(\overline{o_1}+m_2\overline{p_1}+n_2\overline{q_1})+\overline{f_2'(-\infty i)}(\overline{o_2}+m_2\overline{p_2}+n_2\overline{q_2}) \quad (25)$$
$$+\overline{f_3'(-\infty i)}(\overline{o_3}+m_2\overline{p_3}+n_2\overline{q_3})]$$

$$\gamma_3=\frac{1}{2(o_3+m_3p_3+n_3q_3)}[f_3'(-\infty i)(o_3+m_3p_3+n_3q_3)$$
$$+\overline{f_1'(-\infty i)}(\overline{o_1}+m_3\overline{p_1}+n_3\overline{q_1})+\overline{f_2'(-\infty i)}(\overline{o_2}+m_3\overline{p_2}+n_3\overline{q_2}) \quad (26)$$
$$+\overline{f_3'(-\infty i)}(\overline{o_3}+m_3\overline{p_3}+n_3\overline{q_3})]$$

The basic assumption of stress and displacement are for the period of $a\pi$, and by lemma,we can know $g_1'(t), g_2'(t), g_3'(t)$ for the period of $a\pi$ too,and has

$$f_1'(-\infty i)=\gamma_1, f_2'(-\infty i)=\gamma_2, f_3'(-\infty i)=\gamma_3 \quad (27)$$

So, when $z=-\infty i$,equations (3)-(5) can be changed into

$$\sigma_{33}(-\infty i)=2\,\mathrm{Re}\{f_1'(-\infty i)+f_2'(-\infty i)+f_3'(-\infty i)\} \quad (28)$$

$$\sigma_{23}(-\infty i)=-2\,\mathrm{Re}\{u_1f_1'(-\infty i)+u_2f_2'(-\infty i)+u_3f_3'(-\infty i)\} \quad (29)$$

$$H_3(-\infty i)=-2\,\mathrm{Re}\{\eta_1f_1'(-\infty i)+\eta_2f_2'(-\infty i)+\eta_3f_3'(-\infty i)\} \quad (30)$$

We will substitute (27) with (28)–(30), and use balance conditions $\sigma_{33}(-\infty i)=\frac{Y}{a\pi}, \sigma_{23}(-\infty i)=\frac{X}{a\pi}, H_3(-\infty i)=\frac{Z}{a\pi}$ can get

$$\mathrm{Re}\{\gamma_1+\gamma_2+\gamma_3\}=\frac{Y}{2a\pi} \quad (31)$$

$$\mathrm{Re}\{u_1\gamma_1+u_2\gamma_2+u_3\gamma_3\}=\frac{-X}{2a\pi} \quad (32)$$

$$\mathrm{Re}\{\eta_1\gamma_1+\eta_2\gamma_2+\eta_3\gamma_3\}=\frac{-Z}{2a\pi} \quad (33)$$

Then,we will substitute (27) with (24)–(26), and trim them

$$
\begin{aligned}
(o_1 + m_1 p_1 + n_1 q_1)\gamma_1 &= (\bar{o}_1 + m_1 \bar{p}_1 + n_1 \bar{q}_1)\bar{\gamma}_1 \\
&+ (\bar{o}_2 + m_1 \bar{p}_2 + n_1 \bar{q}_2)\bar{\gamma}_2 + (\bar{o}_3 + m_1 \bar{p}_3 + n_1 \bar{q}_3)\bar{\gamma}_3
\end{aligned} \tag{34}
$$

$$
\begin{aligned}
(o_2 + m_2 p_2 + n_2 q_2)\gamma_2 &= (\bar{o}_1 + m_2 \bar{p}_1 + n_2 \bar{q}_1)\bar{\gamma}_1 \\
&+ (\bar{o}_2 + m_2 \bar{p}_2 + n_2 \bar{q}_2)\bar{\gamma}_2 + (\bar{o}_3 + m_2 \bar{p}_3 + n_2 \bar{q}_3)\bar{\gamma}_3
\end{aligned} \tag{35}
$$

$$
\begin{aligned}
(o_3 + m_3 p_3 + n_3 q_3)\gamma_3 &= (\bar{o}_1 + m_3 \bar{p}_1 + n_3 \bar{q}_1)\bar{\gamma}_1 \\
&+ (\bar{o}_2 + m_3 \bar{p}_2 + n_3 \bar{q}_2)\bar{\gamma}_2 + (\bar{o}_3 + m_3 \bar{p}_3 + n_3 \bar{q}_3)\bar{\gamma}_3
\end{aligned} \tag{36}
$$

In equations (31)-(36), because of the coefficients determinant about γ_1, γ_2, γ_3 is not zero, so, the six equations can be determined $\gamma_1, \gamma_2, \gamma_3$.

So, We will substitute $\gamma_1, \gamma_2, \gamma_3$ with equations (21)(22)(23), we can get $f_1'(z_1), f_2'(z_2), f_3'(z_3)$, then integral,we can obtain $f_1(z_1), f_2(z_2), f_3(z_3)$, finally, we continue $f_1'(z_1), f_2'(z_2), f_3'(z_3), f_1(z_1), f_2(z_2), f_3(z_3)$ to (2)-(8), we can get the stress distribution,so that,the stress function and stress distribution in the elastic body can be determined.

So, the one-dimensional hexagonal quasicrystals and infinite elasticity plane with single periodic of second fundamental problem have the solutions which are existence and uniqueness.

2.2 *Example*

Hypothesis, in infinite elastic plane boundary have period displacement,so as to in the boundary of a periodic segment $-\frac{1}{2}a\pi \le t \le \frac{1}{2}a\pi$,the displacement distribution is

$$
g_1(t) = 0, g_2(t) = \begin{cases} \varepsilon(\frac{|t|}{l} - 1) & |t| \le l \\ 0 & l \le |t| \le \frac{a\pi}{2} \end{cases}, g_3(t) = 0,
$$

and r_0 is from $-l \le t \le l$, and left to right for the line and forward. How to solve elastic balance.

Solution: Known by the boundaries of a periodic segment displacement distribution can get

$$g_1'(t)=0, g_2'(t)=\begin{cases} \dfrac{-\varepsilon}{l} & t\in l_1=[-l,0] \\ \dfrac{\varepsilon}{l} & t\in l_2=[0,l] \\ 0 & t\in l_0-l_1-l_2 \end{cases}, g_3'(t)=0$$

Using previous results, one can obtain the stress function

$$f_1'(z_1)=\frac{m_1}{2a\pi i(o_1+m_1p_1+n_1q_1)}[\int_{-l}^{0}-\frac{\varepsilon}{l}\cot\frac{t-z_1}{a}dt+\int_0^l\frac{\varepsilon}{l}\cot\frac{t-z_1}{a}d]+\gamma_1$$

$$=\frac{m_1\varepsilon}{2\pi i(o_1+m_1p_1+n_1q_1)l}\ln\frac{\sin\frac{l-z_1}{a}}{\sin\frac{-z_1}{a}}\frac{\sin\frac{-l-z_1}{a}}{\sin\frac{-z_1}{a}}+\gamma_1$$

$$f_2'(z_2)=\frac{m_2}{2a\pi i(o_2+m_2p_2+n_2q_2)}[\int_{-l}^{0}-\frac{\varepsilon}{l}\cot\frac{t-z_2}{a}dt+\int_0^l\frac{\varepsilon}{l}\cot\frac{t-z_2}{a}d]+\gamma_2$$

$$=\frac{m_2\varepsilon}{2\pi i(o_2+m_2p_2+n_2q_2)l}\ln\frac{\sin\frac{l-z_2}{a}}{\sin\frac{-z_2}{a}}\frac{\sin\frac{-l-z_2}{a}}{\sin\frac{-z_2}{a}}+\gamma_2$$

$$f_3'(z_3)=\frac{m_3}{2a\pi i(o_3+m_3p_3+n_3q_3)}[\int_{-l}^{0}-\frac{\varepsilon}{l}\cot\frac{t-z_3}{a}dt+\int_0^l\frac{\varepsilon}{l}\cot\frac{t-z_3}{a}d]+\gamma_3$$

$$=\frac{m_3\varepsilon}{2\pi i(o_3+m_3p_3+n_3q_3)l}\ln\frac{\sin\frac{l-z_3}{a}}{\sin\frac{-z_3}{a}}\frac{\sin\frac{-l-z_3}{a}}{\sin\frac{-z_3}{a}}+\gamma_3$$

In which $\gamma_1,\gamma_2,\gamma_3$ can be determined by following six formulas

$$\mathrm{Re}\{\gamma_1+\gamma_2+\gamma_3\}=\frac{Y}{2a\pi},\ \mathrm{Re}\{u_1\gamma_1+u_2\gamma_2+u_3\gamma_3\}=\frac{-X}{2a\pi},$$

$$\mathrm{Re}\{u_1\gamma_1+u_2\gamma_2+u_3\gamma_3\}=\frac{-X}{2a\pi},$$

$$(o_1+m_1p_1+n_1q_1)\gamma_1=(\overline{o}_1+m_1\overline{p}_1+n_1\overline{q}_1)\overline{\gamma}_1$$
$$+(\overline{o}_2+m_1\overline{p}_2+n_1\overline{q}_2)\overline{\gamma}_2+(\overline{o}_3+m_1\overline{p}_3+n_1\overline{q}_3)\overline{\gamma}_3$$

$$(o_2+m_2p_2+n_2q_2)\gamma_2=(\overline{o}_1+m_2\overline{p}_1+n_2\overline{q}_1)\overline{\gamma}_1$$
$$+(\overline{o}_2+m_2\overline{p}_2+n_2\overline{q}_2)\overline{\gamma}_2+(\overline{o}_3+m_2\overline{p}_3+n_2\overline{q}_3)\overline{\gamma}_3$$

$$(o_3+m_3p_3+n_3q_3)\gamma_3=(\overline{o}_1+m_3\overline{p}_1+n_3\overline{q}_1)\overline{\gamma}_1$$
$$+(\overline{o}_2+m_3\overline{p}_2+n_3\overline{q}_2)\overline{\gamma}_2+(\overline{o}_3+m_3\overline{p}_3+n_3\overline{q}_3)\overline{\gamma}_3$$

References

[1] Lu Jian-ke,Cai Hai-tao. The Theory of Plane Elasticity Problem with the Period[M]. Hunan: The Science and Technology of Hunan Press,1986.

[2] Lu Jian-ke,Cai Hai-tao. The Plane Elasticity Problem Outline of the Period[M]. Hubei: The University of Wuhan Press,2008,6.

[3] Lu Jian-ke. The Plane Elasticity of Complex Variables Method[M]. Wuhan: The University of Wuhan Press,2005.

[4] Lu Jian-ke. The Course of Analysis Function with Boundary Value[M]. Wuhan: The University of Wuhan Press,2009.

[5] Cai Hai-tao,Lu Jian-ke. First Fundamental Problems of Anisotropic Elastic Plane Weakened by Periodic Collinear Cracks[J]. Journal of Applied Mathematics and Mechanics,2009,11.

[6] Lu Jian-ke,Cai Hai-tao. New Formulations of the Second Fundamental Problem in Plane Elasticity[J]. Journal of Applied Mathematics and Mechanics,1992.

[7] Cai Hai-tao. Periodic Crack problems of Plane Anisotropic Medium[J]. Journal of Mathematics and Physics,1982.

[8] Cai Hai-tao. The Periodic Cracks of an Infinite Anisotropic Media for Plane Skew-Symmetric Loadings[J]. Journal of Applied Mathematics and Mechanics,1986.

[9] Liu Shi-qiang. On Problems of an Elastic Half-Plane with Arbitrary Periodic Cracks[J]. Annuals of Hunan Mathematics,1987.

[10] Zheng Ke. On the Fundamental Problems in an Infinitely Long Elastic Strip with Periodic Cracks[J]. Journal of Wuhan University, 1998,10.

[11] Huang Ming-hai,Li Jing. On Periodic Fundamental Problem for Anisotropic Elastic Strip[J]. Journal of Beijing Industry University,2002,9.

[12] Yu Jing,Liu Guan-ting. The Study of Defect Problem in One-Dimensional Orthorhombic Quasicrystals by Generalized Complex Variable Function Method[J]. The Master Dissertation of Neimeng Normal School,2010,5.

[13] Pi Jian-dong,Liu Guan-ting. A Research on the Problem of Some Defects about One-Dimensional Quasicrystals Plane Elasticity[J]. The Master Dissertation of Neimeng Normal School,2007,5.

[14] Han Fei,Liu Guan-ting. The Dynamics Research on the Problem of Some Defects about One-Dimensional Hexagonal Quasicrystals[J]. The Master Dissertation of Neimeng Normal School,2008,5.

[15] Liu Guan-ting,He Qing-long,Guo Rui-ping. The Plane Strain Theory for One-Dimensional Hexagonal Quasicrystals in Aperiodical Plane[J]. Journal of Physics,2009.

[16] Zhang Lei,Shi Wei-chen. A Research on the Crack of Quasicrystals Material[J]. The Master Dissertation of Shanghai Maritime University,2005,5.

[17] Fan Tian-you. Mathematical Theory of Elasticity and Defects of Quasicrystals[J]. Journal of Advances in Mechanics,2000.

[18] Zhou Wang-min,Fan Tian-you,Yin Shu-yuan,Wang Nian-peng. Anti-Plane Elasticity Problem and Mode □ Crack Problem of Cubic Quasicrystal [J]. Journal of Beijing Institute of Technology,2001,10.

[19] Fan Tian-you. Mathematical Theory of Elasticity of Quasicrystals and Its Applications[M]. Beijing: Science Press,2010.

[20] Muskhelishvili NI.Some Basic Problems of the Mathematical Theory of Elasticity[M]. Cambridge: Cambridge University Press,1953.

THE OPTIMAL CONVEX COMBINATION BOUNDS FOR THE CENTROIDAL MEAN*

HONG LIU

College of Mathematics and Computer Science, Hebei University, , Address
Baoding, Hebei 071002, China

XIANG-JU MENG

Department of Mathematics, Baoding College, Address
Baoding, Hebei 071002, China

In this paper we derive some optimal convex combination bounds related to Centroidal mean. We find the greatest values α and the least values β such that the double inequalities

$$\alpha T_1(a,b)+(1-\alpha)H(a,b)\le T(a,b)\le \beta T_1(a,b)+(1-\beta)H(a,b),\ \alpha=\frac{3}{\pi},\beta=1$$

and

$$\frac{2}{3}A(a,b)+\frac{1}{3}C(a,b)\le T_1(a,b)$$

hold for all $a,b>0$ with $a\ne b$. Here $A(a,b)$, $C(a,b)$, $H(a,b)$, $T(a,b)$ and $T_1(a,b)$ denote the arithmetic, contraharmonic, harmonic, the second Seiffert's and Centroidal means of two positive numbers a and b, respectively.

1. Introduction

For $a,b>0$ with $a\ne b$, the second Seiffert't mean $T(a,b)$ was introduced by Seiffert [1] as follows:

$$T(a,b)=\frac{a-b}{2\arctan\frac{a-b}{a+b}} \tag{1}$$

* This work is supported by the Youth Foundation of Hebei University (Grant 2010Q24) and partly supported by N S Foundation of Hebei Province(Grant A2011201011) and N S Foundation of Hebei Province(A2012201054).

Recently, the inequalities for means have been the subject of intensive research [2-9]. In particular, many remarkable inequalities for T or can be found in the literature [10,11].

Let $A(a,b)=(a+b)/2$, $C(a,b)=(a^2+b^2)/(a+b)$, $H(a,b)=2ab/(a+b)$ and $T_1(a,b)=2(a^2+ab+b^2)/3(a+b)$ be the arithmetic, contraharmonic, harmonic and Centroidal means of two positive real numbers a and b with $a \neq b$. Then

$$\min\{a,b\} < H(a,b) < A(a,b) < T(a,b) < T_1(a,b) < C(a,b) < \max\{a,b\}. \qquad (2)$$

In [10], Hästö proved $A_p \prec T(a,b)$ if $p \leqslant 1$, for all $a,b>0$ with $a \neq b$. In [11], the authors found the greatest value α and the least value β such that the double inequality

$$\alpha D(a,b)+(1-\alpha)A(a,b) < T(a,b) < \beta D(a,b)+(1-\beta)A(a,b)$$

holds for all $a,b>0$ with $a \neq b$.

The purpose of the present paper is to find the greatest values α and the least values β such that the double inequalities

$$\alpha T_1(a,b)+(1-\alpha)H(a,b) \leq T(a,b) \leq \beta T_1(a,b)+(1-\beta)H(a,b)$$

and the inequality $C(a,b)/3+2A(a,b)/3 \leq T_1(a,b)$ hold for all $a,b>0$ with $a \neq b$.

2. Main Results

Firstly, we present the optimal convex combination bounds of Centroidal and harmonic means for the second Seiffert's mean as follows.

Theorem 1 The double inequality

$$\alpha T_1(a,b)+(1-\alpha)H(a,b) < T(a,b) < \beta T_1(a,b)+(1-\beta)H(a,b)$$

holds for all $a,b>0$ with $a \neq b$ if and only if $\alpha \leqslant 3/\pi$ and $\beta \geqslant 1$.

Proof At first, from inequality (2), we need only to prove that

$$T(a,b) > \frac{3}{\pi}T_1(a,b)+\left(1-\frac{3}{\pi}\right)H(a,b), \qquad (3)$$

for all $a,b>0$ with $a \neq b$.

Without loss of generality, we assume $a>b$. Let $t=a/b>1$ and $p=3/\pi$. Then

$$\begin{aligned} &\{T(a,b)-[pT_1(a,b)+(1-p)H(a,b)]\} \\ = \quad & bT(t,1)-b[pT_1(t,1)+(1-p)H(t,1)] \\ = \quad & \frac{b[4pt^2+(12-8p)t+4p]}{6(t+1)\arctan\frac{t-1}{t+1}} f(t) \end{aligned} \tag{4}$$

where

$$f(t)=\frac{3(t^2-1)}{4pt^2+(12-8p)t+4p}-\arctan\frac{t-1}{t+1}. \tag{5}$$

Simple computations lead to

$$f(1)=0, \quad \lim_{t\to+\infty} f(t)=\frac{3}{4p}-\frac{\pi}{4}. \tag{6}$$

$$f'(t)=\frac{(t-1)^2}{(t^2+1)[4pt^2+(12-8p)t+4p]^2} g(t), \tag{7}$$

where

$$\begin{aligned} g(t)= \quad & (-16p^2-24p+36)t^2+(32p^2-96p+72)t \\ & -16p^2-24p+36. \end{aligned} \tag{8}$$

Because of $p=3/\pi$ then

$$g(1)=144(1-p)=144\left(1-\frac{3}{\pi}\right)>0, \quad \lim_{t\to+\infty} g(t)=-\infty, \tag{9}$$

$$g'(t)=-8(4p^2+6p-9)t+32p^2-96p+72.$$

$$g'(1)=144(1-p)>0, \quad \lim_{t\to+\infty} g'(t)=-\infty, \tag{10}$$

$$g''(t)=-32p^2-48p+72. \tag{11}$$

From (11) we clearly see that $g''(t)<0$ for $p=3/\pi$, $t\geq 1$, hence $g'(t)$ is strictly decreasing in $[1,+\infty)$, which together with (10) implies that there exists $\lambda_1>1$ such that $g'(t)>0$ for $t\in[1,\lambda_1)$ and $g'(t)<0$ for $t\in(\lambda_1,+\infty)$, hence $g(t)$ is strictly increasing in $[1,\lambda_1]$ and strictly decreasing for $[\lambda_1,+\infty)$.

As so on, from (7) and (8) there exists $\lambda_2>1$ such that $f(t)$ is strictly increasing in $[1,\lambda_2]$ and strictly decreasing in $[\lambda_2,+\infty)$. Note that if $p=3/\pi$, then the second equality in (6) becomes

$$\lim_{t\to+\infty} f(t)=0.$$

Thus $f(t)>0$ for all $t>1$. Therefore, inequality (3) follows from (4) and (5).

Secondly, we prove that $3T_1(a,b)/\pi+(1-3/\pi)H(a,b)$ is the best possible lower convex combination bound of Centroidal and harmonic means for the second Seiffert's mean.

If $\alpha>3/\pi$, then from (1) one has

$$\begin{aligned}&\lim_{t\to+\infty}\frac{\alpha T_1(t,1)+(1-\alpha)H(t,1)}{T(t,1)}\\ =\ &\lim_{t\to+\infty}\frac{[\alpha t^2+(3-2\alpha)t+\alpha]4\arctan\frac{t-1}{t+1}}{3(t^2-1)}=\frac{\alpha}{3}\pi>1.\end{aligned}\tag{12}$$

Inequality (12) implies that for any $\alpha>3/\pi$ there exists $X=X(\alpha)>1$ such that $\alpha T_1(t,1)+(1-\alpha)H(t,1)>T(t^2,1)$ for $t\in(X,+\infty)$.

Finally, we prove that $1T_1(a,b)+0H(a,b)$ is the best possible upper convex combination bound of the Centroidal and harmonic means for the second Seiffert's mean..

If $\beta<1$, then (9) (with β in place of p) leads to $g(1)=144(1-\beta)>0$. From this result and the continuity of $g(t)$ we clearly see that there exists $\delta=\delta(\beta)>0$ such that $g(t)>0$ for $t\in(1,1+\delta)$. Then (7) implies $f'(t)>0$ for $t\in(1,1+\delta)$. Thus $f(t)$ is increasing for $t\in(1,1+\delta)$. Since (6), then $f(t)>0$ for $t\in(1,1+\delta)$, which is equivalent to, by (4), that

$$T(t,1)>\beta T_1(t,1)+(1-\beta)H(t,1),$$

for $t\in(1,1+\delta)$.

Secondary, we present the optimal lower convex combination bound of the arithmetic and contraharmonic means for the Centroidal mean as follows.

Theorem 2 The inequality $\gamma C(a,b)+(1-\gamma)A(a,b)\le T_1(a,b)$ holds for all $a,b>0$ with $a\ne b$ if and only if $\gamma\le 1/3$.

Proof Firstly, we prove that

$$T_1(a,b)\ge\frac{1}{3}C(a,b)+\frac{2}{3}A(a,b),\tag{13}$$

for all $a,b>0$ with $a\ne b$.

Without loss of generality, we assume $a>b$. Let $t=a/b>1$ Then

$$\begin{aligned}&\{T_1(a,b)-[pC(a,b)+(1-p)A(a,b)]\}\\ &=bT_1(t,1)-b[pC(t,1)+(1-p)A(t,1)]=\frac{b(1-3p)(t-1)^2}{6(t+1)}.\end{aligned}\tag{14}$$

Let

$$h(t) = \frac{b(1-3p)(t-1)^2}{6(t+1)}, \tag{15}$$

simple computations lead to $h(t)=0$ for $p=1/3$ and $h(t)>0$ for $p<1/3$.

If $\gamma>1/3$, then (15) (with γ in place of p) leads to $h(t)<0$. Such we prove that $C(a,b)/3+2A(a,b)/3$ is the best possible lower convex combination bound of the arithmetic and contraharmonic means for the Centroidal mean.

References

1. H. J. Seiffert, *Aufgube β16. Die Wurzel.* 29 (1995),221-222.
2. E. Neuman and J. Sándor, *On the Schwab-Borchardt mean. Mathematica Pannonica.* **1**, 17 (2006),49-59.
3. A. Witkowski, *Convexity of weighted stolarsky means. J. Int. in Pure and Applied Math..* **2**, 7 (2006),article 73.
4. M. Y. Shi, Y. M. Chu and Y. P. Jiang, *Optimal inequalities among various means of two arguments . Abstract and Applied Analysis.* 2009, Article ID 694394(2009), 10 pages.
5. Y. M. Chu, Y. F. Qiu, M. K. Wang and G. D. Wang, *The optimal convex combination bounds of arithmetic and harmonic means for the Seiffert's mean. Journal of Inequalities and Applications.* 2010, Article ID 436457 (2010), 7 pages.
6. S. S. Wang and Y. M. Chu, *The best bounds of the combination of arithmetic and harmonic means for the Seiffert's mean. Int. J. Math. Anal.* 22, Article 4(2010), 1079-1084.
7. C. Zong and Y. M. Chu, *An inequality among identric, geometric and Seiffert's means . Int. Math. Forum .* 26, 5(2010), 1297-1302.
8. H. Liu and X. J. Meng, *The optimal convex combination bounds for Seiffert's mean. Journal of Inequalities and Applications.* 2011, Article ID 686834 (2011), 9 pages.
9. Y. M. Chu, C. Zong and G. D. Wang, *Optimal convex combination bounds of Seiffert and geometric means for the arithmetic mean. J. Math. Inequalities.* 3, 5(2011), 429-434.
10. P. A. Hästö, *A monotonicity property of ratios of symmetric homogeneous means. J. Int. in Pure and Applied Math.* 5, 3(2002), article 71.
11. X. J. Meng H. Liu and H. Y. Gao, *Optimal convex combination bounds for the second Seiffert's mean. Journal of NingXia University(NSE).* 1, 33(2012), 14-16.

THE METHOD OF FUNDAMENTAL SOLUTION FOR A CLASS OF ELLIPTICAL PARTIAL DIFFERENTIAL EQUATIONS WITH COORDINATE TRANSFORMATION AND IMAGE TECHNIQUE

LI-NA WU[1], QUAN JIANG[1,2] *
1 *College of Civil Engineering, Nantong University, Nantong, China*
2 *College of Science, Nantong University, Nantong, China*

The method of fundamental solution (MFS) has high accuracy and fast speed in solving a class of elliptical partial differential equations (PDEs). The coordinate transformation technique is used to convert the elliptical PDEs to biharmonic equations, which can extend the application areas of MFS. In the symmetrical problem, the fundamental functions are reconstructed to increase the computational efficiency. From the examples, it is concluded that the numerical results have high accuracy in the bending problem.

Keywords: Method of fundamental solution; Image technique; Coordinate transformation

1. Introduction

The elliptical PDEs have been widely used in the science and engineering fields. These kinds of PDEs attract the scientists and engineers for a long period. Gilbarg [1] has given the further theoretical research of the second order elliptical PDEs.

MFS is a true boundary type meshless method for elliptical boundary value problems. It depends on the fundamental solution as BEM does. Kupradze and Aleksidze [2-6] have introduced MFS in 60s of last century. With the help of MFS, Steinbigler [7] has given the solution of the symmetry electrical field. Furthermore, the problem with 3-D arbitrary smooth bodies [8], static analysis of plates arid shells [9,10], and some elastical boundary equations [11] have been researched. Fairweather and Karageorghis [12-14] employed the MFS to solve the ellptical boundary value problems of biharmonic functions. Katasurada and Furthermore, Chen [15,16] has given further studies for the computation of MFS in his papers.

* Corresponding author. Tel: +86-513-85012655; fax: +86-513-85012187. E-mail addresses: jiang.q@ntu.edu.cn

It is our purpose to implement MFS for these general PDEs to share the advantages of this method with the potential problems. The numerical experiments and conclusions are also obtained in the last two sections of this paper.

2. Two types of elliptical PDEs and coordinate transformation

A class of elliptical PDEs are considered as follows

$$L_N L_N \phi = q \qquad \text{in domain } \Omega \tag{1}$$

where q are the prescribed functions, and Ω is the bounded domain of two dimension ($N = 2$) or dimension ($N = 3$). The elliptical partial differential operator in Eq.(1) is defined by

$$L_N = \sum_{i=1}^{N} \sum_{j=1}^{N} A_{ij} \frac{\partial^2}{\partial X_i \partial X_j} \tag{2}$$

where $[A_{ij}]$ which consists of the coefficients in Eq.(2), is positive definite matrix. And, the variable X_i is the coordinate for domain Ω. In particular, we have

$$L_2 = A_{11} \frac{\partial^2}{\partial X_1^2} + (A_{12} + A_{21}) \frac{\partial^2}{\partial X_1 \partial X_2} + A_{22} \frac{\partial^2}{\partial X_2^2} \tag{3}$$

for the operator when $N = 2$.

On the boundary $\partial\Omega$ of the domain Ω, The boundary conditions for Eq.(1) are also defined by

$$\phi = f_1, \ \frac{\partial \phi}{\partial n} = f_2 \qquad \text{on } \partial\Omega \tag{4c}$$

or

$$\phi = f_3, \ \sum_{i=1}^{N} \sum_{j=1}^{N} \alpha_{ij} \frac{\partial^2 \phi}{\partial X_i \partial X_j} = f_4 \quad \text{on } \partial\Omega \tag{4d}$$

where $f_i (i = 1, 2, \ldots, 6)$ is the given function on the boundary, α_{ij} is the prescribed function, and n is the unit normal to the boundary.

It is feasible to convert the value problems of the elliptical PDEs to the ones of harmonic or biharmonic equations in consideration that those equations have been further investigated. Therefore, the coordinate transformation is written by

$$x_i = \sum_{j=1}^{N} b_i l_{ij} X_j \tag{5}$$

where x_i is the coordinate after transformation. Furthermore, b_i is the constant about the magnification of the coordinates, and l_{ij} is the constant concerning with the rotation of the coordinates. Furthermore, the matrix l_{ij} related to the rotation of the coordinate can be expressed by Euler angles of ψ, θ and φ. For the two dimensional problem, the angles θ and φ degenerate to zero. Thus, one can get

$$[l_{ij}] = \begin{pmatrix} \cos\psi & \sin\psi \\ -\sin\psi & \cos\psi \end{pmatrix} \tag{6}$$

The operator $L_2^{'}$ can be written by

$$L_2^{'} = B_{11}\frac{\partial^2}{\partial x_1^2} + B_{22}\frac{\partial^2}{\partial x_2^2} + B_{12}\frac{\partial^2}{\partial x_1 \partial x_2} \tag{7}$$

where

$$B_{11} = b_1^2[A_{11}l_{11}^2 + A_{22}l_{12}^2 + (A_{12} + A_{21})l_{11}l_{12}] \tag{8a}$$

$$B_{22} = b_2^2[A_{11}l_{21}^2 + A_{22}l_{22}^2 + (A_{12} + A_{21})l_{21}l_{22}] \tag{8b}$$

$$B_{12} = b_1 b_2[2A_{11}l_{11}l_{21} + 2A_{22}l_{12}l_{22} + (A_{12} + A_{21})(l_{11}l_{22} + l_{12}l_{21})] \tag{8c}$$

The partial differential operator $L_N^{'}$ in Eq.(7) will be degenerate to Laplace operator , when the constants B_{ij} satisfy the relations below

$$B_{ij} = 0 \, (i \neq j, i = 1, \ldots, N, j = 1, \ldots, N) \tag{9a}$$

$$B_{ii} = B \, (i = 1, \ldots, N) \tag{9b}$$

Therefore, one can convert the general PDEs of Eq.(1) to the harmonic and biharmonic equations as follows

$$\nabla^2 \nabla^2 \phi = \frac{q}{B^2} \qquad \text{in domain } \Omega^{'} \tag{10}$$

where, we have $\nabla^2 = \sum_{i=1}^{N} \partial^2 / \partial x_i^2$ for the Laplace operator.

The boundary conditions of Eqs.(4c) and (4d) should be rewritten as

$$\phi = f_1, \quad \sum_{i=1}^{N}\sum_{k=1}^{N}\sum_{j=1}^{N} b_k l_{ki} l_{ji} n_{x_j} \frac{\partial \phi}{\partial x_k} = f_2 \qquad \text{on } \partial\Omega_1^{'} \tag{11a}$$

and

$$\phi = f_3, \quad \sum_{i=1}^{N}\sum_{j=1}^{N}\sum_{m=1}^{N}\sum_{n=1}^{N}\alpha_{ij}b_m b_n l_{mi} l_{nj}\frac{\partial^2\phi}{\partial x_m \partial x_n} = f_4 \quad \text{on } \partial\Omega_2' \tag{11b}$$

3. The formulations of MFS for biharmonic functions

The approximate solution $\phi(P)$ with MFS for Eq.(10) reads [13]

$$\phi(P) = \sum_{i=1}^{n} E_i G_1(P, R_i) + \sum_{i=1}^{n} F_i G_2(P, R_i) + \phi^*(P) \tag{12}$$

where E_i and F_i are the coefficients to be determined by the prescribed boundary conditions. The function $\phi^*(P)$ is the particular solution of Eq.(10). When q is constant, we have

$$\phi^*(P) = \frac{q_2}{64B^2}(x_1^2 + x_2^2)^2 \qquad (N = 2) \tag{13a}$$

$$\phi^*(P) = \frac{q_2}{120B^2}(x_1^2 + x_2^2 + x_3^2)^2 \qquad (N = 3) \tag{13b}$$

In approximate solution (12), $G_1(P, Q_i)$ is the fundamental solution for the harmonic equations. Consequently, we have

$$G_1(P, Q) = -\frac{1}{2\pi}\ln r \qquad (N = 2) \tag{14a}$$

$$G_1(P, Q) = -\frac{1}{4\pi \mathrm{r}} \qquad (N = 3) \tag{14b}$$

And $G_2(P, Q_i)$ is the fundamental solution for the biharmonic equations, which are written as

$$G_2(P, R_i) = -\frac{1}{8\pi}r^2(\ln r - 1) \qquad (N = 2) \tag{15a}$$

$$G_2(P, R_i) = -\frac{r}{8\pi} \qquad (N = 3) \tag{15b}$$

where r is the distance between P and R_i.

Therefore, the coefficients E_i and F_i can be determined by the equations below

$$\sum_{i=1}^{n} E_i G_1(Q_k, R_i) + \sum_{i=1}^{n} F_i G_2(Q_k, R_i) = f_3(Q_k) - \phi^*(Q_k) \tag{16a}$$

$$\sum_{i=1}^{n} E_i \frac{\partial}{\partial n} G_1(Q_k, R_i) + \sum_{i=1}^{n} F_i \frac{\partial}{\partial n} G_2(Q_k, R_i) = f_4(Q_k) - \frac{\partial \phi^*(Q_k)}{\partial n} \quad (16b)$$

4. The image technique for the MFS in symmetrical problem

The domain to be treated is generally regular in science and technique computing. If the transformed domain and boundary $\partial\Omega'$ are symmetrical, the calculation of MFS will be mostly simplified. In general, the boundary conditions on the symmetry of axis or plane is

$$\partial G_1(P,Q)/\partial n = 0, \partial G_2(P,Q)/\partial n = 0 \quad (17)$$

One can reduce the number of the equations in computation, if the fundamental solutions which satisfy the symmetrical conditions are given. Based on the idea of image technique employed in electrostatic problem, the fundamental solutions are reconstructed for the symmetrical problems in the following cases.

When the domain Ω' and its boundary are symmetry about axis x_1, we can condense the computational domain into the half of this region, e.g. the upper plane $(x_2 \geq 0)$. If the fundamental solution which satisfies the condition on the axis of symmetry is found, the numerical computation will be greatly reduced.

Considering the positive unit sources be situated on the point Q and its image Q_2 (see Fig. 1(a)), the fundamental solutions of Eq.(10) can be reconstructed by

$$G_1(P,Q) = -\frac{1}{2\pi} \ln(r_1 r_2) \quad (18a)$$

$$G_2(P,Q) = -\frac{1}{8\pi}[r_1^2(\ln r_1 - 1) + r_2^2(\ln r_2 - 1)] \quad (18b)$$

where r_1 and r_2 are the distance between the points P, Q and Q_2,

$$r_1 = \left[(x_1 - x_1^Q)^2 + (x_2 - x_2^Q)^2\right]^{1/2} \quad (19a)$$

$$r_2 = \left[(x_1 - x_1^Q)^2 + (x_2 + x_2^Q)^2\right]^{1/2} \quad (19b)$$

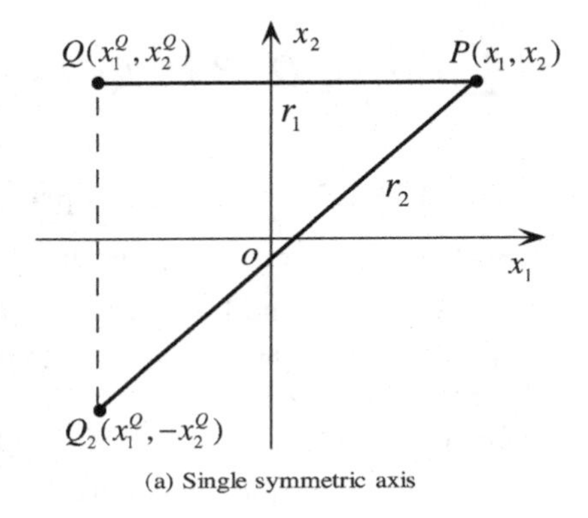

(a) Single symmetric axis

$Q_3(-x_1^Q, x_2^Q)$ x_2 $Q(x_1^Q, x_2^Q)$ r_1 r_3 $P(x_1, x_2)$ o r_2 x_1 r_4 $Q_4(-x_1^Q, -x_2^Q)$ $Q_2(x_1^Q, -x_2^Q)$

(b) Double symmetric axises

Fig. 1. The distribution of source and images in two dimensional domain

Case 1: Single symmetrical axis of x_1

If the domain Ω' is symmetry about the axis x_2, we can also recognize the solutions without any difficulties.

Case 2: Double symmetrical axises of x_1 and x_2

When the domain Ω' has double symmetrical axises of x_1 and x_2, which is shown in Fig.1(b), the computational region can be condensed into the quarter of Ω', e.g. the part of the plane of $x_1 \geq 0, x_2 \geq 0$. Similar to the procedure of the reconstruction of case 1, we consider the positive unit sources are placed on the point Q and its images, which are shown in Fig.1 (b). Therefore, the fundamental solutions should be rewritten as below

$$G_1(P,Q) = -\frac{1}{2\pi}\ln(r_1 r_2 r_3 r_4) \tag{20a}$$

$$G_2(P,Q) = -\frac{1}{8\pi}[r_1^2(\ln r_1 - 1) + r_2^2(\ln r_2 - 1) + r_3^2(\ln r_3 - 1) + r_4^2(\ln r_4 - 1)] \tag{20b}$$

where the distances of r_1 and r_2 can be expressed by Eqs.(190a) and (19b). And the distances of r_3 and r_4, which is shown in Fig. 1(b), are

$$r_3 = \left[(x_1 + x_1^Q)^2 + (x_2 - x_2^Q)^2\right]^{1/2} \tag{21a}$$

$$r_4 = \left[(x_1 + x_1^Q)^2 + (x_2 + x_2^Q)^2\right]^{1/2} \tag{21b}$$

5. Applications of MFS for the bending of orthotropic plate

The bending problem of orthotropic plate is considered here as an application of MFS. The control equation of the discussed problem can be written by [17,18].

$$D_1 \frac{\partial^4 w}{\partial X_1^4} + 2D_3 \frac{\partial^4 w}{\partial X_1^2 \partial X_2^2} + D_2 \frac{\partial^4 w}{\partial X_2^4} = q \tag{22}$$

where w is the bending deflection of the plate, q is the uniform load per unit area. It is emphasized that the equation of (31) is defined in the space (X_1, X_2) before coordinate transforming. Furthermore, the coefficients of D_1, D_2 and D_3 in Eq.(22) are

$$D_1 = D_{11}, D_3 = D_{12} + 2D_{66}, D_2 = D_{22} \tag{23}$$

where D_{11}, D_{22}, D_{12} and D_{66} are principal elastic stiffness.

The moments of bending according to the deflection w can be given by

$$M_{X_1} = -D_{11}\left(\frac{\partial^2 w}{\partial X_1^2} + v_{21}\frac{\partial^2 w}{\partial X_2^2}\right) \tag{24a}$$

$$M_{X_2} = -D_{22}\left(\frac{\partial^2 w}{\partial X_1^2} + v_{12}\frac{\partial^2 w}{\partial X_2^2}\right) \tag{24b}$$

$$M_{X_1X_2} = -2D_{66}\frac{\partial^2 w}{\partial X_1 \partial X_2} \tag{24c}$$

where v_{12} and v_{21} are the Poisson ratios along the different direction.

For simplicity, we only give the two boundary conditions for the deflection w as follows

$$w=0,\frac{\partial w}{\partial n}=0 \qquad \text{(Clamped boundary)} \qquad (25a)$$

$$w=0, M_n=0 \qquad \text{(Simple supported boundary)} \qquad (25b)$$

where

$$M_n = M_{X_1} n_{X_1}^2 + M_{X_2} n_{X_2}^2 + 2M_{X_1X_2} n_{X_1} n_{X_2} \qquad (26)$$

For a special orthotropic plate, e.g. a kind of reinforced concrete plate, we have the relationship of $D_3=(D_1D_2)^{1/2}$ [26]. Therefore, Eq.(31) can be rewritten in the form of

$$(D_1^{1/2}\frac{\partial^2}{\partial X_1^2}+D_2^{1/2}\frac{\partial^2}{\partial X_2^2})^2 w=q \qquad (27)$$

It is obvious that Eq.(36) can be converted to biharmonic function by using the coordinate transformation of Eq.(5). Therefore, one can get the coordinate relation as follows

$$x_1=\left(\frac{D_2}{D_3}\right)^{1/4} X_1,\ x_2=\left(\frac{D_1}{D_3}\right)^{1/4} X_2 \qquad (28)$$

Consequently, the biharmonic equation reads

$$\nabla^2\nabla^2 w = q/D \qquad (29)$$

where $D=(D_1D_2)^{1/2}$.

6. Numerical experiments

In the numerical experiment, the elastic constants are taken as: $D_1=1\times10^6$, $D_2=4\times10^6$ and $D_3=2\times10^6$. The lengths are taken as: $a=40$ and $h=40$. The uniform load is taken as $q=100$. The units of the constants in numerical experiments are omitted for simplicity.

When the plate is simple supported on the boundary $X_1=0,a$, and clamped on the boundary $X_2=0,h$, the numerical computing for the deflection can also be treated by the image technique.

(1) Numerical result for single symmetrical axis x_1

In Fig.2(a), the boundary conditions in the transformed plane are shown as follows

$$w(Q_k)=0,\ \frac{\partial w}{\partial x_2}=0 \quad \text{on } x_2=\left(\frac{D_1}{D_3}\right)^{1/4}\frac{h}{2},\ 0\le x_1\le\left(\frac{D_2}{D_3}\right)^{1/4} a \qquad (30)$$

for the clamped boundaries.

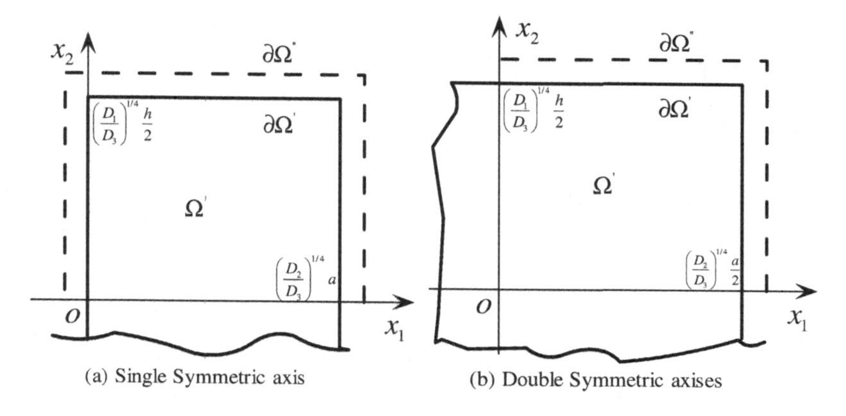

Fig. 2. The domain and boundaries of the symmetrical problems in transformed plane for numerical experiments

The corresponding boundary condition can be written by

$$\begin{aligned}&\sum_{i=1}^{n} E_i G_1(Q_k, R_i) + \sum_{i=1}^{n} F_i G_2(Q_k, R_i) = -\phi^*(P),\\&\sum_{i=1}^{n} E_i \frac{\partial}{\partial x_2} G_1(Q_k, R_i) + \sum_{i=1}^{n} F_i \frac{\partial}{\partial x_2} G_2(Q_k, R_i) = -\frac{\partial \phi^*(P)}{\partial x_2}\end{aligned} \tag{31}$$

We also have

$$w(Q_k) = 0,\ \frac{\partial w}{\partial x_1} = 0 \text{ on } x_1 = 0,\ 0 \le x_2 \le \left(\frac{D_1}{D_3}\right)^{1/4} \frac{h}{2} \text{ and}$$

$$x_1 = \left(\frac{D_2}{D_3}\right)^{1/4} a,\ 0 \le x_2 \le \left(\frac{D_1}{D_3}\right)^{1/4} \frac{h}{2} \tag{32}$$

Therefore, one has expanded the boundary conditions with the coefficients E_i and F_i as

$$\begin{aligned}&\sum_{i=1}^{n} E_i G_1(Q_k, R_i) + \sum_{i=1}^{n} F_i G_2(Q_k, R_i) = -\phi^*(P),\\&\sum_{i=1}^{n} E_i \frac{\partial}{\partial x_1} G_1(Q_k, R_i) + \sum_{i=1}^{n} F_i \frac{\partial}{\partial x_1} G_2(Q_k, R_i) = -\frac{\partial\, \phi^*(P)}{\partial x_1}\end{aligned} \tag{33}$$

From the above linear equations, the coefficients E_i and F_i can be determined to achieve the bending deflection w, which are shown in Tab.2.

(2) Numerical result with double symmetrical axises x_1 and x_2

For the problem with double symmetrical axises. It is obvious that the computing area reduces to the quarter of the domain Ω' (See Fig.5(b)).

We have

$$w(Q_k)=0, \frac{\partial w}{\partial x_2}=0 \quad \text{on } x_2=\left(\frac{D_1}{D_3}\right)^{1/4}\frac{h}{2}, 0\le x_1\le\left(\frac{D_2}{D_3}\right)^{1/4}\frac{a}{2} \tag{34}$$

and

$$w(Q_k)=0, \frac{\partial w}{\partial x_1}=0 \quad \text{on } x_1=\left(\frac{D_2}{D_3}\right)^{1/4}\frac{a}{2}, 0\le x_2\le\left(\frac{D_1}{D_3}\right)^{1/4}\frac{h}{2} \tag{35}$$

Consequently, the bending deflection w can be calculated out after the coefficients E_i and F_i are determined.

Table 1 The bending deflection w along X_1 when $X_2=h/2$

X_1	Numerical solutions of single symmetrical axis	Numerical solutions of double symmetrical axis	Analytical solution [17,18]
5	0.0727043	0.0725055	0.0724765
10	0.1220031	0.1224484	0.1217815
15	0.1476927	0.1485438	0.1474120
20	0.1553392	0.1563706	0.1550783
25	0.1476873	0.1485438	0.1474120
30	0.1219323	0.1224484	0.1217815
35	0.0725536	0.0725055	0.0724965

7. Conclusion

MFS with coordinate transformation and image technique is introduced to the numerical solution of a class of the elliptical PDEs in this paper. The coordinate transformation is used for converting the general PDEs into the harmonic function. For the symmetrical problems, the improved fundamental solutions are reconstructed by using the special boundary conditions on the symmetrical axises.

From the numerical experiments, it is seen that the results in this paper have good accuracy and high efficiency. MFS with coordinate transformation technique can solve more elliptical PDES compared with the classic field biharmonic equations. In particular, the programming, date preparing and the linear equations for computing can be greatly reduced by using the reconstructed fundamental solutions with image technique.

Acknowledgements

The authors would like to express their gratitude for the support from the National Natural Science Foundation of China (No. 10902055).

References

[1] D. Gilbarg, N.S. Trudinger, Elliptic Partial Differential Equations of Second Order, Springer-Verlag, Berlin, 1983.
[2] M.A. Aleksidze, On approximate solutions of a certain mixed boundary value problem in the theory of harmonic functions, Differential Equations 2(1966) 515–518.
[3] V.D. Kupradze, A method for the approximate solution of limiting problems in mathematical physics, Comput. Math. Math. Phys 4(1964) 199-205.
[4] V.D. Kupradze, Potential Methods in the Theory of Elasticity, Israel Program for Scientific Translations, Jerusalem, 1965.
[5] V.D. Kupradze, On the approximate solution of problems in mathematical physics, Russian Math. Surveys 22(1967) 58–108.
[6] V.D. Kupradze, M.A. Aleksidze, The method of functional equations for the approximate solution of certain boundary value problems, Comput. Math. Math. Phys 4(1964) 82-126.
[7] H. Steinbigler, Digitale Berechnung elektrischer Felder, Elektrotechnishe Zeitschrift, Ausgabe A. 90(1969) 663-666.
[8] W.C. Webster, The flow about arbitrary, three-dimensional smooth bodies, Journal of Ship Research 19(1975) 206-218.
[9] N. Simos, A.M. Sadegh, An indirect BIM for static analysis of spherical shells using auxiliary boundaries, International Journal for Numerical Methods in Engineering 32(1991) 313-325.
[10] Y. Wang, W. Wu, Isotropicalized outside singular point method for bending problems of reinforced concrete plate, Mechanics in Engineering 15(1998) 65-68.
[11] U. Heise, Application of the singularity method for the formulation of plane elastostatical boundary value problems as integral equations, Acta Mechanica 31(1978) 33-69.
[12] G. Fairweather, A. Karageorghis, The method of fundamental solutions for elliptic boundary value problems, Advances in Computational Mathematics 9(1998) 69-95.
[13] A. Karageorghis, G. Fairweather, The method of fundamental solutions for the numerical solution of the biharmonic equation, Journal of Computational Physics 69(1987) 434-459.
[14] A. Karageorghis, G. Fairweather, The simple layer potential method of fundamental solutions for certain biharmonic problems, International Journal for Numerical Methods in Fluids 9(1989) 1221-1234.

[15] W. Chen, F.Z. Wang, A method of fundamental solutions without fictitious boundary, Engineering Analysis with Boundary Elements 34(2010) 530-532.

[16] W. Chen, F.Z. Wang, A method of fundamental solutions without fictitious boundary, Engineering Analysis with Boundary Elements 34(2010) 530-532.

[17] S.G. Lekhnitskii, Anisotropic Plates, Gordon and Breach, New York, 1968.

[18] Z.D. Luo, S.J. Li, Mechanics of anisotropic materials, Shanghai Jiaotong University, Shanghai, 1994.

VARIOUS WAVELET METHODS FOR SOLVING FRACTIONAL FREDHOLM-VOLTERRA INTEGRAL EQUATIONS*

PENG-PENG SHI, XIU LI, XING LI

School of Mathematics and Computer Science, Ningxia University, Yinchuan,750021 People's Republic of China

This paper presents the computational techniques for fractional Fredholm-Volterra integral equations. Various rationalized wavelet functions approximation together with collocation method is utilized to reduce this form of integral equations into a system of algebraic equations. Moreover, through illustrative example, a comparison of numerical solutions by using Harr wavelets, Legendre wavelets, Chebyshev wavelets, confirms the expected accuracy. Specially, this method is computationally attractive while the equations have not been solved analytically.

Keywords: Fredholm-Volterra Integral equations; Fractional order; Rationalized wavelets; Operational matrix; Collocation method

1. Introduction

In recent years, many practical problems of continuum mechanics, theoretical physics and other discipline lead to a wide variety of equations regarding fractional order [1][2][3]. There has recently been much attention devoted to the search for more efficient methods for determining the numerical solution, then several numerical approaches have been proposed[4][5][6]. Wavelets have proved to be a wonderful mathematical tool, and have been applied extensively to find the approximate solution of variational problems, differential and partial differential functions, integral equations and integro-differential equations [7][8][9][10][11].

The paper is organized as follows: Section 2 introduces some basic definitions and mathematical basis of fractional calculus, and concerns with the necessary definitions of wavelets and the Block-pluse function. Section 3 introduces

---

* This work is supported by the National Natural Science Foundation of China (10962008; 11261045; 51061015) and Research Fund for the Doctoral Program of Higher Education of China (20116401110002).

Corresponding author: Xing Li, Tel.: +86 951 2061405. E-mail address: li_x@nxu.edu.cn

procedure of three numerical methods to solve the equation by taking advantage of the nice properties of Harr wavelets, Legendre wavelets, Chebyshev wavelets. The procedure involves (i) expanding the admissible functions by wavelets functions with coefficients to be determined; (ii) establishing an operational matrix for performing fractional calculus and integral term; In Section 4, by reducing integral equation into a liner system of algebraic equations, we solve the algebraic equations obtained from the previous steps to evaluate unknown coefficients. Applying the wavelets methods to the fractional Fredholm-Volterra integral equations, considering numerical examples and making a comparison of numerical solutions by using three approximate methods, we demonstrate the accuracy of the proposed numerical schemes.

2. Notations, definitions and preliminaries

In this section, we present some basic notations, definitions and preliminary facts that will be used further in this paper.

2.1. Fractional calculus

Since Liouville attempted seriously to give logical definition, fractional calculus has been researching for 300 years. Numerous definitions of fractional integrals and derivatives have been proposed, sunch as the Riemann–Liouville, the Caputo, the Weyl, the Hadamard, the Marchaud, the Riesz, etc [12].

In this work, the fractional derivative of order $\alpha > 0$ is defined by

$$D^{\alpha}u(t) = \frac{d^n}{dt^n} \mathrm{I}^{n-\alpha}u(t)$$

where

$\alpha \in R, n-1 < \alpha \leq n, n \in N$

$u:(0,\infty) \to R$ be continuous,

$$\mathrm{I}^{\alpha}u(t) = \frac{1}{\Gamma(\alpha)} \int_0^t (t-s)^{\alpha-1} u(s)ds$$

is the Riemann–Liouville fractional integral operator of order $\alpha > 0$ and Γ is the gamma function[13].

In the paper， we consider Fredholm-Volterra integral equations involves fractional integral term as:

$$\sum_{n=1}^{m} \lambda_n \mathrm{I}_n^{\alpha_n} y(t) + \delta_1 \int_0^1 k_1(s,t) y(s)ds + \delta_2 \int_0^t k_2(s,t) y(s)ds = f(t, y(t)),$$

where

$0 \leq \alpha_n, 0 \leq s,t \leq 1, \lambda_n, \delta_1, \delta_2 \in R$.

2.2. Some properties of wavelets

Recently, wavelets have proved to be a wonderful mathematical tool, and have been applied extensively to find the approximate solution of problems. They constitute a group of functions constructed from dilation and transformation of a single function called the mother wavelets function. When the dilation parameter and the translation parameter vary continuously, we have various families of wavelets.

2.2.1. Some properties of Haar wavelets

The orthogonal set of Rationalized Harr functions is a family of square waves with magnitude of ± 1 in some intervals and zeros elsewhere. The first wavelets function is $h_0(t)$. The mother wavelets function is $h_1(t)$ which constructs a group of functions[14]:

$$h_0(t)=1,0\leq t<1$$

$$h_1(t)=\begin{cases}1,0\leq t<1/2\\0,1/2\leq t<0\end{cases}$$

$$h_n(t)=h_1(2^j t-k),n=2^j+k,j\geq 0,0\leq k<2^j .$$

2.2.2. Some properties of Legendre wavelets

The Legendre polynomials of order m , denoted by $L_m(t)$ are defined on the interval $[-1,1]$ and can be determined with the aid of following recurrence formulae[15]:

$$L_0(t)=1,L_1(t)=t,L_{m+1}(t)=\frac{2m+1}{m+1}tL_m(t)+\frac{m}{m+1}L_{m-1}(t).$$

The Legendre wavelets are defined on interval $[0,1)$ by

$$\psi_{n,m}(t)=\begin{cases}(2m+1)^{\frac{1}{2}}2^{\frac{k}{2}}L_m(2^k t-\hat{n}),\frac{\hat{n}-1}{2^k}\leq t<\frac{\hat{n}+1}{2^k}\\0, \qquad elsewhere\end{cases}$$

where $k=2,3,...,\hat{n}=2n-1,n=1,2,3,...,2^{k-1},m=0,1,2,...,M-1$. m is the order of the Legendre polynomials and M is a fixed positive integer.

2.2.3. Some properties of Chebyshev wavelets

Chebyshev wavelets $\phi_{n,m}(t)=\phi(k,n,m,t)$ have four arguments, $n=1,2,3,...,2^{k-1},k$, can assume any positive integer, m is the degree of Chebyshev polynomials of the first kind and t denotes the time[16].

$$\phi_{n,m}(t)=\begin{cases} 2^{\frac{k}{2}}\tilde{T}_m(2^k t-2n+1), \frac{n-1}{2^{k-1}}\leq t<\frac{n}{2^{k-1}} \\ 0, \qquad otherwise \end{cases}.$$

where

$$\tilde{T}_m(t)=\begin{cases} \frac{1}{\sqrt{\pi}}, m=0 \\ \sqrt{\frac{2}{\pi}}T_m(t), m>0 \end{cases}$$

and $m=0,1,2,...,M-1, n=1,2,3,...,2^{k-1}$. In equation, the coefficients are used for orthonormality. Here $T_m(t)$ are Chebyshev polynomials of the first kind of degree m, on the interval $[-1,1]$, and satisfy the following recursive formula: $T_0(t)=1, T_1(t)=t, T_{m+1}(t)=2tT_m(t)-T_{m-1}(t), m=1,2,...$.

2.3. Block-pulse function

The m-set of Block Pulse Functions is defined on $[0,1)$ as follows:

$$b_i(t)=\begin{cases} 1, \frac{i-1}{m}\leq t<\frac{i}{m} \\ 0, otherwise \end{cases}$$

where $i=1,2,...,m$. The functions b_i are disjoint and orthogonal[17]. That is, for $t\in[0,1)$ and $i, j=1,2,...$, we have

$$b_i(t)b_j(t)=\begin{cases} b_i(t), i=j \\ 0, \quad i=j \end{cases}$$

and

$$\int_0^1 b_i(t)b_j(t)dt=\begin{cases} \frac{1}{m}, i=j \\ 0, i=j \end{cases}.$$

The orthogonality property of block-pulse function is obtained from the disjointness property.

Furthermore, $B_m(t)$ is denoted $[b_0(t), b_1(t), ..., b_{m-1}(t)]$.

3. Function approximations and operational matrices

3.1. Function approximation

Any function $f(t)$, which is square integrable in the interval $[0,1)$, can be expanded into a family of wavelets functions series

$$f(t)=\sum_{n=0}^{\infty} c_n\gamma_n(t).$$

This series contains an infinite number of terms. If the infinite series is truncated, then equation can be written as

$$f(t)=\sum_{n=0}^{m-1} c_n\gamma_n(t)=C^T\Upsilon(t)=\hat{f}(t)$$

where $\hat{f}(t)$ denotes the truncated sum. C and $\Upsilon(t)$ are $m\times 1$ matrices, given by

$$C=[c_0,c_1,\ldots,c_{m-1}]^T$$

$$\Upsilon=[\gamma_0(t),\gamma_1(t),\ldots,\gamma_{m-1}(t)]^T$$

where the superscript T indicates transposition.

Furthermore, taking the collocation points as following

$$t_i=\frac{2i-1}{m}, i=1,2,\ldots m.$$

Every numerical method to solve the equation by taking advantage of the nice properties of Harr wavelets, Legendre wavelets, Chebyshev wavelets, we need define the square wavelets matrix of dimensions $m\times m$ denoted $\Phi_{m\times m}$.

3.1.1 Harr wavelets functions approximation

The function $f(t)$ is developed into the finite series as[14]

$$f(t)=\sum_{n=0}^{m-1} a_n h_n(t)=A_m^T H_m(t)=\hat{f}(t)$$

where $m=2^j$. The Haar coefficient vector A_m and Haar function vector $H_m(t)$ are defined as

$$A_m=[a_0,a_1,\ldots,a_{m-1}]^T$$

$$H_m(t)=[h_0(t),h_1(t),\ldots,h_{m-1}(t)]^T.$$

We defined the m-square Haar matrix $\Phi_{m\times m}$ as:

$$\Phi_{m\times m}=[H_m(\frac{1}{2m})\ H_m(\frac{3}{2m})\ \cdots\ H_m(\frac{2m-1}{2m})].$$

Correspondingly, we have

$$\hat{f}_m=[\hat{f}(\frac{1}{2m})\ \hat{f}(\frac{3}{2m})\ \cdots\ \hat{f}(\frac{2m-1}{2m})]=A_m^T\Phi_{m\times m},$$

the m -square Haar matrix $\Phi_{m\times m}$ is invertible, so the Haar coefficient vector A_m^T can be gotten by

$$A_m^T=\hat{f}_m\Phi_{m\times m}^{-1}.$$

3.1.2 Legendre wavelets functions approximation

The function $f(t)$ is developed into the finite Legendre series as[15]

$$f(t)=\sum_{n=1}^{2^{k-1}}\sum_{m=0}^{M-1}b_{nm}\psi_{n,m}(t)=B_{\hat{m}}^T\Psi_{\hat{m}}(t)=\hat{f}(t)$$

where $\hat{m}=2^{k-1}\times m$. The Legendre coefficient vector $B_{\hat{m}}$ and Legendre function vector $\Psi_{\hat{m}}(t)$ are defined as

$$B_{\hat{m}}=[b_{10},b_{11},\dots,b_{1M-1,}b_{20},b_{21},\dots,b_{2M-1,},\dots,b_{2^{k-1}0},b_{2^{k-1}1},\dots,b_{2^{k-1}M-1}]^T$$

$$\Psi_{\hat{m}}(t)=[\psi_{1,0},\psi_{1,1},\dots,\psi_{1,M-1,}\psi_{2,0},\psi_{2,1},\dots,\psi_{2,M-1,},\dots,\psi_{2^{k-1},0},\psi_{2^{k-1},1},\dots,\psi_{2^{k-1},M-1}]^T$$

We defined the m -square Legendre matrix $\Phi_{m\times m}$ as:

$$\Phi_{m\times m}=[\Psi_m(\frac{1}{2m})\ \Psi_m(\frac{3}{2m})\ \cdots\ \Psi_m(\frac{2m-1}{2m})].$$

Correspondingly, we have

$$\hat{f}_m=[\hat{f}(\frac{1}{2m})\ \hat{f}(\frac{3}{2m})\ \cdots\ \hat{f}(\frac{2m-1}{2m})]=B_m^T\Phi_{m\times m}$$

and

$$B_m^T=\hat{f}_m\Phi_{m\times m}^{-1}.$$

3.1.3 Chebyshev wavelets functions approximation

The function $f(t)$ is developed into the finite Chebyshev series as[16]

$$f(t)=\sum_{n=1}^{2^{k-1}}\sum_{m=0}^{M-1}d_{nm}\varphi_{n,m}(t)=D_{\hat{m}}^T\Theta_{\hat{m}}(t)=\hat{f}(t)$$

where $\hat{m}=2^{k-1}\times m$. The Legendre coefficient vector $D_{\hat{m}}$ and Legendre function vector $\Theta_{\hat{m}}(t)$ are defined as

$$D_{\hat{m}}=[d_{10},d_{11},\dots,d_{1M-1,}d_{20},d_{21},\dots,d_{2M-1,},\dots,d_{2^{k-1}0},d_{2^{k-1}1},\dots,d_{2^{k-1}M-1}]^T$$

$$\Theta_{\hat{m}}(t)=[\varphi_{1,0},\varphi_{1,1},\ldots,\varphi_{1,M-1},\varphi_{2,0},\varphi_{2,1},\ldots,\varphi_{2,M-1},\ldots,\varphi_{2^{k-1},0},\varphi_{2^{k-1},1},\ldots,\varphi_{2^{k-1},M-1}]^T$$

We defined the m -square Legendre matrix $\Phi_{m\times m}$ as:

$$\Phi_{m\times m}=[\Theta_m(\frac{1}{2m})\ \ \Theta_m(\frac{3}{2m})\ \ \cdots\ \ \Theta_m(\frac{2m-1}{2m})].$$

Correspondingly, we have

$$\hat{f}_m=[\hat{f}(\frac{1}{2m})\ \ \hat{f}(\frac{3}{2m})\ \ \cdots\ \ \hat{f}(\frac{2m-1}{2m})]=D_m^T\Phi_{m\times m}$$

and

$$D_m^T=\hat{f}_m\Phi_{m\times m}^{-1}.$$

3.2 operational matrices

3.2.1. Wavelets operational matrix for fractional integration

The fractional integral of block-pulse function vector can be written as following form[17]:

$$\frac{1}{\Gamma(\alpha)}\int_0^t(t-\tau)^{\alpha-1}B_m(\tau)d\tau\simeq F_m^\alpha B_m(\tau)$$

where

$$F_m^\alpha=\frac{1}{m^\alpha}\frac{1}{\Gamma(\alpha+2)}\begin{bmatrix}1 & \xi_2 & \xi_3 & \cdots & \xi_m\\ 0 & 1 & \xi_2 & \cdots & \xi_{m-1}\\ 0 & 0 & 1 & \cdots & \xi_{m-2}\\ 0 & 0 & 0 & \ddots & \vdots\\ 0 & 0 & 0 & 0 & 1\end{bmatrix}$$

$$\xi_k=(k+1)^{\alpha+1}-2k^{\alpha+1}+(k-1)^{\alpha+1}.$$

Denote the wavelets operational matrix for fractional integration with P_α, then

$$\frac{1}{\Gamma(\alpha)}\int_0^t(t-\tau)^{\alpha-1}\Upsilon_m(\tau)d\tau=P_m^\alpha\Upsilon_m(\tau)$$

by considering

$$\Upsilon_m(t)=\Phi_{m\times m}B_m(t)$$

For simplicity, we have

$$P_m^\alpha=\Phi_{m\times m}F_m^\alpha\Phi_{m\times m}^{-1}$$

where I_m is the identity matrix of order $m\times m$.

3.2.2. Operation for integral term
Considering the equations,

$$f(t)=\sum_{n=0}^{m-1} c_n\gamma_n(t)=C^T\Upsilon(t)=\hat{f}(t)$$

we obtain that

$$\int_0^1 k_1(s,t)y(s)ds=\int_0^1 k_1(s,t)\sum_{n=0}^{m-1} c_n\gamma_n(s)ds=\sum_{n=0}^{m-1} c_n\int_0^1 k_1(s,t)\gamma_n(s)ds=C^T K_1^F$$

$$\int_0^t k_2(s,t)y(s)ds=\int_0^t k_2(s,t)\sum_{n=0}^{m-1} c_n\gamma_n(s)ds=\sum_{n=0}^{m-1} c_n\int_0^t k_2(s,t)\gamma_n(s)ds=C^T K_2^V$$

where

$$K_1^F=[\int_0^1 k_2(s,t)\gamma_0(s)ds \quad \int_0^1 k_2(s,t)\gamma_1(s)ds \quad \cdots \quad \int_0^1 k_2(s,t)\gamma_{m-1}(s)ds]^T$$

$$K_2^V=[\int_0^t k_2(s,t)\gamma_0(s)ds \quad \int_0^t k_2(s,t)\gamma_1(s)ds \quad \cdots \quad \int_0^t k_2(s,t)\gamma_{m-1}(s)ds]^T$$

in here, substituting function of kernels into it, K_1^F, K_2^V are known vectors consisting of functions.

4. Illustrative examples

In order to show the effectiveness of wavelets methods for solving fractional integral equations, we present some numerical examples. Consider Numerical solution for Fredholm-Volterra integral equations involving single term fractional differential as

$$aI^\alpha y(t)+cy(t)+\delta_1\int_0^1 k_1(s,t)y(s)ds+\delta_2\int_0^t k_2(s,t)y(s)ds=f(t) \quad ,$$

$$0\le t\le 1, 0\le\alpha\le 1, a,c,\delta_1,\delta_2\in R$$

$$y(0)=0$$

Let

$$y(t)=C^T\Upsilon(t)$$

together with the initial states, then we have

$$I^\alpha y(t)=C^T P_m^\alpha\Upsilon(t)$$

Similarly, the input signal $f(t)$ may be expanded by the wavelets functions as follows:

$$f(t)=f_m^T\gamma_m(t)$$

where g_m^T is a known constant vector.

Substituting equations into linear fractional integral equation, we get

$$aC^T P_m^\alpha\Upsilon(t)+cC^T\Upsilon(t)+\delta_1 C^T K_1^F+\delta_2 C^T K_2^V=f_m^T\gamma_m(t)$$

Thus, the fractional integral equation has been transformed into a system of algebraic equations. Substituting values of a,c into it and solving the system of algebraic equations, we can obtain the coefficients C^T. Then using equation, we can get the output response $y(t)$.

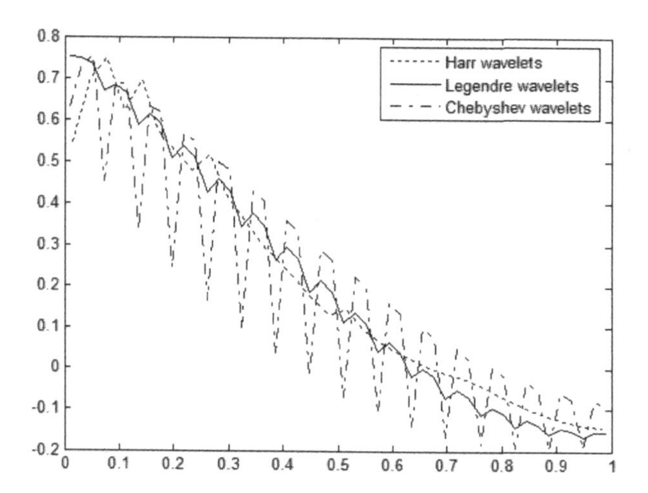

Fig. 1. the Numerical Solutions for equations where free term $f_1(t)=1$.

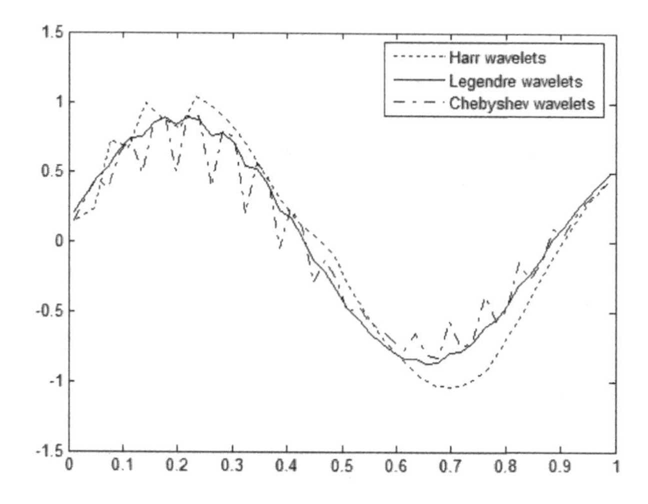

Fig. 2. the Numerical Solutions for equations where free term $f_2(t)=\sin(2\pi t)$.

We assume that $a=c=\delta_1=\delta_2=1, \alpha=0.5, k_1(s+t)=k_2(s+t)=e^{s+t}$ and consider $f_1(t)=1,\ f_2(t)=\sin(2\pi t),\ f_3(t)=\cos(2\pi t)$. Then the numerical solutions by using Harr wavelets, Legendre wavelets, Chebyshev wavelets are in perfect agreement with each other. Taking into account the convenience of process, when Harr wavelets functions approximation is utilized to solve the equation, we assignment $m=32$, otherwise, $m=48$. The numerical solutions by using Harr wavelets, Legendre wavelets, Chebyshev wavelets are shown in figs 1-3, and in perfect agreement with each other. Numerical results for different values of free term $f_i(t)$ are shown in the following Figs 1-3.

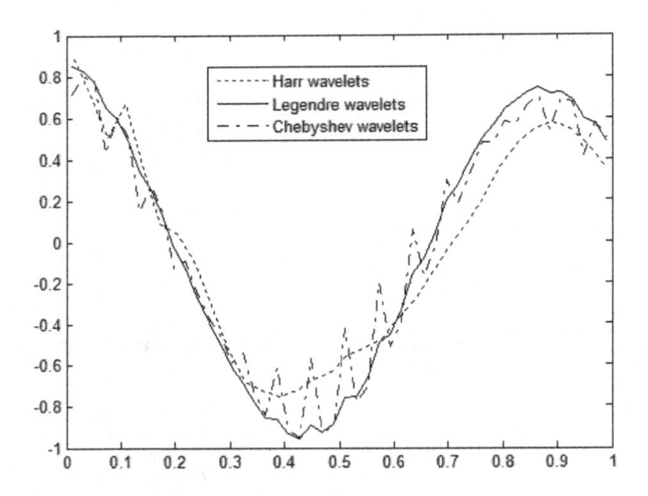

Fig. 3. the Numerical Solutions for equations where free term $f_3(t)=\cos(2\pi t)$.

5. Conclusion

We construct numerical methods for the fractional integral equations based on various wavelets operational matrices of the fractional order integration. General procedures of wavelets method are summarized. Several examples are given to confirm the accuracy of the proposed method.

References

[1] S. Samko, A. Kilbas, O. Marichev, Fractional integrals and derivatives: theory and applications, Gordon and Breach, London, 1993.

[2] A. Carpinteri, F. Mainardi, Fractals and fractional calculus in continuum mechanics, Springer-Verlag, New York, 1997.

[3] L.P. Kholpanov, S.E. Zakiev, Fractional integro-differential analysis of heat and mass transfer, Journal of Engineering Physics and Thermophysics. 78 (1) (2005) 33-46.

[4] V.Daftardar-Gejji, H.Jafari, Solving a multi-order fractional differential equation using adomian decomposition, Applied Mathematic and Computation. 189 (2007) 541-548.

[5] S. Momani, M.A. Noor, Numerical methods for fourth-order fractional integro-differential equations, Applied Mathematic and Computation. 182 (2006) 754-760.

[6] B. Ahmad, S. Sivasundaram, Existence of solutions for impulsive integral boundary value problems of fractional order, Nonlinear Analysis: Hybrid Systems. 4(2010)134-141.

[7] C.H. Hsiao, Haar wavelet direct method for solving variational problems, Mathematics and Computers in Simulation. 64 (2004) 569–585.

[8] ü. Lepik, Numerical solution of differential equations using Haar wavelets, Mathematics and Computers in Simulation. 68 (2005) 127–143.

[9] ü. Lepik, Haar wavelet method for nonlinear integro-differential equations, Applied Mathematics and Computation. 176 (2006) 324-333.

[10] ü. Lepik, Solving PDEs with the aid of two-dimensional Haar wavelets, Computers and Mathematics with Applications. 61 (2011) 1873-1879.

[11] L. Yuanlu, Z. Weiwei, Haar wavelet operational matrix of fractional order integration and its applications in solving the fractional order differential equations, Applied Mathematics and Computation. 216 (2010) 2276–2285.

[12] K.B. Oldham, J. Spanier, The fractional calculus, Academic Press. New York, 1974.

[13] T.F. Nonnenmacher, R. Metzler, On the Riemann-Liouville fractional calculus and some recent applications, Fractals. 3(3) (1995) 557-566.

[14] M.H. Reihani, Z.Abadi, Rationalized harr functions method for solving fredholm and volterra integral equations, Journal of Computational and Applied Mathematics. 200 (2007) 12-20.

[15] S. Yousefi, M. Razzaghi, Legendre wavelets method for the nonlinear volterra–fredholm integral equations, Mathematics and Computers in Simulation. 70 (2005) 1–8.

[16] E. Babolian, F. Fattahzadeh, Numerical computation method in solving integral equations by using Chebyshev wavelet operational matrix of integration, Applied Mathematics and Computation. 188 (2007) 1016–1022.

[17] A. Kilicman, Z. Abdel Aziz Al Zhour, Kronecker operational matrices for fractional calculus and some applications, Applied Mathematics and Computation. 187 (2007) 250–265.